Geodynamics of the Alps 3

SCIENCES

Geoscience, Field Director – Yves Lagabrielle

Dynamics of the Continental Lithosphere,
Subject Head – Sylvie Leroy

Geodynamics of the Alps 3

Collisional Processes

Coordinated by
Claudio L. Rosenberg
Nicolas Bellahsen

ISTE

WILEY

First published 2023 in Great Britain and the United States by ISTE Ltd and John Wiley & Sons, Inc.

ISTE Ltd
27-37 St George's Road
London SW19 4EU
UK

www.iste.co.uk

John Wiley & Sons, Inc.
111 River Street
Hoboken, NJ 07030
USA

www.wiley.com

Library of Congress Control Number: 2022941470

British Library Cataloguing-in-Publication Data
A CIP record for this book is available from the British Library
ISBN 978-1-78945-118-4

ERC code:
PE10 Earth System Science
PE10_4 Terrestrial ecology, land cover change
PE10_5 Geology, tectonics, volcanology
PE10_13 Physical geography

Contents

Introduction

Nicolas BELLAHSEN and Claudio L. ROSENBERG
ISTeP, Sorbonne Université,
UMR 7193, CNRS, Paris, France

The European Alps represent an archetypal mountain belt, mainly because its orogenic wedge is the most studied one. It results from a period of Mesozoic divergence forming the Tethyan Ligurian oceanic realm, linking the Neo-Tethys in the east with the Central Atlantic in the west. This followed the Variscan orogenic event affecting the European lithosphere and preceded a period of convergence starting during upper Cretaceous times with the northward movement of the Adria/Africa plates. This convergence was first accommodated by oceanic subduction, and later followed by continental subduction and eventually collision *sensu stricto*. The transition between oceanic and continental subduction is characterized by exhumation of high-pressure (HP) nappes of oceanic origin during early Eocene times, followed by HP metamorphism and exhumation of continental units during middle to late Eocene (Briançonnais and European distal continental margin). Collision started at the Eocene–Oligocene transition and is characterized by barrovian metamorphism, crustal shortening, high topographic relief and molasse-type peri-orogenic basins. During the whole convergence history, the amount of magmatism was very small. Today, convergence has slowed down, varying from almost zero in the Western Alps to approximately 2 mm/yr in the Eastern Alps, although vertical movements, extensive and compressive earthquakes are recorded, even in the Western Alps.

In spite of the latter observations, the European Alps may be considered as simpler than many other mountain belts that also result from the closure of oceanic domains. Indeed, in many other mountain belts, subduction and collision do not follow a simple sequence. For example, in the middle Paleozoic, during the Caledonian cycle and the Iapetus closure and more recently, during Cenozoic times and the Tethys closure between India and Asia, HP metamorphism events occurred long after the onset of collision. This is possibly due to the subduction of larger oceanic domains, and to longer and more vigorous convergence dynamics, which probably did not affect the Alpine realm.

The book *Geodynamics of the Alps* consists of three volumes. The first includes the recent and present-day structure and tectonic setting of the Alpine chain, from lithospheric mantle to brittle crust and surface topography, but also a historical overview of Alpine research and includes two chapters covering specific Alpine regions (Corsica and Eastern Alps), through all phases of Alpine history.

The second volume presents the pre-collisional history of the Alps. It starts with the Variscan orogeny in the Alpine realm, and continues with the inferred paleogeography, structure and extensional processes affecting continental margins, and their mantle structure in the pre-orogenic Alpine realm. This volume finishes with two chapters describing oceanic and continental subduction processes.

The third volume is entirely dedicated to Alpine collision. It describes and interprets the following elements of the Alpine chain: (1) Alpine magmatism, which is mainly contemporaneous to the very first stages of collision; (2) the formation of the external massifs, which represent a peculiarity in the Alpine chain, rarely occurring in other collisional orogens; (3) the foreland (Molasse) basin, in relation to the lithospheric processes that controlled its evolution; (4) exhumation of the internal part of the chain as constrained from the thermo-chronological record of the foreland basin sediments, and finally (5) the northern deformation front of the Alps, the youngest out of the large-scale collisional structures of the orogen.

This book will not provide a single, consistent and unique view on Alpine geodynamics, but rather create a space for experts on Alpine research to present their state of the art of specific subjects, not withdrawing from proposing their own interpretations, whether or not they are entirely consistent with those suggested in other chapters.

Volume 1

The first volume starts with a chapter on the history of Alpine cross-sections. This chapter does not review the historical evolution of tectonic and geodynamic processes in the Alps, but it shows how the style and format of geological cross-sections have evolved through time in relationship and as a consequence of major tectonic discoveries and debates. It also underlines changes through time in the scientific approach of scientists, showing how the very close link to observations in the mid-19th century evolved toward more conceptual interpretations, eventually lacking an observational base at the beginning of Plate Tectonics, when entirely new concepts of Alpine orogeny were formulated. Only in the last decade of the 20th century, have these conceptual models regained a close link to structural and geophysical observable data.

Chapter 2 of Volume 1 describes the present-day deep structure of the Alps, in which several slabs and slab segments can be identified. The dip-direction and the presence or absence of break-offs in the European slab are discussed in this chapter, strongly suggesting continuous, vertical slabs. In addition, the surface of the Alpine Mohos and the first-order structures of the deep crust are reassessed. Finally, based on the integration of receiver function, Vp/Vs seismic tomography and local earthquake tomography, the geometry and orientation of the west-Alpine subduction channel is imaged.

Chapter 3 of Volume 1 provides a quantitative description of the present-day surface topography of the Alps and its inferred uplift history. Several processes affecting the Alps since the Pliocene, such as sedimentation in the foreland basins, cooling ages and geochemically inferred erosion rates are integrated in the discussion of the present-day topographic structure, suggesting that changes in both climatic and tectonic processes must be invoked to explain the uplift history of the Alps.

Chapter 4 of Volume 1 reviews stress patterns inferred by the analysis of the youngest Alpine brittle fault systems, in addition to inferred deformation fields based on seismic and GPS data. These analyses show that the topographically highest portions of the Alpine chain in the Western and Central Alps were affected by orogen-perpendicular extension during Pliocene time, and this state of deformation is still recorded by seismic activity. The areas affected by such orogen-perpendicular extension coincide with high uplift rates in the northern and central parts of the Alpine arc. The possible collisional and post-collisional large-scale geodynamic causes of this deformation and uplift patterns are discussed.

Volume 1 concludes with two chapters which are both regionally focused. Because of their broad approach, providing the geologic al history of the Alpine realm throughout its entire history and up to the present, they are included within the first volume. Chapter 5 of Volume 1 presents the geology of the Eastern Alps discussing its underlying geodynamic processes. It shows that the stacking of nappes within the upper plate of the Alpine orogen can be inferred to represent a syn-collisional Cretaceous event, hence leading to the conclusion that two distinct orogenies, each of them consisting of a subduction/collision process, affected the Eastern Alps. The younger (Cenozoic) history is also described, with its peculiar style of collision, very different from that in the Western and Central Alps. Indeed, syn-collisional extrusion-related faults accommodate a significant part of collisional convergence in the Eastern Alps.

Chapter 6 of Volume 1 gives an overview of the geology of Corsica. Because the Sardo–Corsican block rifted away from the rest of the Alpine chain at the beginning of collision, it is not affected by the latter process, and it is ideally suited to study subduction and Cenozoic extension, and in particular, to relate the Alpine to the Appeninic chain. This chapter describes and illustrates how the Eocene-aged Alpine/Corsican east-directed subduction is followed by slab break-off and the initiation of a new, west-directed subduction system, which eventually generated the Appenninic chain.

Volume 2

Chapter 1 of Volume 2 deals with the Paleozoic evolution of basement rocks in the Alps. The oldest events recorded correspond to the formation of a Neoproterozoic peri-Gondwanian arc before its accretion to Gondwana, presumably during an Ordovician collisional event. The Cambrian–Ordovician period was marked by tectonic subsidence and widespread magmatic activity, prior to the opening of the Rheic Ocean. Convergence between Laurussia and Gondwana started in the Devonian, which marked the closure of Rheic Ocean, followed by the Variscan collision during the Carboniferous. Evidence of this convergence is provided by Devonian HP-LT eclogites preserved in the Austroalpine units, and Devonian–Tournaisian arc-related magmatism in the External Crystalline Massifs (Western Alps). The main period of nappe stacking and crustal thickening took place during the Mississipian and it was followed by strike-slip tectonics and extension from Pennsylvanian to Permian times. The orogen-scale dextral strike-slip east-Variscan shear zone was active from the Late Carboniferous to the Permian. Because of the intense reorganization during the Alpine orogeny, replacing the Alpine basement in the European Variscan puzzle remains a challenge to be solved.

Chapter 2 of Volume 2 provides an overview of the continental margins involved in Alpine orogeny. The reconstructed stratigraphic sequence and structure of both the European and Adriatic margin are presented based on a literature review showing numerous case studies from all over the Alpine chain. The questions on how many ocean basins and continental blocks existed before the onset of Alpine subduction, and their inferred paleo-geographic positions are reviewed, taking into account the nature of the oceanic substratum, that is, exhumed mantle versus oceanic crust. In addition to describing numerous interpretations of this complex paleo-geographic realm, a new Jurassic paleo-geographic map and a Jurassic-to-Cretaceous evolution are proposed. It suggests the existence of a main oceanic axis, which propagated into different parts of adjacent, pre-fractured continental blocks. A similar pattern is inferred to characterize the North Atlantic ocean in the Early Eocene, where numerous oceanization attempts, leading to hyper-extended crust and exhumed mantle, initiated from one major, stable central axis of accretion.

Chapter 3 of Volume 2 summarizes the petrology, the geochemistry and the age data of mantle and oceanic crustal remnants in the Alps and Appennines, and discusses these data in the frame of rifting and seafloor spreading preceding Alpine subduction. The comparison of these data with those of peridotites worldwide, including mantle samples dredged offshore, points to their origin from passive margins and ultra-slow spreading ridges. Their association with lower crustal granulites and mafic intrusions, in addition to their sedimentary cover typical of continental margins, suggest that they originate from the mantle lithosphere beneath a continental crust. The small volumes of gabbroic dikes recording melt extraction and magmatism during formation of the Piemonte–Liguria Ocean, and the similarity between the REE signature of their basalts, and that of MORB from ultra-slow spreading ridges, point to a likely origin from an ultra-slow spreading ridge.

Chapter 4 of Volume 2 is the first of two chapters on Alpine subduction. Based on a large synthesis of P-T conditions and PTt paths determined from HP rocks, this chapter reconstructs the 3D architecture of oceanic subduction in the Alpine realm and draws several conclusions on the dynamics of Alpine subduction. The consistent alignment of compiled P/T data along a P-T gradient corresponding to that of mature subduction zones suggests that Alpine HP rocks record the P-T conditions of subduction without being significantly affected by overpressure. The magmatic age of subducted and exhumed oceanic HP rocks points to the fact that only oceanic domains next to the continental margin and/or OCT were exhumed, whereas the remaining oceanic crust disappeared during the first stages of subduction. Integrating observations on syn-subduction sedimentary deposits all along the Alpine arc to petro-chronological data on HP rocks shows that an along-strike segmentation of the subduction zone both in terms of crustal-lithospheric structure and processes existed, with subduction accretion dominating in the Western Alps and subduction erosion dominating in the Central Alps. These differences in the nature of the subduction processes are discussed in relationship to the inherited structure of the continental margins and the oceanic crust itself. Finally, the Alpine case study is compared with larger scale subduction zones of the Mediterranean and Pacific.

Chapter 5 of Volume 2 reviews the available data about the subduction of continental crust in the European Alps. The type of crustal material that was involved in the subduction process is characterized (lower vs. upper crust, tilted blocks vs. extensional allochthons) and the data on pressure, temperature and age of metamorphic units are summarized. Finally, a discussion of the present geometry of the subducted continental units allows the setting of some constraints on the kinematic evolution of the mountain belt, and on the exhumation of HP-UHP rocks in particular.

Volume 3

Chapter 1 of Volume 3 focuses on Alpine magmatism, reviewing petrological, geochemical and geochronological data of Alpine magmatic bodies in relationship to their distribution all along the Alpine chain. Based on their spatial occurrence, age and geochemical signature, Alpine magmatic bodies are shown to be generated by different geodynamic processes, which are inferred to be (1) subduction of both continental and oceanic crust, whose melting produced magmas that were emplaced in the upper plate in the Middle and Late Eocene; (2) slab break-offs at the transition from subduction to collision, allowing for the rise of asthenosphere and its direct contact to the lower Alpine crust, whose melting and mixing with asthenospheric melts generated the largest volumes of Alpine magmas, forming the Periadriatic Plutons; (3) extensional processes related to mantle dynamics, limited to the marginal areas of the Alpine chain (the Alpine/Apennine junction and the Alps/Dinarides/Pannonian junction); (4) ascent along inherited lithospheric-scale faults, possibly explaining intraplate magmatism as observed in the Veneto magmatic province, which is located relatively far from the Alpine front.

Chapter 2 of Volume 3 summarizes the present-day knowledge on the external crystalline massifs, basement units deformed during the collisional stage of the mountain belt. In these massifs, the amount of shortening increases from south to north, as well as the P-T conditions of barrovian metamorphism. Each massif experienced a complex sequence of shortening: after slicing off the Dauphinois cover, ductile internal shortening affected the basement, before localization of deformation along basement thrusts resulted in "en bloc" exhumation of the massifs. Specificities between different massifs are described in detail and different hypotheses on the driving forces leading to the formation of the massifs are synthetized.

Chapter 3 of Volume 3 reviews the structure, age and stratigraphy of the northern foreland basin of the Alps, and it discusses the large-scale geodynamic processes underlying its stratigraphic architecture, its facies relationships and the provenance of its clastic material. The along-strike similarity versus change of the stratigraphy through time is synthetized and related to the drainage pattern of the basin. The historical evolution of inferred geodynamic processes is reconstructed over the past decades, thus showing that the stratigraphy and structure of the Molasse Basin may not only reflect thickening of the more internal part of the orogen, but rather the slab dynamics of the lower plate. Hence, subduction loads at deeper crustal levels rather than surface loads may exert the first-order control on basin subsidence.

Chapter 4 of Volume 3 reviews the currently available detrital thermochronological record of exhumation for the European Alps. It allows tracing and discussing temporal and spatial patterns of exhumation across the Alpine chain. The higher temperature white mica 40Ar/39Ar ages, if considered as cooling ages, depict phases of very rapid exhumation, on the order of 3 ± 1 km/Myr, in the Western Alps during the early Oligocene, and somewhat slower exhumation of ~1 km/Myr in the Central Alps during the late Oligocene–early Miocene. Detrital zircon fission-track data indicate a relatively steady exhumation signal since the late Oligocene for the Central Alps, whereas detrital apatite fission-track data emphasize regional differences in exhumation rates since at least Miocene times.

Chapter 5 of Volume 3 focuses on the northern deformation front of the Alps, that is, on the youngest, large-scale structure accommodating collisional shortening by the relative displacement of the orogen on its foreland basin. It shows that the structural style of this fault system varies along strike, with triangle zones being replaced by duplexes and basal décollements changing depth and the stratigraphic level. These variations are correlated to lateral changes in the mechanical stratigraphy, to differences in the links between these faults and their roots in the more internal parts of the orogen, or even to differences in the mechanical properties of the entire European Plate.

Acknowledgments

Romain Bousquet generously provided the vectorized version of his Tectonic Map of the Alps, used in most of the chapters in the three volumes of this book.

1

Cenozoic Magmatism in the Alps and their Surroundings: Relating Magmatic and Geodynamic Events

Alfons BERGER[1], Claudio L. ROSENBERG[2] and Giancarlo MOLLI[3]

[1]*Institute of Geological Sciences, University of Bern, Switzerland*

[2]*ISTeP, Sorbonne Université,*
UMR 7193, CNRS, Paris, France

[3] *Department of Earth Sciences, University of Pisa, Italy*

1.1. Introduction

Present-day Alpine magmatic rocks occur in rather limited volumes, areal extension and time length. This is a main difference, compared to the older and larger orogens that shaped the European continent, such as the Caledonian and the Variscan continents (e.g. Jacob et al. this book). The magmatic rocks discussed in this chapter are part of the Central–Western Mediterranean magmatic province (e.g. Wilson and Bianchini 1999; Savelli 2002; Alagna et al. 2010; Lustrino et al. 2011). However, this contribution only discusses the part of this magmatism related to the building of the Alps (Figure 1.1) and the immediate surroundings involved in geodynamic events, related with the Apennine and Dinarides orogens, the opening of the Ligurian-Provençal and the Pannonian Basins. The magmatic events of the Alps spread over ~25–30 million years as a consequence of distinct processes that

Geodynamics of the Alps 3,
coordinated by Claudio L. ROSENBERG and Nicolas BELLAHSEN.

affected and formed the Alpine Chain. These are mainly the subduction of the Piemont-Ligurian Ocean (PLO; Agard 2021), which is followed by subduction of the Briançonnais and Valaisan units, and finally the southernmost paleo-European margin. Subduction varies in time and space as seen in the crystallization ages of HP–LT metamorphic minerals (e.g. Berger and Bousquet 2008; Chapter 5, Volume 2 of this book by Manzotti and Ballèvre; Agard 2021) and in plate reconstructions (e.g. Schmid et al. 2004; Handy et al. 2010). Subduction of oceanic crust and the Briançonnais continental sliver, the Valaisan and the hyper-extended continental European margin (e.g. Decarlis et al. 2017) are followed by collisional shortening of the entire nappe stack (e.g. Schmid et al. 1996). The subduction history and the composition of the down-going plate affect the fluid input into the mantle wedge, hence the generation of melts. The Adriatic Plate represents the upper plate of the Alpine subduction, but also the lower plate of the Apenninic subduction since the Late Eocene–Early Oligocene (e.g. Lustrino et al. 2009, 2017; Chapter 6 in Volume 1 by Molli and Malavieille; Schmid et al. 2017). Most plutons and dykes are exposed along the axial part of the chain (Figure 1.1) and they are sometimes termed Periadriatic plutons (see next section for more information). A smaller number of magmatic bodies also occurs in the foreland (e.g. Esterél magmatism) and the retroforeland of the Alps (e.g. Veneto volcanic province; Figure 1.1).

Moreover, traces of magmatic products (Elter et al. 1964) are found in some tectonic units of the Apenninic nappe stack. This developed after the subduction reversal occurred during Late Eocene–Early Oligocene when the former retroside of the Alpine prism was partially reworked as the backstop of the newly established Apenninic prism (Malavieille et al. 2016; Schmid et al. 2017). The development of the Apenninic prism was related to the westward subduction of a relict of Ligurian Tethys oceanic basement and the distal Adriatic continental margin (Molli and Malavieille 2011; Malavieille et al. 2016; Malavieille and Molli this book). The Oligocene–Miocene westward subduction of the Adria plate was associated with a rear-wedge (Liguro-Provencal) and later intra-wedge (Tyrrheninan) extension in the wake of the Apenninic slab, a tectonic frame still active today (e.g. Jolivet et al. 1998; Bennett et al. 2012; Carminati and Doglioni 2012).

Within the Apenninic nappe stack Cenozoic magmatic products are found in the so-called Sub-Ligurian units, that is, the oldest remants of the Apenninic foredeeps and now observable below the units of the Ligurian prism (Elter et al. 1999; Mattioli et al. 2012; Di Capua et al. 2014, 2020). In addition, they are included in some top-wedge basins (called Epiligurian units) overlaying the Ligurian prism (Molli et al. 2010; Remitti et al. 2011) and within the Tertiary Piemont Basin (Mutti et al. 1995; Ghibaudo et al. 2014).

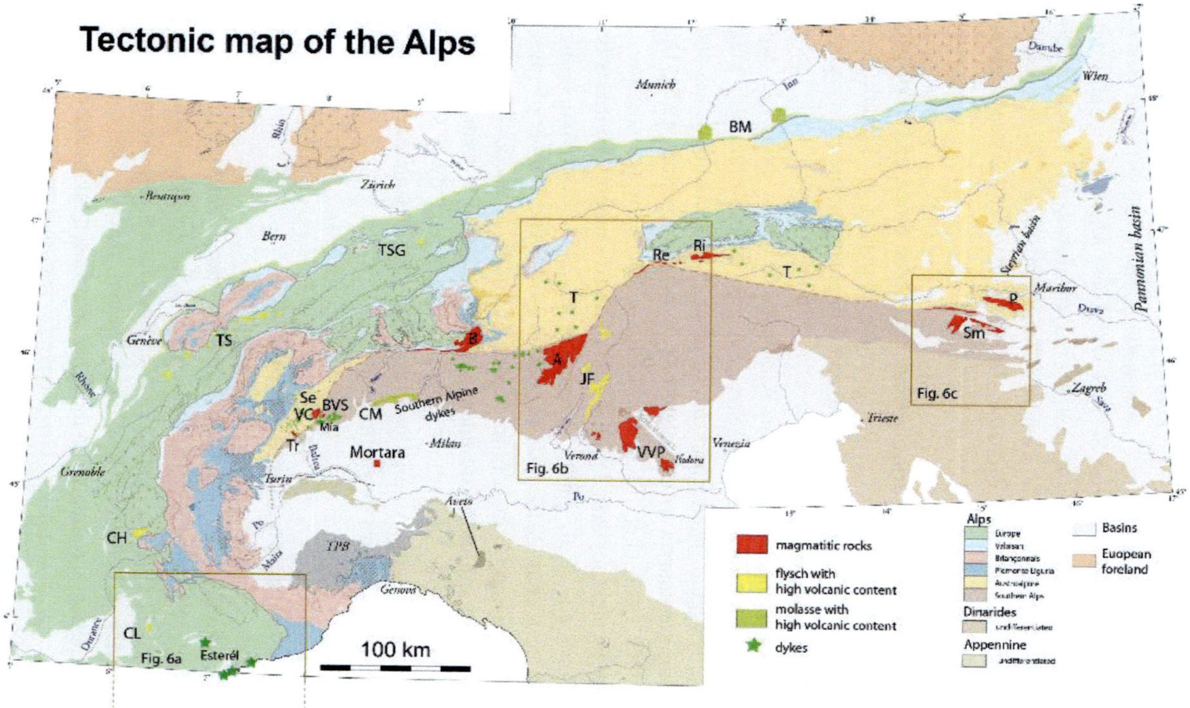

Figure 1.1. *Tectonic map of the Alps (after Bousquet et al. 2012) with the main Cenozoic igneous rock outcrops, buried bodies and volcaniclastic formations*

COMMENT ON FIGURE 1.1.– *Sediments: TS = Taveyannaz sandstone; TSG = Taveyannaz sandstone Glarus; CH = Champsaur sandstone; CL = Clumanc. JF: Judicarien flysch. BM: Bavarian Molasse. CM: Comer Molasse. Aveto: some localites of the Aveto-Petrignacola–M. Senario formation; TPB = Tertiary Piemont Basin. In situ magmatic rocks: Tr = Travesella pluton; VC = Valle del Cervo pluton; BVS = Biella volcanic suite; Se = Sesia dykes; Mia = Miagliano pluton; B = Bergell pluton; A = Adamello pluton; T = Tyrol dykes; Re = Rensen pluton; Ri = Rieserferner pluton; VVP = Veneto volcanic province; Sm = Smrekovec volcano; P = Pohoje pluton; Mortara volcano; Esterél = Esterél magmatic province. Detailed information is given in Table 1.1.*

Magmatism occurred mainly during Eocene/Oligocene times, and its distribution is linked to the coeval tectonic activity, in particular that of the Periadriatic Fault System (Figure 1.1). The latter crosscuts the entire thickness of the crust and forms a major crustal-scale boundary (e.g. Handy et al. 2005). In order to assess a link between Cenozoic magmatic events and Alpine geodynamic evolution, we have summarized the data on age, geochemistry and area of occurrence of magmatic bodies in the Alpine region. We restrict this summary to selected data sets and refer to already existing summaries to complete our review (e.g. Exner 1976; Dal Piaz and Venturelli 1983; Laubscher 1983; Waibel 1994; von Blanckenburg and Davies 1995; Wilson and Bianchini 1999; Tipolo et al. 2002; Rosenberg 2004; Lustrino and Wilson 2007; Conticelli et al. 2009; Lustrino et al. 2011; Carminati et al. 2012; Bergomi et al. 2015; Ji et al. 2019; Müntener et al. 2011). The number of studies and related geochemical datasets on Alpine magmatic bodies is unequally distributed. Some of the main Alpine plutons are intensively investigated in terms of chemistry and age (e.g. Bergell and Adamello), whereas others (e.g. dyke swarms of the Eastern Alps) are well known and documented in geological maps, but poorly investigated in terms of modern geochemical data.

1.2. History of research on Alpine magmatism

Magmatic rocks in the Alps have been known for more than a century (e.g. Suess 1875, Salomon 1897; Steinmann 1913; Cornelius 1915, 1908; Dal Piaz 1926). Suess (1875) noted a spatial alignment between the Rieserferner, the Adamello and the Iffinger-Brixner Massif (Figure 1.1; note, however, that the Brixner granite is Permian, hence genetically unrelated to the other two, based on modern knowledge). He coined the term "Südtiroler Granitbogen" (South-Tyrolean granitic arc). Salomon (1897, 1898) noted the alignment of the aforementioned plutons with several others located further east (Eisenkappel, Schwarzenbach, Pohorje; Figure 1.1) and he defined the "Periadriatischer Randbogen" (Periadriatic marginal arc). He suggested that all these plutons forming an arc around the Adriatic Sea intruded at the same time due to the same process. The term Periadriatic magmatism or Periadriatic plutons was later expanded to include the magmatic bodies of the Central and Western Alps (Steinmann 1913). Salomon (1898) systematically reviewed the literature to discuss maximum and minimum age of the intrusive bodies based on the age and structural/metamorphic relationships to dated sedimentary enclosing rocks. In addition, he also considers the inferred age of faults interpreted to have served as ascent pathways for the magmas. Based on the observations available at that time, Salomon (1898) concludes that the intrusion age is surely Post-Jurassic and most probably bracketed between Neocomian (Lower Cretaceous) and Mid-Eocene. The same author also suggested a geodynamic conceptual model to explain the alignment of the intrusions in map view: the collapse above a "Magma Basin" located at depth would cause part of the magma to intrude upward. Novarese (1901) provided detailed observations on the Traversella pluton, describing that the syenite (tonalite in the modern nomenclature) crosscuts the sedimentary beds and is oriented perpendicularly to these beds, hence inferring its intrusive nature and confirming a previous interpretation by Traverso (1893). Novarese (1901) also described the transformation of the carbonate sediments in contact with the "syenite" into Kalkhornfels by contact metamorphism. Finally, noting that the "syenite" is undeformed ("absence of pressure phenomena, lamination"), he suggested that it cannot be older than the last Alpine deformation, hence "not older than Miocene".

Later field geologists (e.g. Hammer 1903) recognized several dykes in Tyrol and the eastern Southern Alps crosscutting the basement, the Mesozoic sediments and the Alpine structures. They were correctly interpreted as dykes of Alpine age (e.g. Hammer 1903; Andreatta 1953; Reinhard 1964).

Steinmann (1913) noted that the Bergell intrusion and its dykes, in particular, crosscut the Alpine nappe stack. He, therefore, concluded that the age of intrusion must be Tertiary. His interpretations were confirmed by Staub (1918) on the basis of many and more detailed field observations. Argand (1916) followed the interpretations of Novarese for the age, and those of Salomon (1898) for the geodynamic process, considering that due to its increased weight, the Po plain and the Adriatic margin exerted a pressure on the underlying magma, which ascended along the less compressed margin, thus forming the Periadriatic plutons.

The age of the Alpine magmatites was very critical for all interpretations of the geodynamics of the Alps, because the latter rocks were used as relative time markers. Hence, the suggestion that magmatism in the Canavese and Sesia zone was of Permian age (Bianchi and Dal Piaz 1961) instead of the nowadays-determined

Oligocene age (Kapferer et al. 2012) which led to the misinterpretation that the Sesia HP rocks were of Pre-Alpine age. This issue needs the contribution of isotope dating of the volcanics of the Canavese zone (Ahrendt 1969) to be definitely solved.

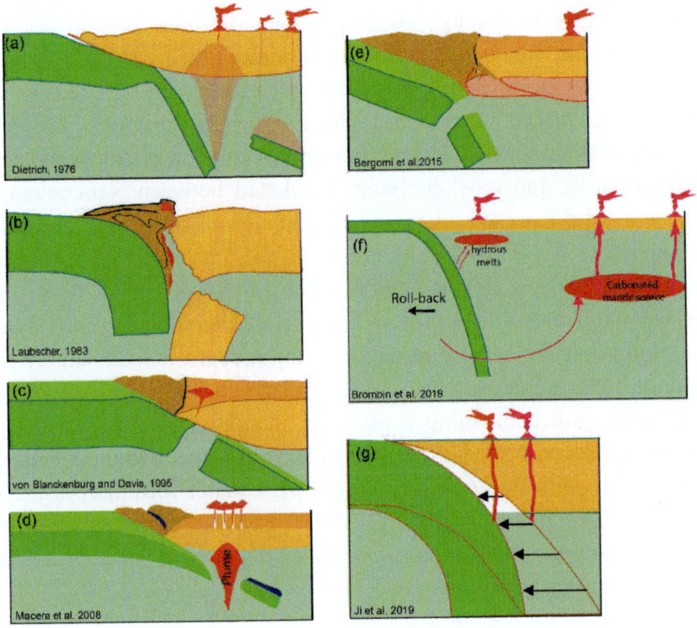

Figure 1.2. *Sketches of the proposed geodynamic setting for Alpine magmatism. All sketches infer a roughly N–S profile through the Alps. The different scenarios are discussed in the text. a) Summary figure of Dietrich (1976); b) redrawn sketch from Laubscher (1983); c) model of von Blanckenburg and Davies 1995; d) model of Macera et al. (2008); e) model of Bergomi et al. (2014); f) model of Brombin et al. 2019; g) model of Ji et al. (2019)*

New geodynamic interpretations of Alpine magmatism were proposed after the first radiometric ages of the plutons were provided in the 1960s (e.g. Grünenfelder and Stern 1960; Ferrara 1962; Chessex 1964), contemporaneously to the establishment of plate tectonics. The first model linking magmatism to a plate tectonic scenario (Dietrich 1976) explained both the Veneto volcanic province and the Periadriatic magmatism (Figure 1.2(a)). Dietrich (1976) suggested that the Paleocene–Eocene Veneto volcanic province (Figure 1.1) resulted from a first phase of melting of the oceanic crust subducted below Adria between the Upper Cretaceous and the Eocene. This phase was followed by slab breakoff and by the melting of the broken off segment, located in a more southern position (Figure 1.2(a)).

The latter would generate the melts of the Veneto volcanic province, whereas the still attached segment would steepen and continue sinking, eventually generating melts in the Late Eocene–Oligocene, emplaced along the Periadriatic Fault (Figure 1.2(a)). Laubscher (1983) also considered a slab breakoff, but of the Adriatic slab, that he associated with a phase of extension in the Oligocene, thus explaining the creation of a space required for the ascent and emplacement of the Periadriatic plutons (Figure 1.2(b)). Davies and von Blanckenburg (1995) suggested that breakoff occurred at the transition between oceanic and continental crust. This breakoff would allow for an asthenospheric rise, causing the increase of heat needed for the melting of the crust (Figure 1.2(c)). This model was inferred to explain the linear distribution of the main plutons along the central axis of the orogen (Figure 1.1) and the inferred asthenospheric geochemical signature interpreted on the base of the isotope geochemistry of the plutons. von Blanckenburg et al. (1998) suggested that such geochemical signature, and the mixed mantle-crustal source of the magmas could alternatively be explained by convective removal of a thickened and cold lithospheric root. The latter would be formed by syn-collisional thickening and would be replaced by hot asthenospheric material (e.g. Dewey 1988). Rosenberg (2004) pointed out that the alignment of the plutons along the Periadriatic Fault System indicates a tectonic control, channeling the ascent of melts along the only steep structure, continuously crossing the entire Alpine crust, but it does not reflect the geometry of the source region. Thus, this alignment cannot be a criterium to discriminate magmatic processes at the base of the crust. Macera et al. (2008) interpreted the Eocene Veneto volcanic magmatism as due to slab breakoff, also taking place at the continent–ocean transition as postulated by Davies and von Blanckenburg (1995), but significantly earlier, namely between 45 and 40 Ma (Figure 1.2(d)). Bergomi et al. (2015), based on new dating of dykes in the Southern Alps, assessed the previously undetected occurrence of Eocene dykes and stocks throughout the Southern Alps, between the Adamello batholith and Lake Como (Figure 1.1). In contrast to all previous models, insisting on the linear distribution of magmatic bodies induced by slab breakoff, they suggested that the entire Alpine magmatism resulted from a very large source region located in the mantle wedge, from the Paleocene to the Oligocene (Figure 1.2(e)). Only in this later period, slab breakoff contributed to extend the source region toward the axial part of the chain, hence generating the Periadriatic magmatism. The study of Brombin et al. (2019) is focused on the Veneto magmatism, for which it proposed that poloidal flow allowed for the rise of asthenospheric material from beneath the slab, hence the temperature increase and melting at the base of the upper plate (Figure 1.2(f)). Finally, Ji et al. (2019) considered that the age data of all magmatic bodies indicate a general northwestward younging (Figure 1.2(g)). They interpreted this inferred trend as a consequence of retreat of the European slab, shifting the ascent pathways of melt in the retreat direction.

1.3. Occurrence of magmatic rocks

We subdivide the Cenozoic magmatic bodies of the Alpine area into different groups, selected by their age and regional position (Figures 1.1 and 1.3; Table 1.1) as follows:

a) Magmatic rocks of the Periadriatic area (calc-alkaline series emplaced between ~42 and 30 Ma).

b) Sesia zone area (ultrapotassic, shoshonitic and calc-alkaline rocks emplaced between 33 and 30 Ma).

c) Southern Alpine dykes (basalts, andesites emplaced between 42 and 33 Ma).

d) Tyrol dykes (basalts, andesites, rhyolites and some plutonic equivalents emplaced ~30 Ma).

e) Esterél magmatic province (basalts, andesites and dacites emplaced between 40 and 20 Ma).

f) Veneto volcanic province (VVP) (nephelinites, basanites and alkali basalts emplaced between 52 and 22 Ma).

g) Pannonian Basin (volcanic and plutonic rocks of different composition younger than 18 Ma).

h) Central Alpine crustally derived granites (continental, peraluminuous melts emplaced between 30 and 25 Ma).

i) Mortara volcano (calc-alkaline volcano hidden in subsurface below the Molasse deposits of the Po plain and the retroforeland sequences; older than 28 Ma).

j) Volcanic-bearing sediments of the northern foreland (eroded products of Alpine magmatic rocks).

k) Volcanic-bearing sediments of the southern foreland (products of Alpine volcanics and other regions).

The age, distribution, mineralogy and geochemistry of these groups are summarized in the following sections. Additional data are found in Lustrino et al. (2011) and Bergomi et al. (2015). The names of units and magmatic bodies are listed in Table 1.1. The following description skips the major chemical data, which are summarized in Lustrino et al. (2011). In order to have first-order insight on the source region, available εNd and $^{87}Sr/^{86}Sr_{(i)}$ are shortly mentioned (Figure 1.4).

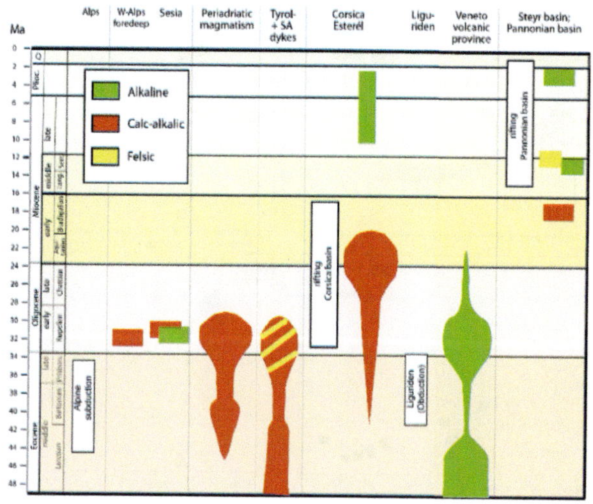

Figure 1.3. *Summary of the timing of the discussed magmatism. The thickness of the signs indicates the relative volumes inferred from present day outcrops. "Tyrol and SA. dykes" refer to all dykes in the Southern Alps and Eastern Alps*

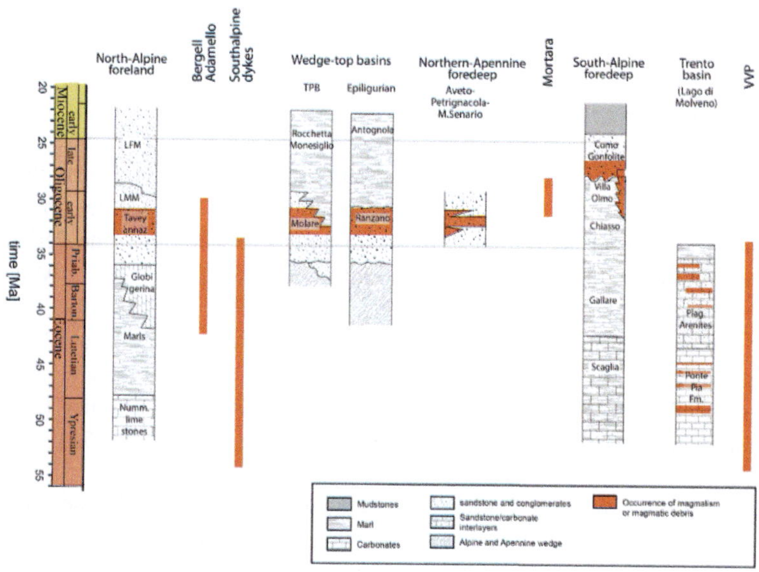

Figure 1.4. *Composite stratigraphic columns compiled by Malusa et al. (2011) and Lu et al. (2019). In addition, the main time intervals of the magmatism of group A, C, F and I are shown. TPB: Tertiary Piemont Basin; VVP: Veneto volcanic province*

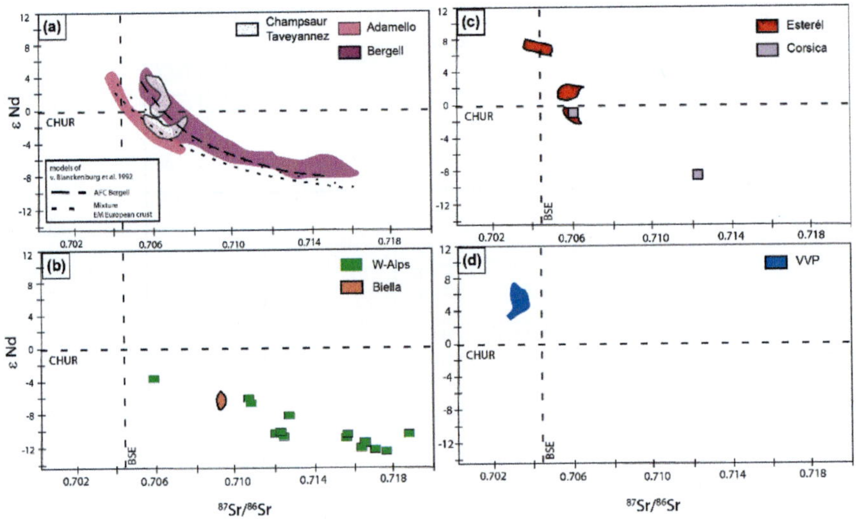

Figure 1.5. *Sr and Nd isotope data of the different magmatism (data from Lustrino et al. 2011 and literature therein, and in addition Lustrino et al. 2017; Conticelli et al. 2009; Milani et al. 1999; von Blanckenburg et al. 1992). a) Periadriatic magmatism, b) Sesia and W Alps magmatism, c) Esterél and Corsica magmatism and d) Veneto Volcanic province*

1.3.1. *Periadriatic magmatic rocks*

The plutons exposed along the Periadriatic Fault System represent the most significant volume of the Cenozoic magmatic rocks of the Alpine Chain (e.g. Laubscher 1983). From west to east, the main bodies located along the Periadriatic Fault are as follows (Figure 1.1): Traversella (42–33 Ma; Ji et al. 2019), Miagliano (33 Ma; Berger et al. 2012), Insubric dykes (32 Ma; Romer et al. 1996), Bergell (33–28 Ma; Samperton et al. 2015; Oberli et al. 2004), Adamello (~42–32 Ma; Schaltegger et al. 2019), Giudicarie lamellae (35–32 Ma; Pomella et al. 2011; Martin et al. 1993), Rensen (32 Ma; Barth et al. 1989), Rieserferner (~32 Ma; Romer and Siegesmund 2003) and Karawanken lamellae (Scharbert 1975).

The Periadriatic magmatic rocks are defined by their calc-alkaline character, which includes fractionation and assimilation processes (e.g. von Blanckenburg et al. 1992; Hürlimann et al. 2016). The dominant rock types are tonalities, granodiorites and granites and their volcanic equivalents. The formation of these magmatic rocks includes mantle- and crustal-derived melts (Bellieni et al. 1991; Kagami et al. 1991; Blundy and Sparks 1992; von Blanckenburg et al. 1992; Tiepolo et al. 2002; Tiepolo and Tribuzio 2005; Tiepolo et al. 2011, Hürlimann et al. 2016; Marxer and Ulmer 2019), as shown by the range of $^{87}Sr/^{86}Sr_{(i)}$ isotopic ratios (~0.704–0.716) and the εNd values (Figure 1.4). As seen in the mixing line between mantle and crustal end-members in Figure 1.4, crustal contamination is high (see also von Blanckenburg et al. 1992). Therefore, high heat input into the crust is required to allow for its partial melting (von Blanckenburg et al. 1992, 1998). Indeed, such heat input in the lower crustal plate is documented by UHT metamorphic rocks inferred to be 34 Ma in some of the nappes derived from the European basement (Oalmann et al. 2019; see also section 1.5.2).

1.3.2. *Sesia zone area*

The magmatic rocks of the Sesia zone include the Valle del Cervo pluton (syenites, monzonites, granites) and dykes (andesites, lamprophyres; Figure 1.1; Bigioggero et al. 1994; Owen 2008; Berger et al. 2012a; Zanoni et al. 2021). We separate these magmatic rocks from the Periadriatic ones of the same region (e.g. Traversella and Miagliano plutons), because of their more alkaline chemistry versus the calc-alkaline evolution of the Periadriatic serie (Venturelli et al. 1984; Owen 2008). Most of the Sesia magmatic rocks occur as dykes, of high-K, shoshonitic and lamprophyric composition (Dal Piaz et al. 1979; Beccaluva et al. 1983; Venturelli et al. 1984; Owen 2008). The main characteristics are radiogenic initial Sr ratios and unradiogenic Nd values (e.g. Owen 2008; Figure 1.4), pointing to an enriched mantle source. Such an enrichment has been often connected with sediment input into the mantle (e.g. Venturelli et al. 1984; Owen 2008; Conticelli et al. 2009). This interpretation is not only based on the radiogenic isotopes, but also on several key trace element ratios (e.g. low Ce/Pb, Nb/U and U/Th). The age of the lamprophyric dykes is poorly constrained, but available K/Ar data indicate emplacement around 30–32 Ma (Dal Piaz et al. 1973). But the high-K Val Cervo pluton is well dated (~30 Ma; Berger et al. 2012a; Romer et al. 1996; Table 1.1).

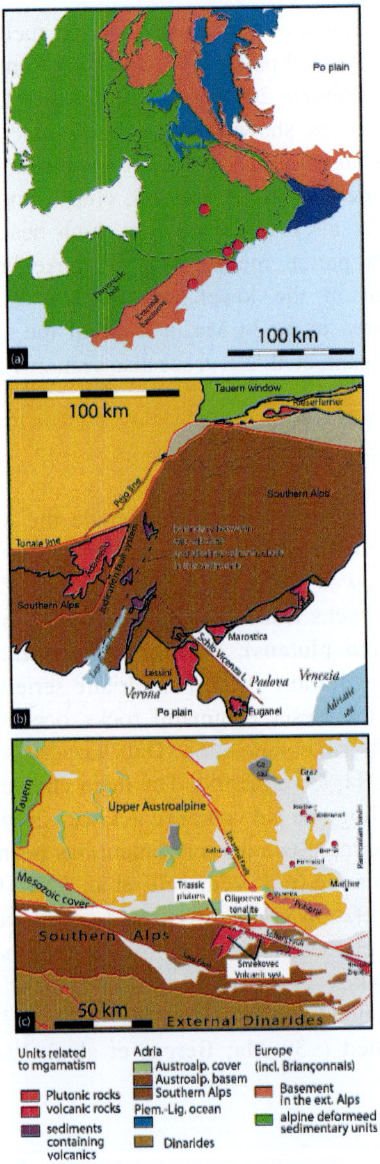

Figure 1.6. *Zoom-in of the tectonic map shown in Figure 1.1 (see there for location). a) Area including the Esterèl magmatism. The main locations of the magmatism (violet circles from Lustrino et al. 2017); b) magmatism between the Adamello and the VVP, including the flysch units containing volcanic pebbles; c) triple junction of the Dinaric Alps and Pannonian Basin with the major magmatic centers*

1.3.3. *Dykes and stocks in the Southern Alps*

Numerous subvolcanic dykes and intrusive stocks are mapped in the Southern Alps, in the area comprised between Ivrea Zone and Adamello pluton (Figure 1.1; Bigi et al. 1992). They are low- to high-K calc-alkaline basalts to andesites (Bergomi et al. 2014 and literature therein). The timing of their emplacement is well constrained by U/Pb data on zircon and it yields ages from 42 to 33 Ma (D'Adda et al. 2011; Bergomi et al. 2014). In addition, K/Ar data indicate that several dykes are as old as 64 Ma (Fantoni et al. 1999; Zanchi et al. 1990). Some of these dykes overlap both in space and time with the southern Adamello plutons, both in time and composition. The dykes in the Southern Alps do not form a volumetrically significant group of magmatic rocks. However, they may represent magma conduits that conducted additional volumes of magmas to subvolcanic bodies and to volcanoes, whose erosion and deposition provided the volcanic clasts preserved in the foreland sediments (see sections 1.3.10 and 1.3.11).

1.3.4. *Dykes and stocks in Tyrol*

Magmatic dykes also occur in the Ortler–Cevedale, Campo area (Figure 1.1; Gatto et al. 1976, Beccaluva et al. 1983, Dal Piaz et al. 1988) and in the Austroalpine units, south and east of the Tauern Window (Deutsch 1984; Neubauer et al. 2018). These dykes dominantly consist of andesites, with minor basaltic andesites, dacites and rhyolites. Several small stocks with similar chemistry are also reported (e.g. Dal Piaz et al. 1988), which crystallized as plutonic equivalents (quartz diorites to granites). These subvolcanic to plutonic bodies are numerous, but their chemical and geochronological data are rare. One larger dataset consists of K/Ar and Rb/Sr ages that spread between 30 and 32 Ma (Deutsch 1984; Dal Piaz et al. 1988). These ages are confirmed by newly dated small bodies of tonalites that resulted in a zircon U/Pb age of 31 Ma (Neubauer et al. 2018).

1.3.5. *Esterél magmatic province*

The magmatic rocks of the "Esterél magmatic province" (e.g. Ivaldi et al. 2003; Lustrino et al. 2017; Figure 1.1), in Provence, mainly occur as clasts within Oligo–Miocene detritic deposits located in the foreland of the Alpine orogen (St. Raphaël, Cap d'Antibes; Figure 1.1), or at its most external deformed margin (St. Antonin; Figure 1.1). The dominant rock types are andestitic to dacitic volcanic to subvolcanic rocks. The whole area includes two age groups, one Late Eocene to Early Miocene (defined by Lustrino et al. 2017 as LEEM) and a second one of Lower Miocene to Pliocene age. In the context of this contribution, the LEEM are of great interest, due to their overlap in time with Alpine magmatism (Figure 1.1; Table 1.1). For this reason, we do not describe the Lower Miocene to Pliocene group in this review. The isotopic composition of the LEEM indicates a mantle source including high εNd values and low Sr(i) values (Figure 1.4; Lustrino et al. 2017). As suggested by their Pb isotopic signature, the parent melts probably derived from a sub-continental lithospheric mantle source.

1.3.6. *Veneto volcanic province (VVP)*

The VVP (Figure 1.1) is subdivided into the Euganian Hills, Berici Hills and Lessini Mountains, located SW of the Schio–Vicenza Line (Figure 1.6(b)), and the Marostica hills, located NE of this tectonic boundary. In addition, some rare dykes occur in the Southern Alps close by (Lucchini et al. 1983). The volcanics are alkali basalts, basanites and nephelinites with a strong alkaline character and mostly a primitive composition (Barbieri et al. 1981; Siena and Coltorti 1989; Milani et al. 1999; Beccaluva et al. 2001; Macera et al. 2008; Gasperini et al. 2006; Brombin et al. 2019). The non-radiogenic isotope values of the volcanic rocks in combination with the trace elements indicate a subcontinental mantle source, whose characteristics are acquired after the interaction with crustal-derived fluids (e.g. Beccaluva et al. 2001; Brombin et al. 2019). The enriched REE pattern in pyroxenes and the composition of fluid inclusions in the same minerals of mantle xenoliths interpreted as possible subduction-related metasomatism (e.g. Cannatelli 2012). The calculated degree of mantle melting from the trace element ratios spread between 3 and 15% (Table 3 in Beccaluva et al. 2007).

The VVP is active at least between 52 and 30 Ma (Visonà et al. 2007). In addition, one magmatic center is dated as young as 22 Ma (Brombin et al. 2019). Despite the long time span of magmatic activity, the spatial distribution of these magmatic rocks is quite punctual, without significant change through time in the nature of the volcanic rock types (Figures 1.1 and 1.6(b)). The older volcanics of the VVP are interbedded with Eocene and Oligocene sedimentary deposits related to the ENE–WSW opening of the Alpone-Agno Graben (Zampieri 1995).

1.3.7. *Pannonian magmatic province*

The Pannonian Basin is not a major topic of this review (an overview of its magmatism is given by Seghedi and Downes 2011 and Lukács et al. 2018), but the spatial overlap between the Alpine, Pannonian and Dinaridian magmatism (Figure 1.1) requires a short notice, especially in relation to the easternmost plutons of the Periadriatic Lineament (Figures 1.1 and 1.6(c)). The most prominent is the Pohorje pluton (Fodor et al. 2008; Trajanova et al. 2008), which is located close to the Periadriatic Lineament and whose composition is similar to that of the Periadriatic plutons, but it has a significantly younger age (18.6 ± 0.1 Ma; Table 1.1). The volcanic activity in the Styrian Basin (Figure 1.1) is of similar age to that of the Pohorje pluton. Moreover, in the Pohorje area, volcanic and plutonic rocks of the same chemical evolution exist in more or less the same area (Figure 1.6(c)). The volcanics of the Styrian Basin are Miocene in age (~15 Ma; e.g. Harangi 2001; Handler et al. 2006; Lukács et al. 2018), hence only slightly younger than the Pohorje pluton. The occurrence of both volcanic and plutonic rocks of similar age and composition at the present-day erosion level points to significant vertical movements and exhumation in the area. Some Oligocene sub-surface plutonic rocks (detected in drill cores) and outcrops of volcanic rocks exist in the Pohorje area, hence they may be related to the Periadriatic group (Benedek 2002, cited in Fodor et al. 2008; Kralj 2012; Figure 1.6(c)).

1.3.8. *Central Alpine crustal melts*

Only minor areas of exhumed migmatites and/or pure crustal-anatectic plutons of Alpine age are exposed in the Alps. The Novate leucogranite (Figure 1.1), located at the western margin of the Bergell pluton, is the largest of such crustal-derived melts. It is a typical two-mica granite emplaced at ~25 Ma (Liati et al. 2000).

The major elements as the peraluminous character, the low Na content as well as the Sr isotopy outside the calc-alkaline evolution, and the peraluminous character of these granites, indicate that they derive from crustal melts (von Blanckenburg et al. 1992). Moreover, the bulk rock REE pattern with the Eu anomaly fits with their continental character (Mottana et al. 1978). The Novate granite is the largest of such crustal melts and the only well-investigated granite, but more such plutonic bodies and veins are known, mainly inside the migmatites (e.g. Burri et al. 2005; Berger et al. 2008; Rubatto et al. 2009). No modern geochemical data are available yet from these small granitic bodies.

1.3.9. *Mortara volcano*

The Mortara volcano is hidden below ~5,500 m of detritic sediments in the Po plain (AGIP drill-site "Mortara-1": Ruffini 1995; Mattioli et al. 2002a; Figure 1.1). Therefore, not many data are available for this magmatic system. It is located in the southern foreland within the Adriatic domain that was barely deformed during collision (Figure 1.1). Two types of volcanic rocks have been recognized in this volcanic body: porphyritic lavas of andesitic and dacitic-rhyodacitic compositions.

The available samples show a calc-alkaline affinity including fractional crystallization and crustal contamination (Mattioli et al. 2002a). Their typical calc-alkaline evolution based on the available geochemical data indicates a possible relation to the Periadriatic magmatism, in spite of their completely different position. The Mortara volcano is located inside the retroforeland basin of the Alps in the upper crust of the Adriatic Plate (Figures 1.1 and 1.6). The estimated age of the sedimentary cover (~28 Ma) defines the upper age limit of the Mortara volcano. Given the difficulty of identifying such magmatic bodies, not exposed at the surface, we should bear in mind that the Mortara volcano may be one of a series of volcanic edifices hidden/buried below the retroforeland sediments.

1.3.10. *Volcanoclastic sediments of the northern foreland*

Terrigenous sediments containing volcanic pebbles are widespread all around the Alpine foreland (Figure 1.1). In the northern foreland, these are the Taveyannaz sandstone, the Champsaur sandstone and some Molasse units in Germany and Austria (Figure 1.5; Table 1.1). The estimated ages and compositions of the volcanic components overlap with those of the calc-alkaline series of the Periadriatic magmatism (Lu et al. 2018). Therefore, the volcanic clasts in the flysch and Molasse units of the northern foreland are often associated with the Periadriatic magmatic rocks (Waibel 1993; Féraud et al. 1995; Ruffini et al. 1997; Brügel et al. 2000; Boyet et al. 2001; Lu et al. 2018, 2019; Figure 1.5). Most of the Taveyannaz pebbles are ~32 Ma in age (Lu et al. 2019), but one member of the Taveyannaz sandstone in the Glarus Alps includes volcanic clasts, whose dominant age is ~42 Ma (Lu et al. 2018).

1.3.11. *Volcanic-bearing sediments of the southern foreland*

In addition to the deposits on the northern border of the Alps, several sedimentary units containing volcanic pebbles also occur in the Southern Alpine retroforeland. These are the flysch deposits of the Giudicarie belt, the Molasse/siliciclastic units of the southern foreland (Gonfolite Lombarda) and the top wedge basins along the transition to the Apennines (Figure 1.5). In the flysch basins of the Giudicarie belt, the volcanic pebbles are subdivided based on their petrology and geochemistry into a calc-alkaline and an alkaline series. The calc-alkaline pebbles have been interpreted as part of the Adamello system, and the alkaline ones as part of the VVP (Martin and Macera 2013). The source of the magmatic fragments of the Southalpine foreland basin and some of the top wedge basins may be the same as that of the Periadriatic magmatism based on their calc-alkaline character (Giger and Hurford 1989; Malusa et al. 2016; Lu et al. 2019; Figure 1.5).

However, U/Pb ages of these pebbles spread between 48 and 30 Ma (Lu et al. 2019), thus overlapping in age with the Southalpine dykes, the Adamello and the VVP magmatism.

In the Apennines, Cenozoic volcanic fragments are found in the Aveto-Petrignacola–M. Senario Fm. (Figure 1.1). This unit has been related to the Sub-Ligurian tectonic domain, within the Apenninic nappe structure (Elter et al. 1964; 1999; Mattioli et al. 2012; Di Capua et al. 2014). It was recognized within the Epiligurian top-wedge basin overlaying the Ligurian units (Molli et al. 2010; Remitti et al. 2011), and within the Tertiary Piemont Basin (Mutti et al. 1995; Ghibaudo et al. 2014). These units consist of turbiditic sequences with pelitic-arenaceous and conglomeratic members, including in their middle and upper part volcanic and volcanoclastic fragments. The volcanic fraction is made up of basaltic andesites, andesites and dacites, with minor rhyolites, basalts and gabbros, found as fragments, clasts and pebbles up to 0.5 m^3 in size. Petrographic, major and trace elements and radiogenic isotope data are compatible with calc-alkaline magmatism (Mattioli et al. 2012; Di Capua 2014). Biostratigraphic age data of these turbiditic deposits (Early Oligocene, ~32 and 29 Ma; Cerrina Feroni et al. 1990; Elter et al. 1999; Plesi et al. 2000) overlap with the Ar/Ar ages of their volcanic clasts (~32–29 Ma; Mattioli et al. 2002b), indicating that volcanism, erosion and sedimentation took place in a narrow time span. The sedimentological and depositional features of the host deposits and the volcano-clastic components indicates a short distance between the volcanic source and the basin (Elter et al. 1999; Mattioli et al. 2012), strongly supporting the idea of volcanos located within the backside of the Alpine orogen or within the Adria-backstop areas. A source region within the Alpine orogeny (volcanos related to Biella dykes or feeded by Bergell intrusion) appears suitable on the base of sediment provenance investigations (Malusà et al. 2011; Anfinson et al. 2016), although a quite large distance (on the order of ca. 300 km) is not fully consisting with the sedimentological characters of the volcanoclastic deposits according to Mattioli et al. (2012). In contrast, although strikingly similar in age, petrographic features and mineral chemistry (Cibin et al. 1998; Mattioli et al. 2012), the volcano-clastic sediments within the Epiligurian deposits (Ranzano Fm.) and Tertiary Piemont Basin (Mutti et al. 1995; Cibin et al. 2003; Ghibaudo et al. 2014) may be related to a depositional basin/sink area located at a larger distance to the volcanic sources. This is in line with the inference previously indicated and the suggestion of a volcanic source located somewhere within the backside of the Alpine orogen or within the Adria-backstop. The alternative interpretation, pointing to the Esterél Province as the source region (see Di Capua et al. 2014, 2020), is associated with the problem of the source to sink transfer of needs to by-pass the accretionary collisional belt represented by the Western Alps/Corsica orogen and the proto-Apenninic accretionary wedge.

1.4. Inferred spatial distribution of Alpine magmatic bodies in the retrodeformed Oligocene Alps

In order to discuss the geodynamic setting of the Cenozoic magmatism in the Alps, we relocated the magmatic bodies to their inferred Early Oligocene positions in map view, by retrodeforming the major collisional displacements, as constrained by tectonic studies (Figure 1.7). With the exception of the Tyrol dykes, all Alpine magmatic bodies are located along the Periadriatic Fault or south of it, therefore retrodeformation of the Miocene thrusts of the Southern Alps and strike-slip movements along the Periadriatic Fault is sufficient to reconstruct the original positions of these bodies at the onset of collision (approximately 32 Ma). In contrast to several other reconstructions, we locate stable reference points in the Po plain, between Apennines and Alps, in an area that was barely affected by deformation of both orogens. Therefore, the Periadriatic Fault is displaced northward after retro-deformation of the South-Alpine thrusts (Figure 1.7). As a result, the northern limit of the Southern Alps, which is the Insubric Line (part of the Periadriatic Fault System), coincides with the northern limit of its lower crust, which is located further north in the present structure of the orogen (Figure 1.8; Schmid et al. 1996; Rosenberg and Kissling 2013). This creates a straight and nearly vertical path for the melts, from below the Adriatic Moho to the surface (Figure 1.8). The amount of strike-slip movements along Insubric Line, Giudicarie Fault, Schio-Vicenza Fault and the Lavantal Faults (see Figure 1.6 for locations; the inferred amounts of shortening used for the retro-deformation are provided in section 1.7 (Appendix 1)).

The restored map shows the following first-order features: A SW–NE oriented alignment of plutons, whose age is between 32 and 30 Ma, strikes parallel to and all along the axis of the Alpine Chain. From east to west, these plutons are the Karawanken, Rieserferner, Rensen, several tonalitic lamellae, the northern Adamello (Presanella and Avio) and the Bergell pluton, and if we include the westernmost bodies in spite of a bend to this linear distribution, the Valle del Cervo, Traversella and Miagliano in the southwest (Figure 1.7). These intrusions form the group of the Periadriatic plutons. The areas defined by the distribution of dykes appear to be distributed on both sides of the Periadriatic Lineament in the Western Alps, whereas they are alternatively located to the north and south of the latter structure in the Central and Eastern Alps (Figure 1.7). The distribution north or south of the Periadriatic Lineament is sharply defined. All the remaining magmatic bodies are located south of the latter linear structure, and, except for the Mortara volcano and Marostica hills, they are systematically older. Although the number of dated samples is limited, a general, Alpine-wide trend of northwestward younging (Ji et al. 2019) cannot be assessed on the basis of the map of Figure 1.7, because the Valle del Cervo, Traversella, Miagliano, and Mortara magmatic bodies are at odds with such a trend.

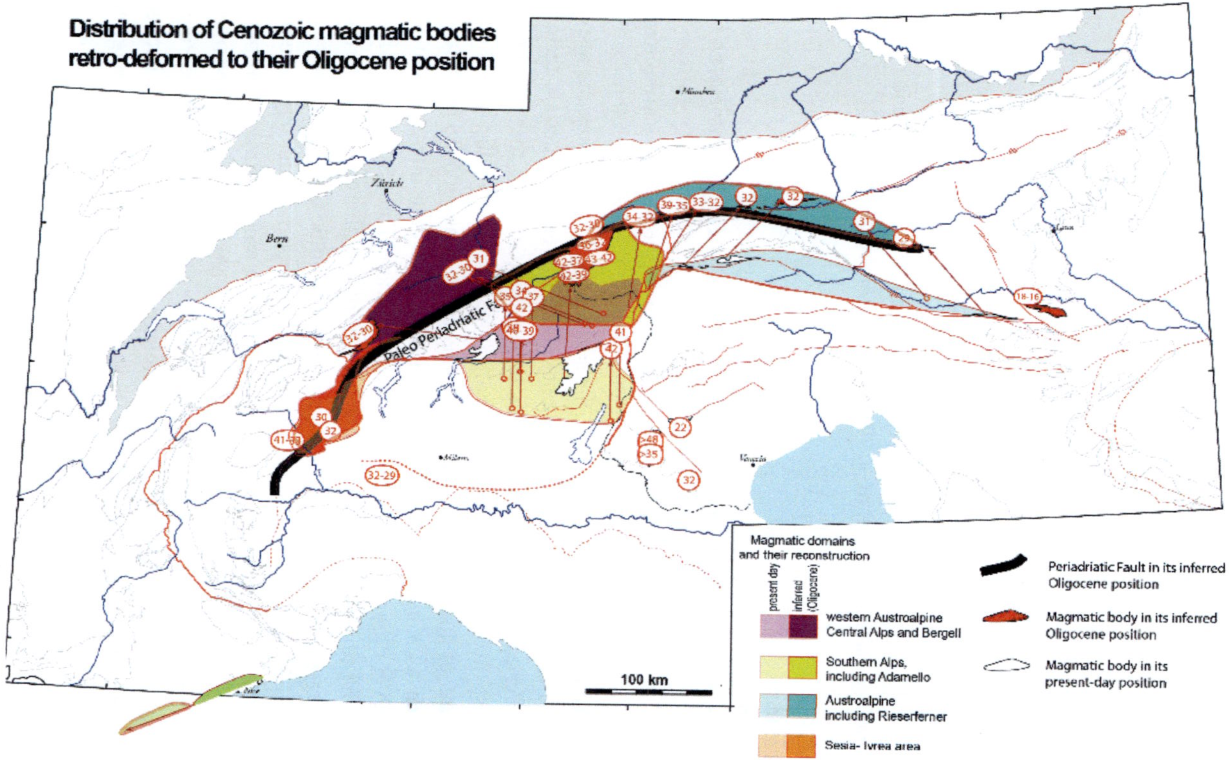

Figure 1.7. *Distribution of Cenozoic magmatic bodies retro-deformed to their Oligocene position. For detailed explanation of the retro-deformation procedure, see Appendix 1*

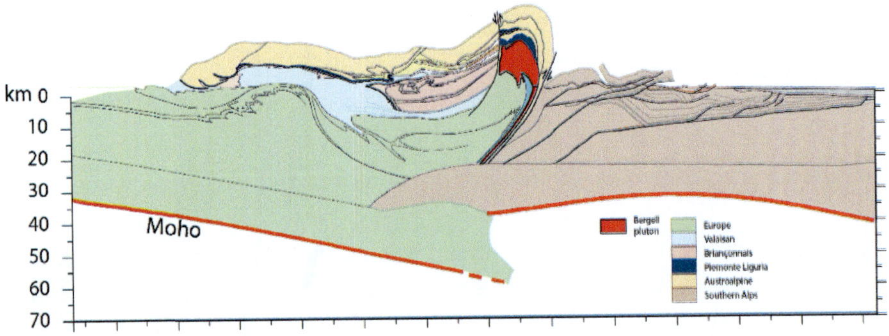

Figure 1.8. *Orogen-scale cross-section along the NFP20 seismic line. Modified after Schmid et al. (1996) and Rosenberg et al. (2015)*

1.5. Discussion

1.5.1. *Consequences of the spatial configuration in Oligocene times*

The map in Figure 1.7 shows a major distinction in the spatial distribution of magmatic bodies as a function of their size: although the larger plutons are all aligned along the Periadriatic Fault System, the smaller stocks and dykes can also be found at great distances to the main lineament. This difference suggests a tectonic control of the Periadriatic Fault System on the amount of melt that can ascend through the crust. Indeed, the Periadriatic Fault System represents the only steeply dipping structure of the Alps that crosscuts the crust continuously from top to bottom, as much as it can be inferred on the basis of seismic interpretation (e.g. Schmid et al. 1996; Figure 1.8). In addition, the Periadriatic Fault System is inferred to be active at the time of ascent and emplacement of the plutons (e.g. Rosenberg et al. 1995; Berger et al. 1996). Therefore, the spatial link between plutons and Periadriatic Fault System appears to be due to the fact that the latter structure with its wide (>1 km) mylonitic belt enables and facilitates the ascent of melts. In spite of the intuitive character of this conclusion based on map observations, understanding the mechanical process allowing for the creation of space in the plane of shearing is not straightforward. Different solutions invoking the interaction between magma-pressure and pre-existing planes of anisotropy (schistosity) were suggested by studies interpreting the ascent of the Periadriatic plutons along the active Periadriatic Fault System (Berger et al. 1996; Rosenberg 2004).

As described above (Figure 1.7), the suggested northwestward trend of younging is only consistent with the area of the Adamello and the dated dykes to its western and southeastern regions. The limited size of this area and the interruption of this trend in the rest of the chain do not allow to confirm any systematic age trend across the chain, in contrast to the conclusions of Ji et al. (2019) based on the present-day distribution of magmatic bodies. The simple conclusion that can be drawn from Figure 1.7 is that magmatic intrusions emplaced between ~34 and 30 Ma do concentrate along a linear structure, coinciding with the newly developed plate interface of the collisional setting (= Periadriatic Fault and its mylonitic belt).

The area of the Tyrol dykes is very peculiar, because it is located quite far north compared to the Periadriatic Fault System. These dykes and stocks are intruded in the Austroalpine nappes, which were thrust above the PLO and the Briançonnais nappes well before the emplacement of the dykes and stocks in the Oligocene. A steep or even subvertical ascent path for these melts would have its source at a depth well in front of the northern termination of the mantle wedge (Figure 1.8). However, the example of the Bergell pluton (Rosenberg et al. 1995; Berger et al. 1996) shows that melts ascend along the Periadriatic Fault, and migrate away from it, to their site of final emplacement, along gently dipping planes, covering distances of ~50 km. The northern margin of the Bergell pluton is located ~25 km north of the Periadriatic Fault System, but the pluton accommodated 21 km of post-emplacement shortening (Schmid et al. 1996; Rosenberg and Kissling 2013). If a pluton of similar size and shape was emplaced east of the Bergell, it could feed swarms of dykes in the overlying Austroalpine units far north of the Periadriatic Fault. Due to post-emplacement tilting (Davidson et al. 1996) an E–W cross-section through the Bergell area represents 10 km difference in the exposed structural level of the Alpine nappe stack. Hence, the area of the Tyrol dykes, NE of the Bergell may well be representative of the eroded hanging wall of the Bergell pluton. If this is the case, such dyke swarms may not be located directly above the primary source region of the magmas.

When restored to their Early Oligocene position, the magmas of the VVP come to lie in a very distant position (some 200 km south) with respect to the Periadriatic Fault (Figure 1.7). Most probably, the thrusts south of the Adamello Batholith did not exist at that time (Schönborn 1992), meaning that the southern deformation front of the Alps was located much further north, far from the VVP, thus suggesting that melting and ascent of these magmas was largely unrelated from Alpine subduction, unless the subduction angle was very flat.

1.5.2. *Subduction magmatism, slab breakoff and syn-collisional magmatism*

Given the long and well-documented subduction history of the Alpine orogen, it is often puzzling to realize that subduction magmatism, that is, Pre-Oligocene magmatism, is not very significant (see McCarthy et al. 2020), at least in terms of the preserved volumes of crystallized melts. However, several magmatic bodies described above were emplaced in the Eocene, at the same time of HP metamorphism of the Liguro-Piemont, Briançonnais and European units, which is bracketed between 46 and 34 Ma (e.g. Berger and Bousquet 2008). The older portion of the Alpine, calc-alkaline magmatism (~50 to 33 Ma) mainly consists of a limited number of dykes and stocks (Bergomi et al. 2015), and no large plutons, except for the southern Adamello. However, it should not be forgotten that the present-day erosion level of the Austroalpine and Southalpine nappes, where the latter dykes and stocks are located, lies below the original level of most Eocene–Oligocene volcanoes. In addition, volcanic clasts with sedimentary rocks are partly dated as Late Eocene to Early Oligocene, as documented in different flysch and molasse units (e.g. Garzanti and Malusa 2005; Malusa et al. 2011; Sharmann et al. 2017; Lu et al. 2018, 2019; Figure 1.4), point to the existence of Late Oligocene–Oligocene volcanic edifices in the Alpine Chain. Therefore, Eocene subduction-related magmatism in the Alps may have been more significant in terms of volumes of melt compared to its present-day distribution of plutons in map view.

The evidence for subduction magmatism also relies on the composition of magmatic rocks. The signature of subduction magmatism is preserved, for example, in the volcanic pebbles of the Taveyannaz sandstone (Reichenwallner 2019), which show the typical calc-alkaline evolution related to fractionation of primitive melts in a hydrated environment. Fluid-fluxed partial melting in suprasubduction mantle wedges, combined with crustal assimilation and fractional crystallization of the magmas produce calc-alkaline series, as commonly observed in the Alps, as summarized in section 1.3 and discussed in the following.

Subduction commonly leads to arc-magmatism. This process is related to changing temperature fields and to the input of fluids from the down-going, dehydrating slab into the overlying mantle wedge (e.g. Hacker et al. 2003a, 2003b; Best 2011; Zheng 2019). The input of aqueous fluids in the subduction system leads to partial melting of the mantle wedge, provided the temperature is sufficiently elevated (see sketched isotherms of the wedge in Figure 1.9). This temperature depends both on the thermal history of the subducting slab, its subduction angle (Figure 1.1(a)), and subduction velocity. Based on experimental results (Müntener et al. 2001), nearly pure fractional crystallization from wet mantle melts is proposed to explain parts of Adamello rocks (e.g. Hürlimann et al. 2016; Müntener et al. 2021) with its characteristic depletion of high field strength elements (e.g. Ti, Nb and Ta) and a typical Sr anomaly. Indeed, even the petrology and geochemistry of several dykes (from high-Mg basalts to andesites) in the Adamello area have been interpreted as a result of pure fractionation of a mantle melt (Hürlimann et al. 2016; Figure 1.9).

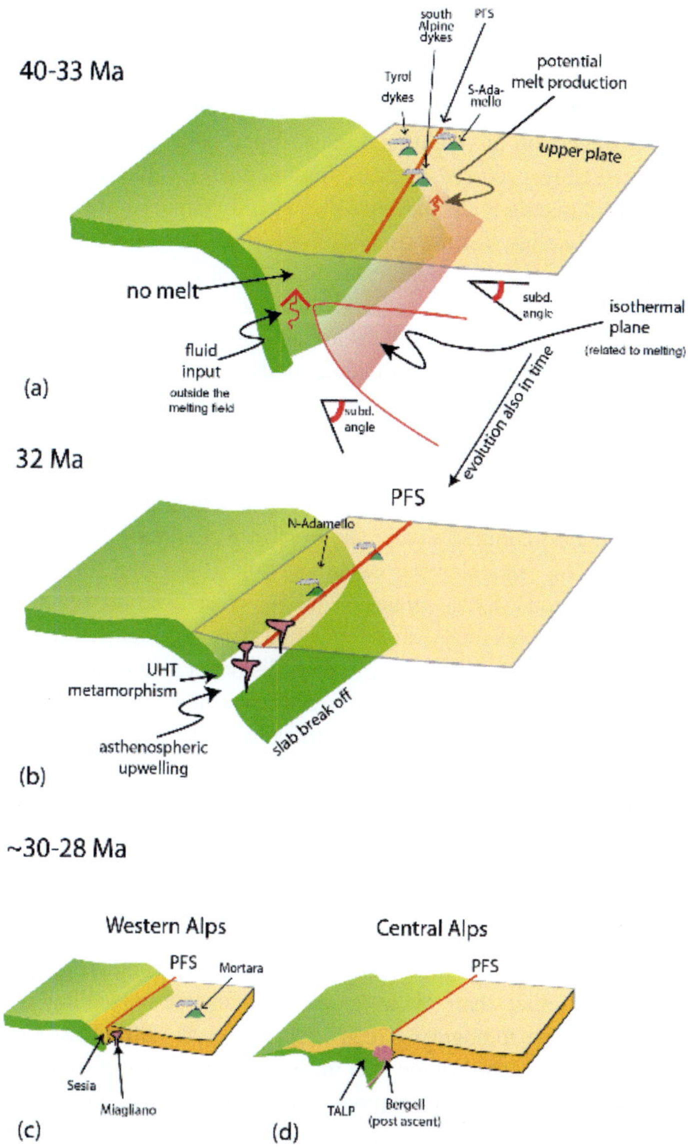

Figure 1.9. *Sketches for the evolution during subduction and collision: a) simplified sketch for subduction including a wide spread magmatism in the upper plate (Tyrol dykes, Southern Alpine dykes, etc.); b) slab breakoff including dominant emplacement in the direct surroundings of the Periadriatic Fault System (PFS); (c and d) two geometries at the collision stage: c) a situation in the Central Alps with strong uplift of the lower plate; d) a situation as in the Western Alps with still a subduction geometry*

However, the composition of the tonalities and granodiorites of the Periadriatic plutons have been also interpreted to result from the assimilation of large amounts of crustal melts (e.g. ~30%), based on O, Nd and Sr isotopes (e.g. Kagami et al. 1991; von Blanckenburg et al. 1992). Melting such large amounts of crustal material requires the existence of an additional heat input. During short-lived or cold subduction, as probably occurred in the Alps, magmatic rocks are inferred to be mainly derived from mantle melts, because these geodynamic settings create thermal conditions which only allow for the generation of low amounts of melt (e.g. 1 wt%) from the subducted crust (e.g. Gerya and Meilick 2011).

It is noteworthy that the Alpine plutons are almost exclusively bracketed between 32 and 30 Ma, hence immediately post-dating subduction, as constrained by the ages of HP metamorphism (Berger and Bousquet 2008). This time window corresponds to the onset of collision. For this and several other reasons discussed below, all these plutons (Periadriatic plutons) were interpreted to be generated by the upwelling of hot, asthenospheric mantle, due to slab breakoff (von Blanckenburg and Davies 1995; von Blanckenburg et al. 1998), initiated when the European continental margin entered the subduction zone. This model was based on the following observations: (1) geochemistry of igneous rocks indicates large crustal components in several Periadriatic plutons; (2) relatively short time interval of magmatism in the Alps, mainly coinciding with the beginning of collision; (3) linear, orogen-parallel distribution of the magmatic bodies, inferred to be the expression of a linear, orogen-parallel slab breakoff; (4) thermal modeling of subduction indicates that heating recorded in a subducting crust is not sufficient for generating large melt volumes. In this respect, the slab breakoff model has the advantage of explaining the high amount of crustal melting in spite of the limited time and amount of continental subduction that occurred in the Alpine Chain. In addition, the observed UHT metamorphic conditions attained at approximately 34 Ma in the down-going plate (Oalmann et al. 2019) are an independent evidence for the occurrence of an anomalous, large heat input in the crustal rocks of the downgoing plate, which is difficult to interpret other than by asthenospheric mantle upwelling due to slab breakoff. Finally, seismic tomography (e.g. Lippitsch et al. 2003; Kästle et al. 2020) indicates that the length of the Alpine slab in the Central Alps is compatible with the amount of syn-collisional shortening estimated in the lower plate, therefore suggesting that the imaged slab may consist entirely of the European Plate. As a result, the oceanic slab is not imaged by tomography, hence it must have broken off.

Furthermore, the transition from subduction to collision in the Alps may be set at 34–32 Ma, which is the age of the oldest sediments of the Molasse Basin. This age coincides with the UHT metamorphism (Oalman et al. 2019) and immediately precedes the most frequently determined age of emplacement of the Periadriatic plutons. Only after changing from subduction to collision geometry, the thickened European Plate underwent Barrovian type of metamorphism. The peak temperature of this metamorphism occurred at 20–25 Ma (Berger et al. 2011; Boston et al. 2017), that is, several million years after emplacement of all Periadriatic plutons, and the only magmatic bodies corresponding to this age window are those of the Novate stock and its dyke swarm. Hence, the thermal conditions attained during collision led to very minor partial melting, forming migmatites and very small, granitic peraluminous stocks (see section 1.3.8).

1.5.3. *Magmatism in the Sesia zone*

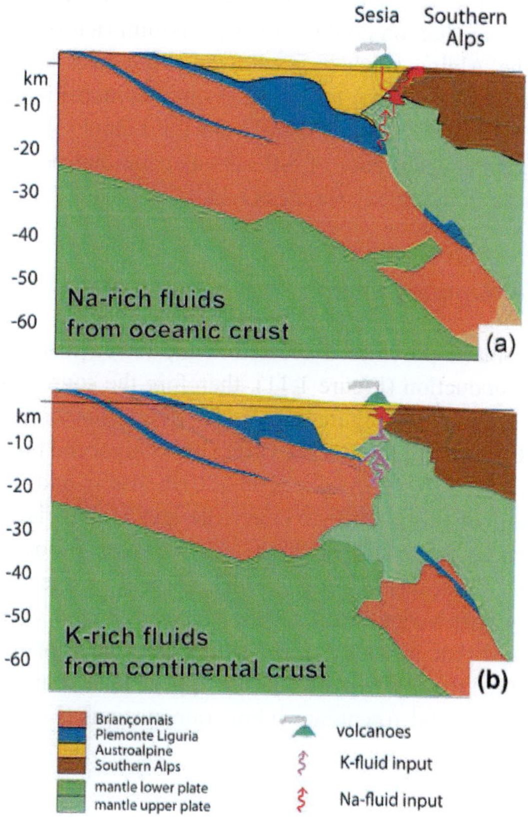

Figure 1.10. *Retrodeformed sketch profiles along the ECORS-CROP section (Schmid and Kissling 2000; Schmid et al. 2017) at ~32 Ma. The sketches show the geodynamic setting across the Periadriatic Line in the Western Alps and the role of the Na-dominated (a) and K-dominated (b) fluids (see text for explanation)*

Although contemporaneous to the other magmatic bodies, the ultrapotassic rocks of the Sesia zone indicate a compositional difference to the other plutonic and volcanic rocks of the Alps (Müntener et al. 2021). These ultrapotassic rocks intruded into the Sesia zone contemporaneously with calc-alkaline magmas (e.g. Traversella pluton, Biella volcanic suite). In order to understand this unusual association, the dynamics of the Sesia zone need to be considered. Its geodynamic evolution includes subduction, uplift and accretion of the Sesia zone on the upper plate, which is followed by subduction of the PLO and the Briançonnais. Subduction of the oceanic lithosphere (PLO) will deliver fluids as elsewhere of oceanic subduction in the Alps (e.g. sodium input; Figure 1.10a). In contrast to the central and eastern part of the Alps, the following subduction of the Briançonnais includes subduction of large proportions of sediments and continental basement units (e.g. Dora Maira, Gran Paradiso). This subduction could deliver potassium rich fluids (Figure 1.10b). The structure of the Alpine subduction allowed these two ingredients (Na- and K-fluid inputs in a mantle wedge) to induce distinct composition of melts (Figure 1.9; see also Campbell et al. 2014). The different fluid input (Na-rich versus K-rich fluids) would explain the occurrence of different associations, pointing to a different type of altered mantle as the source for these rocks.

1.5.4. *The Alps/Apennine junction and its magmatism*

The Esterél magmatic province is located in the lower plate and in an external part of the Alpine subduction (Figure 1.11), therefore the source of its melts is not related to Alpine subduction (Figure 1.11). However, another subduction system developed east of Corsica in Late Eocene or Earliest Oligocene times (e.g. Lacombe and Jolivet 2005; Molli 2008; Molli et al. 2010; Molli and Malavieille 2011; Chapter 6, Volume 1 of this book by Malavieille and Molli; Figure 1.11). It is the Apenninic subduction, associated with the early stages of development of the Apenninic-Ligurian accretionary wedge, which formed on the rear of the former Alpine accretionary retrowedge (Molli 2008; Malavieille et al. 2016; Schmid et al. 2017). This indicates that in Late Eocene/Oligocene times the European and the Adriatic Plates formed the lower plate and the upper plate, respectively, of the two opposite dipping Alpine and Apenninic subduction systems (Figure 1.11; see Molli and Malavieille 2011).

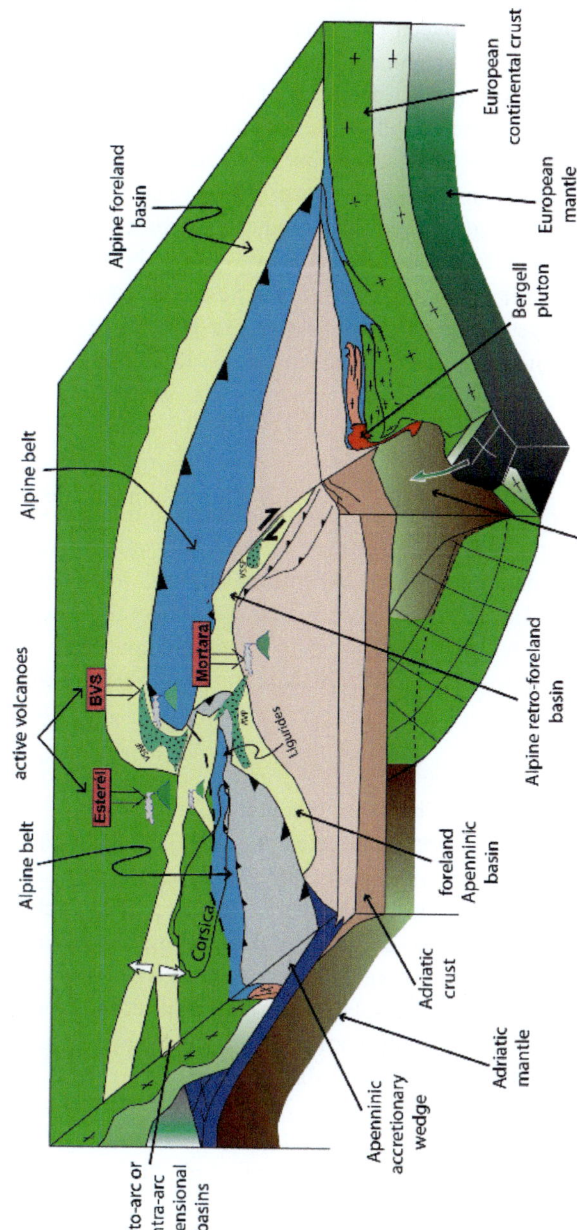

Figure 1.11. *3D sketch of the plate situation in Oligocene times, which include Alpine type subduction and Ligurian accretionary prism. AVP: Aveto Petrignacola; VSNF: Volcanic sediments of the northern foreland; VSSF: Volcanic sediments of the southern foreland; BVS: Biella volcanic suite*

The newly developed Apenninic subduction below the Sardegna-Corsica block includes the inherited configuration of the former Alpine subduction system (Handy et al. 2010; Schmid et al. 2017; see Chapter 6, Volume 1 of this book by Malavieille and Molli). Therefore, the Esterél magmatism may be connected with crustal stretching of a "proto"-arc basin at the inception of the Apenninic subduction (Carminati et al. 2012). The development of such a basin (and related magmatism) may be not exclusively controlled by subduction, but additionally influenced by the early stage of the Corsica-Sardinia rotation (Alvarez 1972). The combination of subduction and small-scale moving/rotating blocks induces asymmetric thinning of the lithosphere, resulting in basin formation and decompression melting. This setting requires very small-scale mantle dynamics between the Sesia zone and the Esterél area. Such small-scale changes were recently observed in relation to rotation of blocks of crustal scale in Papua New Guinea (e.g. Wallace et al. 2005).

1.5.5. *The Veneto volcanic province (VVP)*

Following recent interpretations, the VVP represents an intra-plate type of magmatism (e.g. Brombin et al. 2019 and literature therein). The volcanic rocks are derived from primitive magmas formed in a continental mantle, with minor CO_2 input (Brombin et al. 2019). The relatively close spatial and temporal connection to the Alpine subduction led some investigations to the interpretation of an Alpine, subduction-related magmatism (e.g. Dietrich 1976; Macera et al. 2008; Bartoli et al. 2013).

In the retrodeformation presented in section 1.4 (Figure 1.7), the location of the VVP is stable relatively to the Alpine Chain throughout the Cenozoic, because it lies in the footwall of the Alpine collisional S-directed thrusting. Indeed, this area is inferred to have remained coupled and not offset from its underlying lower crust, in contrast to the rest of the Southern Alps, further north and west (e.g. Schönborn 1992, 1999). However, this area and the entire Adriatic Plate moved northward in the time interval of the VVP magmatism (e.g. Handy et al. 2010; Thorsvik et al. 2012; van Hinsbergen et al. 2020). This displacement is in conflict to the evidence of a protracted magmatic activity, spatially confined to a small upper crustal area (VVP), supposed to be a continuously connected stable source.

Present-day Vp velocities indicate a large negative anomaly in the lithosphere between the Alps and the Adriatic Sea (e.g. Giacomuzzi et al. 2011; Qorbani et al. 2015), which is related either to partial melting of the mantle, hydration of the mantle or changes in the temperature field. Irrespective of such interpretations, the VVP lies along the NW boundary of a large transitional structure of the mantle (Giacomuzzi et al. 2011). The negative Vp velocities in all tomographic models are much larger in spatial extent as the outcropping area of the VVP and these seismic models only visualize the present-day situation.

The geodynamic causes for the formation of these intraplate melts of the VVP remain largely unconstrained. In contrast, the ascent of these melts may well be related to large-scale strike-slip zones in the area as the Proto-Giudicarie Line. Such a precursor has been inferred in order to explain sinistrally offset Cretaceous to Eocene flysch basins, indicating local extensional tectonics in the Paleocene (Castellarin et al. 2006; Castellarin 2009). This interpretation is also required to maintain kinematic compatibility between the Pre-Adamello, N–S shortening that is restricted to the area west of the Giudicarie Fault (Schönborn 1992; Castellarin et al. 2006). As shown by the different basins and the graben structures (Zampieri 1995; Castellarin et al. 2006; Castellarin 2009), the detritic, VVP-bearing sediments were deposited within transtensional basins. The geodynamic connection between this Eocene strike-slip deformation and the contemporaneous Alpine subduction is not clear yet. However, analogous areas characterized by small volumes of alkaline melts ascending along crustal-scale strike-slip zones in the upper plate of a subduction system have been proposed to exist elsewhere (Lustrino and Sharkov 2006; Faccenna et al. 2010).

1.6. Summary and conclusions

Based on their age, chemistry and spatial distribution, we related the generation of the Alpine magmatic bodies to the following geodynamic events: (1 + 2) magmatism related to Alpine subduction followed by slab breakoff; (3) magmatism related to extensional process accommodated along the lateral terminations of the Alps; (4) intraplate magmatism. Syn-collisional magmatism is nearly absent.

1) Subduction of different tectonic units of both continental and oceanic affinity resulted in the generation of melts that were emplaced in the upper plate in the Middle and Late Eocene (Bergomi et al. 2015).

2) Numerous arguments indicate that the Periadriatic plutons were generated by slab breakoff at the transition from subduction to collision: the occurrence of UHT metamorphic rocks in the European basement nappes, with an age (ca. 34 Ma) barely older than that of the Periadriatic plutons; seismic tomography showing that the depth of the subducted slab is not significantly longer than the inferred amount of collisional shortening of the lower plate; the very internal position of the plutons with respect to the Alpine Chain; and the inferred large volume proportion of crustal melts required to generate the latter magmas.

3) The third type of magmatism is connected with the extensional dynamics of plates in the marginal areas of the Alpine Chain. They are related to local changes in the geodynamic setting (Alpine/Apennine junction and Alps/Dinarides/Pannonian junction), which influence (or are caused by) mantle dynamics, which in turn control the magmatic processes. This implies that the magmatic products have different geodynamics causes. For example, the triple junction of Dinarides/Alps/Pannonian Basin may include older subduction-related melts and younger ones due to lithospheric thinning related to the opening of the Pannonian Basin. The Esterél magmatism may be related to extension related to Apenninic subduction (Forearc Basin), but it is contemperanuous with Alpine subduction related magmatism. Such local geodynamic models require a spatial understanding of the mantle structure and kinematics on the small scale, including rotational movements (e.g. Wallace et al. 2005).

4) Intraplate type of magmatism (VVP) only occurred in the Veneto province, which is located far from the Alpine front. However, its position may coincide with that of an inferred lithospheric-scale strike-slip zone, preceding the formation of the Giudicarie Fault System. Hence, intraplate magmatism may be channeled by a steep strike-slip zone at high angle to the axis of the chain, connecting the mantle with volcanoes at the topographic surface.

1.7. Appendix 1: Explanatory notes to the retro-deformation of the magmatic bodies

Dated dykes west of Adamello (D'Adda et al. 2011; Bergomi et al. 2015): the dated dykes of the Southern Alps have been displaced northward by 66 km to account for collisional shortening that affected their enclosing rocks following Schönborn (1992).

Dated dykes south-east of Adamello: the dated dykes of the Southern Alps have been displaced northward by 58 km. In this area, most part of the collisional shortening is accommodated by the Milan Belt (66 km following Schönborn (1992)).

Valle del Cervo, Traversella: these plutons lie almost in contact with the Periadriatic Fault, whose displacement has been mainly west-side up along steeply inclined planes. Therefore, if they should be displaced at all, it is only by very small amounts westward.

Miagliano: this pluton was not displaced by the retro-deformation, because its enclosing rocks in the Western Alps were not affected by shortening during Alpine collision.

Bergell: this pluton was displaced westward along the (retro-deformed) trace of the Insubric Line, to account for 100 km of inferred dextral strike-slip during Alpine collision. We are conscious of the fact that the absolute offset by dextral strike-slip along the Insubric Line is poorly constrained, varying between some 150 km (Handy et al. 2014) and 50 km (Mancktelow et al. 2001). We took a value of 100 km which seems to be a good average between the existing interpretations.

Adamello: the Adamello batholith was displaced northward by 87 km, following the estimates of Post-Adamello shortening by Schönborn (1992).

Tonalitic Lamellae: these magmatic bodies crop out along the Giudicarie Fault, a structure that is generally inferred to form contemporaneously (Schneider et al. 2015), or significantly later compared to the age of these magmatic bodies (Schmid et al. 2013). Therefore, the map of Figure 1.7, which depicts the early collisional stage of the Alps, does not show the Giudicarie Fault. Hence, the tonalitic lamellae are located between the Rensen (in the east) and the Adamello (in the west). A more precise relocation cannot be performed. The idea behind this is that these bodies have been sheared and transported along the Giudicarie Fault from their original site of emplacement.

Southwestern Tyrol dykes: these dykes are in the intermediate area between E-Alpine domain and South-Alpine domain. In the retrodeformed map, these dykes are in the westernmost part of the E-Alpine area, north of the Southalpine magmatic domain, thus showing that some dextral displacement took place before indentation along the Giudicarie Fault System.

Rensen and Rieserferner: these plutons are emplaced north of and adjacent to the Defereggen-Antholz-Vals (DAV) Fault in the Eastern Alps. The Fault is inferred to be sinistral transpressive (Wagner et al. 2006); however, no marker exists to allow for a quantification of this sinistral offset. We displaced these magmatic bodies by 20 km eastward to show their lateral displacement along the DAV, but we are aware of the fact that the absolute value of this displacement is not constrained.

Reifnitz: this pluton was emplaced along the eastern part of the Periadriatic Fault (the Gailtal Fault). The amount of dextral displacement along the latter Fault is not known. If the pluton was affected by the entire amount of orogen-parallel extension inferred for the Eastern Alps (Favaro et al. 2017; Rosenberg et al. 2018), then 75 km of westward displacement are needed to relocate the Refnitz pluton to its original position at the onset of collision.

Pohorje: this pluton was emplaced adjacent to the dextral Lavantal Fault in the Eastern Alps. Based on the retro-deformation of orogen-parallel extension in the Eastern Alps (Favaro et al. 2017; Rosenberg et al. 2018), it is moved westward by 75 km and 12 km northwestward to account for dextral displacements along the Lavantal Fault.

VVP: the VVP were not displaced by the retro-deformation because they are in the footwall of all Alpine and Apenninic structures.

Mortara Volcano: this volcano was not displaced by the retro-deformation, because it is in the footwall of all Alpine and Apenninic structures.

Western Alpine area: this area, exposed on both sides of the Canavese Fault in the Western Alps, was barely displaced during retro-deformation. Its South-Alpine area is not inferred to have been shortened during collision (e.g. Pieri and Groppi 1981) and the area immediately west of the Canavese Fault was not affected by significant collisional displacement either.

Central Alpine area: this magmatic domain has been translated westward to counterbalance collisional, dextral strike-slip along the Insubric Line. The southwestern part of the area has been slightly extended northwards to account for collisional shortening.

Southern Alpine area between Lake Garda and Lake Como: this magmatic domain has been displaced northward by 87 km in the east and 56 in the west based on the amounts of shortening calculated by Schönborn (1992). In addition to this displacement, the surface of the entire domain has been increased in N–S direction to account for its collisional shortening.

Eastern Alpine area: this magmatic domain has been shortened in E–W direction and extended northward in its western part to account for, respectively, orogen-parallel extension, and significant N–S shortening in its westernmost part.

Group	*L11*	Name	Rock type	System	Method[2]	Age	Number of individual analysis[1]	Individual ~analytical error	References	Remarks
A		Biella Volcanic suite	Andesite and others	Zir U-Pb	SG TIMS	32.7–31.7	9 g	0.05	Kapferer et al. 2012	Also called "Volcanic cover of Sesia zone"
A	*2*	Travesella (N)	Tonalite, granodiorite	Zir U-Pb	LA-ICPMS	33	3 s	0.3	Ji et al. 2019	
A		Travesella (S)		Zir U-Pb	LA-ICPMS	42–41	3 s	0.3	Ji et al. 2019	
A		Miaglano	Tonalite	Zir U-Pb	SG TIMS	33	1 s	0.04	Berger et al. 2012	
A		Insubric dykes	Prophyric granodiorite	Tit U-Pb	TIMS	32	1 s		Romer et al. 1996	
A	*5*	Bergell	Tonalite, granodiorite	Zir	SG TIMS	32–28	134 g	0.02	Samperton et al. 2015	See also: von Blanckenbrug 1990; Oberli et al. 2004; Gregory et al. 2009
A	*6*	Adamello	Tonalite, granodiorite	Zir U-Pb	SG TIMS	42–32			Schaltegger et al. 2019	Schoene et al. 2012; Schaltegger et al. 2009

Group	*L11*	Name	Rock type	System	Method[2]	Age	Number of individual analysis[1]	Individual ~analytical error	References	Remarks
A	*9*	Giudicarie lamallae	Tonalite	Zir U-Pb	TIMS	35–32	5 s	1	Pomella et al. 2011	
A	*11*	Rieserferner	Tonalite, granodiorite	Aln U-Th-Pb	TIMS	32.2–31.8			Romer and Siegesmund 2003	
A	*10*	Rensen	Tonalite, granodiorite	Aln Th-Pb	TIMS	31.1	1s	0.3	Barth et al. 1989	
A		Karawanken	Tonalite, granodiorite	Bt Rb/Sr		**28, 29 ± 6**		5	Scharbert 1975	
B	*1*	Sesia dykes	Lamproite, lamprophyres	K-Ar		31.6		1.3	Dal Piaz et al. 1973	
B	*3*	Valle del Cervo pluton	Syenite, monzonite	Zir U-Pb	SG TIMS	30.4		0.05	Berger et al. 2012	
C		Southern Alpine dykes	Andesites, basalts	Zir U-Pb		42–34	9 s	1	D'Adda et al. 2011; Bergomi et al. 2014	
C				K/Ar	WR and amph	64–38	20 s	2	Fantoni et al. 1999; Zanchi et al. 1990	
D	7	Tyrol dykes	Andesites, rhyolites	Amph K-Ar		24–31	20 s	0.5	Deutsch 1984	

Group	*L11*	Name	Rock type	System	Method[2]	Age	Number of individual analysis[1]	Individual ~analytical error	References	Remarks
D				Bt Rb/Sr		31–32	5 s	0.8	Dal Piaz et al. 1988	
D			Tonalite	Zir U/Pb		30.7	1 s	0.3	Neubauer et al. 2018	
E		Esterél magmatic prov.		Ar/Ar		41–20			Lustrino et al. 2017	
F	*15*	Veneto volcanic prov.	Nephelinite, basanite,	Zir U-Pb		52–30	3	0.3	Visonà et al. 2007	
F			Variable volcanics	K-Ar Bt/Wr		42–32	12	1	Borsi et al. 1969	
G	*13*	Pohorje	Tonalite, granodiorite	Zir U-Pb	TIMS	18.64	1s	0.11	Fodor et al. 2008	
G		Steyr Basin	Basalts, etc.	K/Ar		16.0–20.0			Seghedi and Downes 2011	
H	*4*	Novate	Two-mica granite	Zir U-Pb	SIMS	25	1	0.5	Liati et al. 2000	
I	*14*	Mortara	Tonalite, granodiorite	Sediments		28	-	-	Ruffini 1995	
J		Taveyannaz	Mainly andesite clasts	Amph K-Ar Ar/Ar		32			Ruffini et al. 1997; Fischer and Villa 1990	

Group	*L11*	Name	Rock type	System	Method[2]	Age	Number of individual analysis[1]	Individual ~analytical error	References	Remarks
J		Taveyannaz Glarus		Zir U-Pb	LA-ICPMS	41–29	sev.100	n.g.	Lu et al. 2018	
J		Champsau sandstone		Ar-Ar		34–33			Boyet et al. 2001	
J		Eastern Alps Molasse	Andesites-dacites	WR K-Ar		40.5–29.5			Brügel et al. 2000	
J		Eastern Alps Molasse		Zir U-Pb		39–30			Sharmann et al. 2017	
K		= Trento Basin		Zir U-Pb	LA-ICPMS	48.2–33	sev.100	n.g.	Lu et al. 2019	
K		Gonfolite Lombarda		Zir U-Pb	LA-ICPMS	34.5–28			Malusà et al. 2011	

Column "*L11*" shows the numbers used in Lustrino et al. 2011; n.g. = not given.

[1]S: sample; g: grain.

[2]LA-ICPMS: laser ablation ICPMS; SG TIMS: single grain thermal ionistation mass spectrometry; SIMS: secondary ion mass spectrometry; Ar and Sr data are all mineral separates measured with mass spectrometry.

Table 1.1. *Updated list of magmatic rocks in the Alpine larger area (see also Lustrino et al. 2011; Bergomi et al. 2014; Ji et al. 2019). Selected ages are given, which are interpreted to be the most robust, and only selected references are given*

1.8. References

Agard, P. (2021). Subduction of oceanic lithosphere in the Alps: Selective and archetypal from (slow-spreading) oceans. *Earth-Science Reviews*, 214, 103517.

Agard, P. (2023) Oceanic Subduction in the Alps: Lessons from the Rock Record. In *Geodynamics of the Alps 2*, Rosenberg, C.L. and Bellahsen, N. (eds). ISTE Ltd, London, and John Wiley & Sons, New York.

Ahrendt, H. (1969). Tertiärer Vulkanismus in der Canavese-Zone? *Neues Jahr. Geol. Paläont., Mitt.*, 9, 513–516.

Alagna, K.E., Peccerillo, A., Martin, S. (2010). Tertiary to present evolution of Orogenic magmatism, in Italy. *Journal of the Virtual Explorer – The Geology of Italy: Tectonics and Life Along Plate Margins*, 36(18), 1–63. doi: 10.3809/jvirtex.2010.00233.

Altherr, R., Lugovic, B., Meyer, H.P., Majer, V. (1995). Early Miocene post-collisional calc-alkaline magmatism along the easternmost segment of the Periadriatic fault system (Slovenia and Croatia). *Mineral. Petrol.*, 54, 225–247.

Alvarez, W. (1972). Rotation of the Corsica–Sardinia microplate. *Nature Physical Science*, 235, 103–105.

Andreatta, C. (1953). Syntektonische und posttektonische magmatische Erscheinungen der Ortlergruppe in Beziehung zum alpinen Magmatismus. *Tschermaks mineralogische und petrographische Mitteilungen*, 3, 93–114.

Anfinson, O.A., Malusa, M.G., Ottria, G., Dafov, L.D., Stockli, D.F. (2016). Tracking coarse-grained gravity flows by LASS-ICP-MS depth-profiling of detrital zircon (Aveto Formation, Adriatic foredeep, Italy). *Marine and Petroleum Geology*, 77, 1163–1176.

Barbieri, G., De Zanche, V., Medizza, F., Sedea, R. (1981). Considerazioni sul vulcanismo terziario del Veneto occidentale e del Trentino meridionale. *Rend. Soc. Geol. It.*, 4, 267–270.

Barth, S., Oberli, F., Meier, M. (1989). U-Th-Pb systematics of morphologically characterized zircon and allanite: A high resolution isotopic study of the Alpine Rensen pluton (northern Italy). *Earth Planet. Sci. Lett.*, 95, 235–254.

Bartoli, O., Meli, S., Sassi, R., Magaraci, D. (2013). Amphiboles and clinopyroxenes from Euganean (NE Italy) cumulus enclaves: Evidence of subduction-related melts below Adria microplate. *Rendiconti Lincei. Scienze Fisiche e Naturali*, 24, 151–161.

Beccaluva, L., Bigioggero, B., Chiesa, S., Colombo, A., Fanti, G., Gatto, G.O., Gregnanin, A., Montrasio, A., Piccirillo, E.M., Tunesi, A. (1983). Post collisional orogenic dyke magmatism in the Alps. *Mem. Soc. Geol. Ital.*, 26, 341–359.

Beccaluva, L., Bonadiman, C., Coltorti, M., Salvini, L., Siena, F. (2001). Depletion events, nature of metasomatizing agent and timing of enrichment processes in lithospheric mantle xenoliths from the Veneto Volcanic Province. *J. Petrol.*, 42, 173–187.

Bellieni, G., Cavazzini, G., Fioretti, A.M., Peccerillo, A., Poli, G. (1991). Geochemical and isotopic evidence for crystal fractionation, AFC and crustal anatexis in the genesis of the Rensen Plutonic Complex (Eastern Alps, Italy). *Contrib. Mineral. Petrol.*, 92, 21–43.

Berger, A. and Bousquet, R. (2008). Subduction-related metamorphism in the Alps: Review of isotopic ages based on petrology and their geodynamic consequences. In *Tectonic Aspects of the Alpine-Dinaride-Carpathian System*, Siegesmund, S., Fügenschuh, B., Froitzheim, N. (eds). Geological Society, Special Publications, London.

Berger, A., Rosenberg, C.L., Schmid, S.M. (1996). Ascent, emplacement and exhumation of the Bergell pluton within the Southern Steep Belt of the Central Alps. *Schweiz. Mineral. Petrogr. Mitt.*, 76, 357–382.

Berger, A., Burri. T., Alt-Epping, P., Engi, M. (2008). Tectonically controlled fluid flow and water assisted melting in the middle crust: An example from the Central Alps. *Lithos*, 102, 598–615.

Berger, A., Thomsen, T., Ovtcharova, M., Mercolli, I., Kapferer, N., Fügenschuh, B. (2012a). Dating emplacement and evolution of the orogenic magmatism in the internal Western Alps: 1. *The Miagliano Pluton. Swiss J. Geos.*, 105, 49–65.

Berger, A., Mercolli, I., Kapferer, N., Fügenschuh, B. (2012b). Single versus double exhumation in the internal Western Alps: A key to interpret near-surface Oligocene tectonics. *Int. J. Earth Sci.*, 101, 1877–1894.

Bergomi, M.A., Zanchetta, S., Tunesi, A. (2014). The Tertiary dike magmatism in the Southern Alps: Geochronological data and geodynamic significance. *Int. J. Earth Sci.*, 104, 449–473.

Best, M.G. (2011). *Igneous and Metamorphic Petrology*, 2nd edition. Blackwell Publishing. Oxford.

Bianchi, A. and Dal Piaz, G.B. (1962). Gli inclusi di "micascisti eclogitici" nelle porfiriti permiane della zona del Canavese fra Biella ed Oropa. *Rendiconti della Società Mineralogica Italiana*, 18, 299–299.

Bigioggero, B., Colombo, A., DelMoro, A., Gregnanin, A., Macera, P., Tunesi, A. (1994). The Oligocene Valle del Cervo Pluton: An example of shoshoitic magmatism in the Western Italian Alps. *Mem. Soc. Geol. Ital.*, 46, 409–421.

von Blanckenburg, F. (1992). Combined high-precision chronometry and geochemical tracing using accessory minerals: Applied to the Central-Alpine Bergell intrusion (central Europe). *Chem. Geol.*, 100, 19–40.

von Blanckenburg, F. and Davies, J.H. (1995). Slab breakoff: A model for syncollisional magmatism and tectonics in the Alps. *Tectonics*, 14, 120–131.

von Blanckenburg, F., Früh-Green, G., Diethelm, K., Stille, P. (1992). Nd-, Sr-, O-isotopic and chemical evidence for a two-stage contamination history of mantle magma in the Central-Alpine Bergell intrusion. *Contrib. Mineral. Petrol.*, 110, 33–45.

von Blanckenburg, F., Kagami H., Deutsch A., Oberli F., Meier M., Wiedenbeck M., Barth S., Fischer, H. (1998). The origin of Alpine plutons along the Periadriatic Lineament. Schweiz. *Mineral. Petrogr. Mitt.*, 78, 55–66.

Blundy, J.D. and Sparks, R.S.J. (1992). Petrogenesis of mafic inclusions in granitoids of the Adamello massif, Italy. *J. Petrol.*, 33, 1039–1104.

Boyet, M., Lapierre, H., Tardy, M., Bosch, D., Maury, R. (2001). Nature des sources des composants andésitiques des Grès du Champsaur et des Grès de Taveyannaz; implications dans l'évolution des Alpes occidentales au Paleogene. *Bull. Soc. Géol. France*, 172, 487–501.

Brombin, V., Bonadiman, C., Coltorti, M., Fahnestock, M.F., Bryce, J.G., Marzoli, A. (2018). Refertilized mantle keel below the Southern Alps domain (North-East Italy): Evidence from Marosticano refractory mantle peridotites. *Lithos*, 300–301, 72–85.

Brombin, V., Bonadiman, C., Jourdan, F., Roghi, G., Coltorti, M., Webb, L., Callegaro, S., Bellieni, G., De Vecchi, G., Sedea, R. et al. (2019). Intraplate magmatism at a convergent plate boundary: The case of the Cenozoic northern Adria magmatism. *Earth Science Reviews*, 192, 355–378.

Brügel, A., Dunkl, I., Frisch, W., Kuhlemann, J., Balogh, K. (2000). The record of Periadriatic volcanism in the Eastern Alpine Molasse zone and its paleogeographic implications. *Terra Nova*, 12, 42–47.

Burri, T., Berger, A., Engi, M. (2005). Tertiary migmatites in the Central Alps: Regional distribution, field relations, conditions of formation, and tectonic implications. *Schweiz. Mineral. Petrogr. Mitt.*, 85, 215–232.

Callegari, E. and Dal Piaz, G.B. (1973). Field relationships between the main igneous masses of the Adamello intrusive massif, nothern Italy. *Mem. Ist. Geol. Mineral. Univ. Padova*, 29, 1–38.

Callegari, E., Cigoli, C., Medeot, O., D'Antonio, M. (2004). Petrogenesis of calc-alkaline and shoshonitic post-collisional Oligocene volcanics of the Cover-Serie of the Sesia zone, western Italian Alps. *Geodynamica Acta*, 17, 1–29.

Campbell, I.H., Stepanov, A., Liang, H.-Y., Allen, C.M., Norman, M., Zhang, Y.-Q., Xie, Y.-W. (2014). The origin of shoshonites: New insights from the Tertiary high-potassium intrusions of eastern Tibet. *Contrib. Mineral. Petrol.*, 167, 983.

Cannatelli, C. (2012). Multi-stage metasomatism in the lithosphere beneath the Veneto Volcanic Province (VVP, Northern Italy). *Miner. Petrol.*, 104, 177–195.

Carminati, E., Lustrino, M., Doglioni, C. (2012). Geodynamic evolution of the central and western Mediterranean: Tectonics vs. igneous petrology constraints. *Tectonophysics*, 579, 173–192.

Carraro, F. and Ferrara, G. (1968). Alpine "tonalite" at Miagliano, Valle del Cervo (Zona dioritica–kinzigitica): A preliminary note. *Schweiz. Mineral. Petrogr. Mitt.*, 48, 75–80.

Castellarin, A. (2009). Basic stratigraphy and tectonics of the Southern Alps around the Giudicarie Lineament (Southern Alps, Italy). *Italian Journal of Geosciences*, 128, 409–417.

Casterallin, A., Vai, G.B., Cantello, L. (2006). The Alpine evolution of the Southern Alps around the Giudicarie faults: A Late Cretaceous to Early Eocene transfer zone. *Tectonophysics*, 414, 203–223.

Cerrina Feroni, A., Martinelli, P., Perilli, N. (1991). Stratigrafia e struttura dell'Unità di Canetolo in Val Cedra (Appennino Parmense). *Mem. Descr. Carta Geolog. d'Italia*, 46, 301–312.

Cesana, A., Demonte, L., Venegoni, M. (1976). Origin of the Cervo pluton according to the U and Th content and distribution. *The Miagliano Pluton Rendiconti Societa Italiana Mineralogia Petrologica*, 32, 637–646.

Cesare, B. (1992). Metamorfismo di contatto di rocce pelitiche nell'aureola di Vedrette di Ries (Alpi Orientali Italia). *Atti Ticinesi Sci. Terra*, 35, 1–7.

Cibin, U., Tateo, F., Catanzariti, R., Martelli, L., Rio, D. (1998). Composizione, origine ed età del vulcanismo andesitico oligocenico inferiore dell'Appennino settentrionale: le intercalazioni vulcano-derivate nella Formazione di Ranzano. *Bollettino della Società Geologica d'Italia*, 117, 569–591.

Cibin, U., Spadafora, E., Zuffa, G.G., Castellarin, A. (2001). Continental collision history from arenites of episutural basins in the Northern Apennines, Italy. *GSA-Bulletin*, 113, 4–19.

Cibin, U., Di Giulio, A., Martelli, L. (2003). Oligocene–Early Miocene tectonic evolution of the northern Apennines (northwestern Italy) traced through provenance of piggyback basin fill successions. In *Tracing Tectonic Deformation Using the Sedimentary Record*, McCann, T. and Saintot, A. (eds). Geological Society, Special Publication, London.

Conticelli, S., Guarnieri, L., Farinelli, A., Mattei, M., Avanzinelli, R., Bianchini, G., Boari, E., Tommasini, S., Tiepolo, M., Prelević, D. et al. (2009). Trace elements and Sr–Nd–Pb isotopes of potassic, shoshonitic, and calc-alkaline magmatism of the Western Mediterranean region: Genesis of ultrapotassic to calc-alkaline magmatic associations in a post-collisional geodynamic setting. *Lithos*, 107, 68–92.

Cornelius, H.P. (1915). Geologische Beobachtungen in den italienischen Teilen des Albigna-Disgraziamassivs. *Geol. Rundsch.*, 6, 166–177.

D'Adda, P., Zanchi, A., Bergomi, M., Berra, F., Malusà, M.G., Tunesi, A., Zanchetta, S. (2011). Polyphase thrusting and dyke emplacement in the central Southern Alps (Northern Italy). *Int. J. Earth Sci.*, 100, 1095–1113.

Dal Piaz, G.B. (1926). Il confine alpino-dinarico dall'Adamello al massiccio di Monte Croce nell'Alto Adige. *Atti Accad. Sci. Veneto-Trentino-Istriana*, 17, 3–7.

Dal Piaz, G.V. (1934). Studi geologici sull'Alto Adige orientale e regioni limitrofe. *Mem. Ist. Geol. Univ. Padova*, 10, 238.

Dal Piaz, G.V. and Venturelli, G. (1983). Brevi riflessioni sul magmatismo post-ofiolitico nel quadro dell'evoluzione spazio-temporale delle Alpi. *Mem. Soc. Geol. Ital.*, 26, 5–19.

Dal Piaz, G.V., Venturelli, G., Scolari, A. (1979). Calc-alkaline to ultrapotassic postcollisional volcanic activity in the internal northwestern Alps. *Memorie della Società Geologica Italiana*, 32, 1–16.

Dal Piaz, G.V., Del Moro, A., Martin, S., Venturelli, G. (1988). Post-collisional magmatism in the Ortler–Cevedale Massif (Northern Italy). *Jahrb. Geol. Bund.*, 131, 533–551.

Davidson, C., Rosenberg, C.L., Schmid, S.M. (1996). Symmagmatic folding of the base of the Bergell pluton, Central Alps. *Tectonophysics*, 265, 213–238.

Davies, J.H. and von Blanckenburg, F. (1995). Slab breakoff: A model of lithosphere detachment and its test in the magmatism and deformation of collisional orogens. *Earth Planet. Sci. Lett.*, 129, 85–102.

Decarlis, A., Beltrando, M., Manatschal, G., Ferrando, S., Carosi, R. (2017). Architecture of the distal Piedmont-Ligurian rifted margin in NW Italy: Hints for a flip of the rift system polarity. *Tectonics*, 36, 2388–2406.

Del Moro, A., Pardini, G.C., Quercoli, C., Villa, I.M., Callegari, I.E. (1985). Rb/Sr and K/Ar chronology of Adamello granitoids, Southern Alps. *Mem. Soc. Geol. Ital.*, 26, 285–299.

Deutsch, A. (1984). Young Alpine dykes south of the Tauern Window (Austria): A K–Ar and Sr isotope study. *Contrib. Mineral. Petrol.*, 85, 45–57.

Di Capua, A., Groppelli, G., Vezzoli, G. (2014). Volcanism, relative sealevel change, and the stratigraphic record: An Oligocene example. In *STRATI 2013. At the Cutting Edge of Stratigraphy*, Rocha, R., Pais, J., Kullberg, J.C., Finney, S. (eds). Springer International Publishing, Cham.

Dietrich, V.J. (1976). Evolution of the Eastern Alps: A plate tectonics working hypothesis. *Geology*, 4, 147–152.

Drescher-Kaden, F.K. and Storz, M. (1926). Ergebnisse petrographischtektonischer Untersuchungen im Bergeller Granit. *Neues Jahrb. Mineral., Abh.*, 54, 284–291.

Elter, P., Gratziu, C., Labesse, B. (1964). Sul significato dell'esistenza di una tettonica alloctona costituita da formazioni terziarie nell'Appennino Settentrionale. *Bollettino della Societa Geologica Italiana*, 83, 373–394.

Elter, P., Catanzariti, R., Ghiselli, F., Marroni, M., Molli, G., Ottria, G., Pandolfi, L. (1999). L'Unità Aveto (Apennino Settentrionale): caratteristiche litostratografiche, biostratigrafia, petrografia delle arenite ed assetto strutturale. *Bollettino Società Geologica Italiana*, 118, 41–63.

Exner, C. (1971). Geologie der Karawankenplutone östlich Eisenkappel Kärnten. *Mitt. Geol. Ges. Wien*, 64, 1–108.

Exner, C. (1976). Die geologische Position der Magmatite des Periadriatischen Lineaments. *Verh. Geol. Bundesanst. Austria*, 2, 3–64.

Faccenna, C., Becker, T.W., Lallemand, S., Lagabrielle, Y., Funiciello, F., Piromallo, C. (2010). Subduction-triggered magmatic pulses: A new class of plumes? *Earth Planet. Sci. Lett.*, 299, 54–68.

Faninger, E. (1970). Tonalit von Pohorje und seine Differentiate. *Geologija*, 13, 90–104.

Fantoni, R., Bersezio, R., Forcella, F., Gorla, L., Mosconi, A., Picotti, V. (1999). New dating of the Tertiary Magmatic products of the central Southern Alps, bearings on the interpretation of the Alpine tectonic history. *Mem. Sci. Geol.*, 51, 47–61.

Favaro, S., Handy, M.R., Scharf, A., Schuster, R. (2017). Changing patterns of exhumation and denudation in front of an advancing crustal indenter, Tauern Window (Eastern Alps). *Tectonics*, 36, 1053–1071.

Féraud, G., Ruffet, G., Stephan, J.F., Lapierre, H., Delgado, E., Popoff, M. (1995). Nouvelles données géochronologiques sur le volcanisme paléogène des Alpes occidentales : existence d'un événement magmatique bref généralisé. *Magmatisme dans le sud-est de la France*, 38.

Fodor, L.I., Gerdes, A., Dunkl, I., Koroknai, B., Pécskay, Z., Trajanova, M., Horváth, P., Vrabec, M., Jelen, B., Balogh, K., Frisch, W. (2008). Miocene emplacement and rapid cooling of the Pohorje pluton at the Alpine-Pannonian-Dinaridic junction, Slovenia. *Swiss J. Geosci.*, 101(Supplement 1), S255–S271.

Garzanti, E. and Malusà, M.G. (2008). The Oligocene Alps: Domal unroofing and drainage development during early orogenic growth. *Earth Planet. Sci. Lett.*, 268, 487–500.

Gasperini, D., Bosch, D., Braga, R., Bondi, M., Macera, P., Morten, L. (2006). Petrology and Sr–Nd–Pb–O isotope composition of mantle xenoliths from the Veneto Volcanic Province (North-eastern Italy). *Geochem. J.*, 40, 377–404.

Gatto, G.O., Gregnanin, A., Molin, G.M., Piccirillo, E.M., Scolari, A. (1976). Le manifestazioni "andesitiche" polifasiche dell'Alto Adige occidentale nel quadro geodinamico alpino. *Studi Trentini di Scienze Naturali*, 53, 21–47.

Gerya, T.V. and Meilick, F.I. (2011). Geodynamic regimes of subduction under an active margin: Effects of rheological weakening by fluids and melts. *J. Metam. Geol.*, 29, 7–31.

Ghibaudo, G., Massari, F., Chiambretti, I. (2014). Oligo-Miocene tectono-sedimentary evolution of the Langhe Sub-basin: From continental to basinal setting (Tertiary Piedmont Basin – Northwestern Italy). *Journal of Mediterranean Earth Sciences*, 6, 53–144.

Giacomuzzi, G., Chiarabba, C., De Gori, P. (2011). Linking the Alps and Apennines subduction systems: New constraints revealed by high-resolution teleseismic tomography. *Earth Planet. Sci. Lett.*, 301, 531–543.

Giger, M. and Hurford, A.J. (1989). Tertiary intrusives of the Central Alps: Their tertiary uplift erosion redeposition and burial in the southalpine foreland. *Eclogae Geol. Helv.*, 82, 857–866.

Hacker, B., Abes, G., Peacock, S. (2003a). Subduction factory 1. Theoretical mineralogy, densities, seismic wave speeds, and H_2O contents? *J. Geophys. Res.*, B108, 2029.

Hacker, B., Peacock, S., Abes, G., Holloway, S. (2003b). Subduction factory 2. Are intermediate-depth earthquakes in subducting slabs linked to metamorphic dehydration reactions? *J. Geophys. Res.*, B108, 2030.

Hammer, W. (1903). Porphyrite und Diorit aus den Ultenthaler Alpen. *Jahrbuch Geol. Reichsanst.*, 53, 65–90.

Handler, R., Ebner, F., Neubauer, F., Bojar, A.-V., Hermann, S. (2006). 39Ar/40Ar dating of Miocene tuffs from the Styrian part of the Pannonian basin: An attempt to refine the basin stratigraphy. *Geologica Carpathica*, 57, 483–494.

Handy, M.R., Babist, J., Wagner, R., Rosenberg, C.L., Konrad, M. (2005). Decoupling and its relation to strain partitioning in continental lithosphere: Insight from the Periadriatic fault system (European Alps). *Geological Society, London, Special Publications*, 243, 249–276.

Handy, M.R., Schmid, S.M., Bousquet, R., Kissling, E., Bernoulli, D. (2010). Reconciling plate-tectonic reconstructions of Alpine Tethys with the geological–geophysical record of spreading and subduction in the Alps. *Earth-Science Reviews*, 102, 121–158.

Handy, M., Ustaszewski, K., Kissling, E. (2015). Reconstructing the Alps–Carpathians–Dinarides as a key to understanding switches in subduction polarity, slab gaps and surface motion. *Int. J. Earth Sci.*, 104, 1–26.

Harangi, S. (2001) Neogene magmatism at the Alpine-Pannonian transition zone – A model for melt generation in a complex geodynamic setting. *Acta Vulcanologica*, 13, 25–39.

van Hinsbergen, D.J.J., Torsvik, T.H., Schmid, S.M., Maţenco, L.C., Maffione, M., Vissers, R.L.M., Gürer, D., Spakman, W. (2020). Orogenic architecture of the Mediterranean region and kinematic reconstruction of its tectonic evolution since the Triassic. *Gondwana Research*, 81, 79-229.

Hürlimann, N., Müntener, O., Ulmer, P., Nandedkar, R., Chiaradia, M., Ovtcharova, M. (2016). Primary magmas in continental arcs and their differentiated products: Petrology of a post-plutonic dyke suite in the Tertiary Adamello batholith (Alps). *J. Petrol.*, 57, 495–534.

Ivaldi, J.P., Bellon, H., Guardia, P., Mangan, C., Müller, C., Perez, J.L., Terramorsi, S. (2003). Contexte lithostructural, âges 40 K-40 Ar et géochimie du volcanisme calco-alcalin tertiaire de Cap-d'Ail dans le tunnel ferroviaire de Monaco. *Comptes rendus géosciences*, 335, 411–421.

Jacob, J.-B., Guillot, S., Thiéblemont, D., Vanardois, J., Trap, P., Faure, M., Marquer, D. (2023). Paleozoic evolution and Variscan inheritance in the Alps. In *Geodynamics of the Alps 2*, Rosenberg, C.L. and Bellahsen, N. (eds). ISTE Ltd, London, and John Wiley & Sons, New York.

Ji, W., Malusà, M., Tiepolo, M., Langone, A., Zhao, L., Wu, F.-Y. (2019). Synchronous Periadriatic magmatism in the Western and Central Alps in the absence of slab breakoff. *Terra Nova*, 31, 120–128.

Kaganai, H., Ulmer, P., Hansmann, W., Dietrich, V., Steiger, R.H. (1991). Nd-Sr isotopic and geochemical characteristics of the southern Adamello (N Italy) intrusives: Implications for crustal versus mantle origin. *J. Geophs. Res.*, 96, 14331–14346.

Kapferer, N., Mercolli, I., Berger, A., Ovtcharova, M., Fügenschuh, B. (2012). Dating emplacement and evolution of the orogenic magmatism in the internal Western Alps: 2. The Biella Volcanic Suite. *Swiss J. Geos.*, 105, 67–84.

Kästle, E.D., Rosenberg, C.L., Boschi, L., Bellahsen, N., Meier, T., El-Sharkawy, A. (2020). Slab break-offs in the Alpine subduction zone. *Int. J. Earth Sci.*, 109, 587–603.

van Keken, P., Hacker, B., Syracuse, E., Abes, G. (2011). Subduction factory 4. Depth-dependent flux of H_2O from subducting slabs worldwide? *Journ. Geophys. Res.*, 116, B01401.

Kralj, P. (2012). Facies architecture of the upper Oligocene submarine Smrekovec stratovolcano, Northern Slovenia. *Journal of Volcanology and Geothermal Research*, 247–248, 122–138.

Lacombe, O. and Jolivet, L. (2005). Structural relationships between Corsica and the Pyrenees-Provence domain at the time of Pyrennean orogeny. *Tectonics*, 24, TC1003.

Laubscher, H.P. (1983). The late Alpine (Periadriatic) intrusions and the Insubric Line. *Mem. Soc. Geol. Ital.*, 26, 21–30.

Liati, A., Gebauer, D., Fanning, M. (2000). U-Pb SHRIMP dating of zircon from the Novate granite (Bergell, Central Alps): Evidence for Oligocene-Miocene magmatism, Jurassic/Cretaceous continental rifting and opening of the Valais trough. *Schweiz. Mineral. Petrogr. Mitt.*, 80, 305–316.

Lippitsch, R., Kissling, E., Ansorge, J. (2003). Upper mantle structure beneath the Alpine orogen from high-resolution teleseismic tomography. *Journal of Geophysical Research: Solid Earth*, 108, 2376.

Lu, G., Winkler, W., Rahn, M., von Quadt, A., Willett, S.D. (2018). Evaluating igneous sources of the Taveyannaz formation in the Central Alps by detrital zircon U-Pb age dating and geochemistry. *Swiss J. Geosci.*, 111, 399–416.

Lu, G., Di Capua, A., Winkler, W., Rahn, M., Guillong, M., von Quadt, A., Willett, S.D. (2019). Restoring the source-to-sink relationships in the Paleogene foreland basins in the Central and Southern Alps (Switzerland, Italy, France): A detrital zircon study approach. *Int. J. Earth Sci.*, 108, 1817–1834.

Lucchini, F., Simboli, G., Zenatti, A., Barbieri, M., Nicoletti, M., Petrucciani, C. (1983). Petrochimica e dati radiometrici K/Ar e Rb/Sr dei "filoni basici" di Corvara in Badia e dei lamprofiri di Val Fiscalina (Dolomiti Orientali). Revisione classificata ed implicationi genetiche nel quadro del magmatismo filoniano delle Alpi. *Mineralogica Petrographica Acta*, 27, 233–249.

Lukács, R., Harangi, S., Guillong, M., Bachmann, O., Fodor, L., Buret, Y. (2018). Early to Mid-Miocene syn-extensional massive silicic volcanism in the Pannonian Basin (East-Central Europe): Eruption chronology, correlation potential and geodynamic implications. *Earth-Science Reviews*, 179, 1–19.

Lustrino, M. and Sharkov, E. (2006) Neogene volcanic activity of western Syria and its relationship with Arabian plate kinematics. *J. Geodyn.*, 42, 115–139.

Lustrino, M. and Wilson, M. (2007). The circum-Mediterranean anorogenic Cenozoic igneous province. *Earth-Science Reviews*, 81, 1–65.

Lustrino, M., Morra, V., Fedele, L., Franciosi, L. (2009). Beginning of the Apennine subduction system in central western Mediterranan: Constraints from Cenozoic "orogenic" magmatic activity of Sardinia, Italy. *Tectonics*, 28, TC5016.

Lustrino, M., Duggen, S., Rosenberg, C.L. (2011). The central–western Mediterranean: Anomalous igneous activity in an anomalous collisional tectonic setting. *Earth Science Review*, 104, 1–40.

Lustrino, M., Fedele, L., Melluso, L., Morra, V., Ronga, F., Geldmacher, J., Duggen, S., Agostini, S., Cucciniello, C., Franciosi, L. et al. (2013). Origin and evolution of Cenozoic magmatism of Sardinia (Italy). A combined isotopic (Sr-Nd-Pb-O-Hf-Os) and petrological view. *Lithos*, 180, 138–158.

Lustrino, M., Fedele, L., Agostini, S., Di Vincenzo, G., Morra, V. (2017). Eocene-Miocene igneous activity in Provence (SE France): 40Ar/39Ar data, geochemical-petrological constraints and geodynamic implications. *Lithos*, 288–289, 72–90.

Macera, P., Gasperini, D., Ranalli, G., Mahatsente, R. (2008). Slab detachment and mantle plume upwelling in subduction zones: An example from the Italian South-Eastern Alps. *J. Geodyn.*, 45, 32–48.

Malavieille, J. and Molli, G. (2023). Alpine Corsica: From Tethys subduction to Mediterranean extension. In *Geodynamics of the Alps 1*, Rosenberg, C.L. and Bellahsen, N. (eds). ISTE Ltd, London, and John Wiley & Sons, New York

Malavieille, J., Molli, G., Genti, M., Dominguez, S., Beyssac, O., Taboada, A., VitaleBrovarone, A., Chia-Lu, Y., Chic-Tung, Y. (2016). Formation of ophiolite-bearing tectono-sedimentary mélanges in accretionary wedges by gravity driven submarine erosion: Insights from analog models and case studies. *J. Geodyn.*, 100, 87–103.

Malusà, M., Villa, I., Vezzoli, G., Garzanti, E. (2011). Detrital geochronology of unroofing magmatic complexes and the slow erosion of Oligocene volcanoes in the Alps. *Earth Planet. Sci. Lett.*, 301, 324–336.

Malusà, M.G., Anfinson, O.A., Dafov, L.N., Stoeckli, D.F. (2016). Tracking Adria indentation beneath the Alps by detrital zircon U–Pb geochronology: Implications for the Oligocene–Miocene dynamics of the Adriatic microplate. *Geology*, 44, 155–158.

Mancktelow, N.S., Stöckli, D.F., Grollimund, B., Müller, W., Fügenschuh, B., Viola, G., Seward, D., Villa, I. (2002). The DAV and Periadriatic fault systems in the Eastern Alps south of the Tauern window. *Int. J. Earth Sci.*, 90, 593–622.

Manzotti, P. and Ballèvre, M. (2023). Continental subduction in the Alps: From field data to kinematic models. In *Geodynamics of the Alps 2*, Rosenberg, C.L. and Bellahsen, N. (eds). ISTE Ltd, London, and John Wiley & Sons, New York.

Manzotti, P., Bosse, V., Pitra, P., Robyr, M., Schiavi, F., Ballèvre, M. (2018). Exhumation rates in the Gran Paradiso Massif (Western Alps) constrained by in situ U–Th–Pb dating of accessory phases (monazite, allanite and xenotime). *Contrib. Mineral. Petrol.*, 173, 24.

Martin, S. and Macera, P. (2014). Tertiary volcanism in the Italian Alps (Giudicarie fault zone, NE Italy): Insight for double alpine magmatic arc. *Italian Journal of Geosciences*, 133, 63–84.

Martin, S., Prosser, G., Morten, L. (1993). Tectonomagmatic evolution of sheeted plutonic bodies along the north Giudicarie Line (northern Italy). *Geol. Rundsch.*, 82, 51–66.

Marxer, F. and Ulmer, P. (2019). Crystallisation and zircon saturation of calc-alkaline tonalite from the Adamello Batholith at upper crustal conditions: An experimental study. *Contrib. Mineral. Petrol.*, 174, 84.

Mattioli, M., Di Battistini, G., Zanzucchi, G. (2002a). Geochemical features of the Tertiary buried Mortara volcanic body (Northern Apennines, Italy). *Bollettino della Società Geologica Italiana*, 1, 239–249.

Mattioli, M., Di Battistini, G., Zanzucchi, G. (2002b). Petrology, geochemistry and age of the volcanic clasts from the Canetolo Unit (Northern Apennines, Italy). *Bollettino della Società Geologica Italiana*, 1, 399–416.

Mattioli, M., Lustrino, M., Ronca, S., Bianchini, G. (2012). Alpine subduction imprint in Apennine volcaniclastic rocks. Geochemical–petrographic constraints and geodynamic implications from Early Oligocene Aveto-Petrignacola Formation (N Italy). *Lithos*, 134–135, 201–220.

McCarthy, A., Tugend, J., Mohn, G., Candioti, L., Chelle-Michou, C., Arculus, R., Schmalholz, S., Müntener, O. (2020). A case of Ampferer-type subduction and consequences for the Alps and the Pyrenees. *Am. J. Sci.*, 320, 313–372.

Milani, L., Beccaluva, L., Coltorti, M. (1999). Petrogenesis and evolution of the Euganean Magmatic Complex, Veneto Region, North-East Italy. *Eur. J. Mineral.*, 11, 379–399.

Molli, G. (2008). Northern Apennine–Corsica orogenic system: An updated overview. In *Tectonic Aspects of the Alpine-Dinaride-Carpathian System*, Siegesmund, S., Fügenschuh, B., Froitzheim, N. (eds). Geological Society London Special Publications, London.

Molli, G. and Malavieille, J. (2011). Orogenic processes and the Corsica/Apennines geodynamic evolution: Insights from Taiwan. *Int. J. Earth Sci.*, 100, 1207–1224.

Molli, G., Crispini, L., Mosca, P., Piana, P., Federico, L. (2010). Geology of the Western Alps-Northern Apennine junction area: A regional review. *Journal of the Virtual Explorer*, 36/3, 10. doi: 10.3809/jvirtex.2010.00215.

Mottana, A., Morten, L., Brunfelt, A.O. (1978). Distribuzionc delle terre rare nel massiccio Val Masino – Val Bregaglia (Alpi Centrali). *Rend. Soc. Ital. Mineral Petrol.*, 34, 485–497.

Müntener, O., Kelemen, P.B., Grove, T.L. (2001). The role of H_2O during crystallization of primitive arc magmas under uppermost mantle conditions and genesis of igneous pyroxenites: An experimental study. *Contrib. Mineral Petrol.*, 141, 643.

Müntener, O., Ulmer, P., Blundy, J. (2021). Superhydrous arc magmatism in the Alpine context. *Elements*, 17, 35–40.

Mutti, E., Papani, L., Di Biase, D., Davoli, G., Mora, S., Segadelli, S., Tinterri, R. (1995). Il bacino terziario epimesoalpino e le sue implicazioni sui rapporti Alpi-Appennino. *Mem. Sci. Geologi. Padova*, 47, 217–244.

Neubauer, F., Heberer, B., Dunkl, I., Liu, X., Bernroider, M., Dong, Y. (2018). The Oligocene Reifnitz tonalite (Austria) and its host rocks: Implications for Cretaceous and Oligocene–Neogene tectonics of the south-eastern Eastern Alps. *Geologica Carpathica*, 69, 237–253.

Nussbaum, C. (2000). Neogene tectonics and thermal maturity of sediments of the easternmost Southern Alps (Friuli area, Italy). PhD Thesis, University of Neuchâtel.

Oalmann, J., Duesterhoeft, E., Möller, A., Bousquet, R. (2019). Constraining the pressure–temperature evolution and geodynamic setting of UHT granulites and migmatitic paragneisses of the Gruf Complex, Central Alps. *Int. J. Earth Sci.*, 108, 911–930.

Oberli, F., Meier, M., Berger, A., Rosenberg, C.L., Gieré, R. (2004). U-Th-Pb and 230Th/238U disequilibrium isotope systematics: Precise accessory mineral chronology and melt evolution tracing in the Alpine Bergell intrusion. *Geochimica et Cosmochimica Acta*, 68, 2543–2560.

Owen, J.P. (2008). Geochemistry of lamprophyres from the western Alps, Italy: Implications for the origin of an enriched isotopic component in the Italian mantle. *Contrib. Mineral. Petrol.*, 155, 341–362.

Pieri, M. and Groppi, G. (1981). Subsurface geological structure of the Po Plain. Progetto Finalizzato Geodinamica, Sottoprogetto "Modello Strutturale". *Publication Italian CNR*, 144, 1–13.

Plesi, G., Chicchi, S., Daniele, G., Palandri, S. (2000). La struttura dell'alto Appennino reggiano-parmense, fra Valditacca, il Passo di Pradarena e il M. Ventasso. *Bollettino della Societa Geologica Italiana*, 119, 267–296.

Pomella, H., Klotzli, U., Scholger, R., Stipp, M., Fugenschuh, B. (2011). The Northern Giudicarie and the Meran-Mauls fault (Alps, Northern Italy) in the light of new paleomagnetic and geochronological data from boudinaged Eo-/Oligocene tonalites. *Int. J. Earth Sci.*, 100, 1827–1850.

Qorbani, E., Bianchi, I., Bokelmann, G. (2015). Slab detachment under the Eastern Alps seen by seismic anisotropy. *Earth Planet. Sci. Lett.*, 409, 96–108.

Réhault, J.P., Honthaas, C., Guennoc, P., Bellon, H., Ruffet, G., Cotten, J., Sosson, M., Maury, R.C. (2012). Offshore Oligo-Miocene volcanic fields within the Corsica-Liguria Basin: Magmatic diversity and slab evolution in the western Mediterranean Sea. *Journal of Geodynamics*, 58, 73–95.

Reichenwallner, S. (2019). Die vulkanische Aktivität in den Westalpen hergeleitet aus detritischem Amphibol und Klinopyroxen im Taveyannaz-Sandstein. Master's Thesis, Universität Bern.

Remitti, F., Vannucchi, P., Bettelli, G., Fantoni, L., Panini, F., Vescovi, P. (2011). Tectonic and sedimentary evolution of the frontal part of an ancient subduction complex at the transition from accretion to erosion: The case of the Ligurian wedge on the northern Apennines, Italy. *GSA Bullettin*, 123, 51–70.

Romer, R.L. and Siegesmund, S. (2003). Why allanite may swindle about its true age. *Contrib. Mineral. Petrol.*, 146, 297–307.

Romer, R.L., Schärer, U., Steck, A. (1996). Alpine and pre-Alpine magmatism in the root-zone of the western Central Alps. *Contrib. Mineral. Petrol.*, 123, 138–158.

Rosenberg, C.L. (2004). Shear zones and magma ascent: A model based on a review of the tertiary magmatism in the Alps. *Tectonics*, 23, TC3002.

Rosenberg, C.L. and Kissling, E. (2013). Three-dimensional insight into Central-Alpine collision: Lower-plate or upper-plate indentation. *Geology*, 41, 1219–1222.

Rosenberg, C.L., Berger, A., Schmid, S.M. (1995). Observations from the floor of a granitoid pluton: Inferences on the driving force of final emplacement. *Geology*, 23, 443–446.

Rosenberg, C.L., Schneider, S., Scharf, A., Betrand, A., Hammerschmidt, K., Rabaute, A., Brun, J.-P. (2018). Relating collisional kinematics to exhumation processes in the Eastern Alps. *Earth Science Reviews*, 176, 311–344.

Rubatto, D., Hermann, J., Berger, A., Engi, M. (2009). Protracted fluid-induced melting during Barrovian metamorphism in the Central Alps. *Contributions of Mineralogy and Petrology*, 158, 703–722.

Ruffini, R. (1995). Evidenze di attività vulcanica terziaria nelle Alpi Occidentali: Problemi ed ipotesi. PhD Thesis, Università di Torino.

Ruffini, R., Polino, R., Calegari, E., Hunziker, L.C., Pfeiffer, H.R. (1997). Volcanic clast rich turbidites of the Taveyanne sandstone from the Thones syncline (Savoie, France): Records for a Tertiary postcollisional volcanism. *Schweiz. Mineral. Petrogr. Mitt.*, 77, 161–174.

Salomon, W. (1897). Über Alter, Lagerungsform und Entstehungsart der periadriatischen granitisch-körnigen Massen, Tschermaks Mineral. *Petrogr. Mitt.*, 17, H 2/3, 109–284.

Salomon, W. (1908). Die Adamellogruppe: Ein alpines Zentralmassiv und seine Bedeutung für die Gebirgsbildung und unsere Kenntniss von dem Mechanismus der Intrusionen, Teil 1. *Abh. Geol. Bundesanst. Austria*, 21, 1–433.

Salomon, W. (1910). Die Adamellogruppe: Ein alpines Zentralmassiv und seine Bedeutung für die Gebirgsbildung und unsere Kenntniss von dem Mechanismus der Intrusionen, Teil 2. *Abh. Geol. Bundesanst. Austria*, 22, 435–603.

Samperton, K.M., Schoene, B., Cottle, J.M., Keller, C.B., Crowley, J.L., Schmitz M.D. (2015). Magma emplacement, differentiation and cooling in the middle crust: Integrated zircon geochronological–geochemical constraints from the Bergell Intrusion, Central Alps. *Chem. Geol.*, 417, 322–340.

Savelli, C. (2002). Time–space distribution of magmatic activity in the western Mediterranean and peripheral orogens during the past 30 Ma (a stimulus to geodynamic considerations). *J. Geodyn.*, 34, 99–126.

Savelli, C. (2015). Fast episodes of West Mediterranean-Tyrrhenian oceanic opening and revisited relations with tectonic setting. *Scientific Reports*, 5, 14271.

Schabert, S. (1975). Radiometrische Altersdaten von Intrusivgesteinen im Raum Eisenkappel (Karawanken, Kärnten). *Verh. geol. Bundesanstalt*, 1975, 301–304.

Schaltegger, U., Brack, P., Ovtcharova, M., Peytcheva, I., Schoene, B., Stracke, A. Marocchi, M., Bargossi, G.M. (2009). Zircon and titanite recording 1.5 million years of magma accretion, crystallization and initial cooling in a composite pluton (southern Adamello batholith, northern Italy). *Earth Planet. Sci. Lett.*, 286, 208–218.

Schaltegger, U., Nowak, A., Ulianov, A., Fisher, C.M., Gerdes, A., Spikings, R., Whitehouse, M.J., Bindeman, I., Hanchar, J.M., Duff, J., Vervoor, J., Sheldrake, T., Caricchi, L., Brack, P., Müntener, O. (2019). Zircon petrochronology and 40Ar/39Ar thermochronology of the Adamello intrusive suite, N. Italy: Monitoring the growth and decay of an incrementally assembled magmatic system. *J. Petrol.*, 60, 701–722.

Schmid, S.M. and Kissling, E. (2000). The arc of the Western Alps in the light of geophysical data on deep crustal structure. *Tectonics*, 19, 62–85.

Schmid, S.M., Aebli, H.R., Heller, F., Zingg, A. (1989). The role of the Periadriatic line in the tectonic evolution of the Alps. In *Alpine Tectonics*, Coward, M.P., Dietrich, D., Park, R.G. (eds). Geol. Soc. London Spec. Publ. No. 45, London.

Schmid, S.M., Pfiffner, O.A., Froitzheim, N., Schönborn, G., Kissling, E. (1996). Geophysical geological transect and tectonic evolution of the Swiss-Italian Alps. *Tectonics*, 15, 1036–1064.

Schmid, S.M., Fügenschuh, B., Kissling, E., Schuster, R. (2004). Tectonic map and overall architecture of the Alpine orogen. *Eclogae Geolog. Helv.*, 97, 93–117.

Schmid, S.M., Bernoulli, D., Fügenschuh, B., Matenco, L., Schefer, S., Schuster, R., Tischler, M., Ustaszewski, K. (2008). The Alpine–Carpathian–Dinaridic orogenic system: Correlation and evolution of tectonic units. *Swiss J. Geosci.*, 101, 139–183.

Schmid, S.M., Scharf, A., Handy, M.R., Rosenberg, C.L. (2013). The Tauern Window (Eastern Alps, Austria): A new tectonic Map, with cross-sections and a tectonometamorphic synthesis. *Swiss J. Geosci.*, 106, 1–32.

Schmid, S.M., Kissling, E., Diehl, T., van Hinsbergen, D.J.J., Molli, G. (2017). Ivrea mantle wedge, arc of the Western Alps, and kinematic evolution of the Alps–Apennines orogenic system. *Swiss J. Geosci.*, 110, 581–612.

Schneider, S., Hammerschmidt, K., Rosenberg, C.L., Gerdes, A., Frei, D., Bertrand, A. (2015). U–Pb ages of apatite in the western Tauern Window (Eastern Alps): Tracing the onset of collision-related exhumation in the European plate. *Earth Planet. Sci. Lett.*, 418, 53–65.

Schoene, B., Schaltegger, U., Brack, P., Latkoczy, C., Stracke, A., Günther, D. (2012). Rates of magma differentiation and emplacement in a ballooning pluton recorded by U–Pb TIMS-TEA, Adamello batholith, Italy. *Earth Planet. Sci. Lett.*, 355–356, 162–173.

Schönborn, G. (1992). Alpine tectonics and kinematic models of the central Southern Alps. *Mem. Sci. Geol. Padova*, 44, 229–293.

Schönborn, G. (1999). Balancing cross sections with kinematic constraints: The Dolomites (northern Italy). *Tectonics*, 18, 527– 545.

Seghedi, I. and Downes, H. (2011). Geochemistry and tectonic development of Cenozoic magmatism in the Carpathian–Pannonian region. *Gondwana Research*, 20, 655–672.

Séranne, M. (1999). The Gulf of Lion continental margin (NW Mediterranean) revisited by IBS: An overview. In *Mediterranean Basins*, Durand, B., Jolivet, L., Horvàth, F., Séranne, M. (eds). Geol. Soc. Lond. Spec. Publ. No. 156, London.

Sharman, G.R., Hubbard, S.M., Covault, J.A., Hinsch, R., Linzer, H.-G., Graham, S.A. (2017). Sediment routing evolution in the North Alpine foreland basin Austria: Interplay of transverse and longitudinal sediment dispersal. *Basin Res.*, 30, 426–447.

Siena, F. and Coltorti, M. (1989). Lithospheric mantle evolution: Evidences from ultramafic xenoliths in the Lessinian volcanics (northern Italy). *Chem. Geol.*, 77, 347–364.

Staub, R. (1918). Geologische Beobachtungen am Bergellermassiv. *Vjschr. Natf. Ges. Zürich*, 63, 1–18.

Steenken, A., Siegesmund, S., Heinrichs, T. (2000). The emplacement of the Rieserferner Pluton (Eastern Alps, Tyrol): Constraints from field observations, magnetic fabrics and microstructures. *J. Struct. Geol.*, 22, 1855–1873.

Steinmann, G. (1913). Die Bedeutung der jüngeren Granite in den Alpen. *Geol. Rundschau*, 4, 220–224.

Stipp, M., Fügenschuh, B., Gromet, L.P., Stunitz, H., Schmid, S.M. (2004). Contemporaneous plutonism and strike-slip faulting: A case study from the Tonale fault zone north of the Adamello pluton (Italian Alps). *Tectonics*, 23, TEC3004.

Thompson, A.B., Matile, L., Ulmer, P. (2002). Some thermal constraints on crustal assimilation during fractionation of hydrous, mantle-derived magmas with examples from central Alpine Batholiths. *J. Petrol.*, 43, 403–422.

Tiepolo, M. and Tribuzio, R. (2005). Slab-melting during Alpine orogeny: Evidence from mafic cumulates of the Adamello batholith (Central Alps, Italy). *Chem. Geol.*, 216, 271–288.

Tiepolo, M., Tribuzio, R., Vannucci, R. (2002). The compositions of mantle-derived melts developed during the Alpine continental collision. *Contrib. Mineral. Petrol.*, 144, 1–15.

Tiepolo, M., Tribuzio, R., Langone, A. (2011). High-Mg andesite petrogenesis by amphibole crystallization and ultramafic crust assimilation: Evidence from Adamello hornblendites (Central Alps, Italy). *J. Petrol.*, 52, 1011–1045.

Torsvik, H.T., van der Voo, R., Preeden, U., Conall, M.N., Steinberger, B., Doubrovine, P.V., van Hinsbergen, D.J.J., Domeier, M., Gaina, C., Tohver, E. et al. (2012). Phanerozoic polar wander, palaeogeography and dynamics. *Earth-Science Reviews*, 114, 325–368.

Trajanova, M., Pécskay, Z., Itaya, T. (2008). K-Ar geochronology and petrography of the Miocene Pohorje Mountains batholith (Slovenia). *Geologica Carpathica*, 59, 247–260.

Trommsdorff, V. and Evans, B.W. (1972). Progressive metamorphism of antigorite schists in the Bergell Tonalite aureole. *Am. J. Sci.*, 272, 423–137.

Trommsdorff, V. and Nievergelt. P. (1983). The Bregaglia (Bergell) Iorio intrusive and its field relations. *Mem. Soc. Geol. Ital.*, 26, 55–68.

Ulmer, P., Callegari, E., Sonderegger, U.C. (1985). Genesis of the mafic and ultramafic rocks and their genetical relations to the tonalitic–trondhjemitic granitoids of the southern part of the Adamello Batholith (Northern Italy). *Mem. Soc. Geol. It.*, 26, 171–222.

Ustaszewski, K., Schmid, S.M., Fügenschuh, B., Tischler, M., Kissling, E., Spakman, W. (2008). A map-view restoration of the Alpine–Carpathian–Dinaridic system for the Early Miocene. *Swiss J. Geosci.*, 101, 273–294.

Venturelli, G., Thorpe, R.S., Dal Piaz, G.V., Del Moro, A., Potts, P.J. (1984). Petrogenesis of calc-alkaline shoshonitic and associated ultrapotassic Oligocene volcanic rocks from the Northwestern Alps, Italy. *Contrib. Mineral. Petrol.*, 86, 209–220.

Visonà, D., Caironi, V., Carraro, A., Dallai, L., Fioretti, A.M., Fanning, M. (2007). Zircon megacrysts from basalts of the Venetian Volcanic Province (NE Italy): U-Pb ages, oxygen, isotopes and REE data. *Lithos*, 94, 168–180.

Wagner, R., Rosenberg, C.L., Handy, M., Möbus, C., Albertz, M. (2006). Fracture-driven intrusion and upwelling of a mid-crustal pluton fed from a transpressive shear zone – The Rieserferner Pluton (Eastern Alps). *GSA Bulletin*, 118, 219–237.

Waibel, A.F. (1993). Nature and plate-tectonic significance of orogenic magmatism in the European Alps: A review. *Schweiz. Mineral. Petrogr. Mitt.*, 73, 391–405.

Wallace, L., McCaffrey, R., Beavan, J., Ellis, S. (2005). Rapid microplate rotations and backarc rifting at the transition between collision and subduction. *Geology*, 33, 857–860.

Wilson, M. and Bianchini, G. (1999). Tertiary–quaternary magmatism within the Mediterranean and surrounding regions. In *The Mediterranean Basins: Tertiary Extension within the Alpine Orogen*, Durand, B., Jolivet, L., Horvath, F., Sèranne, M. (eds). Geol. Soc. London, London.

Zampieri, D. (1995). Tertiary extension in the southern Trento platform, Southern Alps, Italy. *Tectonics*, 14, 645–657.

Zanchi, A., Chiesa, S., Gillot, P.Y. (1990). Tectonic evolution of the Southern Alps in the Orobic chain: Structural and geochronological indications for pre-Tertiary compressive tectonics. *Mem. Soc. Geol. It.*, 45, 77–82.

Zanoni, D. (2010). Structural and petrographic analysis at the north-eastern margin of the Oligocene Traversella pluton (Internal Western Alps, Italy). *Italian Journal of Geosciences*, 129, 51–68.

Zanoni, D., Corti, L., Roda, M. (2022). Cooling history of the Biella pluton and implication for Oligocene to Miocene tectonics of the Sesia-Lanzo zone, Austroalpine, Western Alps. *International Geology Review*, 64, 203–232.

Zheng, Y.-F. (2019) Subduction zone geochemistry. *Geosci. Front.*, 10, 1223–1254.

2

Evolution of the External Crystalline Massifs of the European Alps: From Massif to Lithosphere Scale

Marco HERWEGH[1], Alfons BERGER[1], Nicolas BELLAHSEN[2], Yann ROLLAND[3] and Edi KISSLING[4]

[1]*Institute of Geological Sciences, University of Bern, Switzerland*

[2]*ISTeP, Sorbonne Université, UMR 7193, CNRS, Paris, France*

[3]*EDYTEM, Université Savoie Mont Blanc, CNRS, UMR 5204, Le Bourget-du-Lac, France*

[4]*Department of Earth Sciences, Institute of Geophysics, ETH Zürich, Switzerland*

2.1. Introduction

This contribution summarizes the Alpine evolution of the so-called "external crystalline massifs" (ECM), which are characterized in map view by lens-shaped basement cores surrounded by sediments (Figure 2.1). In 3D view, these geometries result from dome-like basement culminations. Extending from the Western Alps near Nice to the Swiss Central Alps, the ECM align in an arcuate

Geodynamics of the Alps 3,
coordinated by Claudio L. ROSENBERG and Nicolas BELLAHSEN.

manner, following the trend of the relatively external part of the Western to the Central Alps (Figure 2.1). ECM expose remnants of the proximal part of the former southern and southeastern European continental margin, which was only weakly extended during Mesozoic rifting (Figure 2.2; see also Burkhard 1988; Haenni and Pfiffner 2001; Spillmann 2011; Bellahsen et al. 2014; Cardello and Mancktelow 2014). Toward the internal parts of the orogen, Penninic units follow, with the sediments of the Valaisan trough, the units of the Briançonnais swell forming nowadays the internal crystalline massifs (ICM) and the subsequent units of the former Piemont-Ligurian ocean (Figure 2.2). From the Cretaceous till late Oligocene, convergence between the Adriatic and the European plates occurred, resulting in subduction of the Penninic (Piemont–Valaisan) units underneath Adria. Generally, in the past, a classical subduction of the European plate under the Adriatic plate with an indentation of the latter into the European crust during continent-continent was proposed (e.g. Schmid et al. 1996; Pfiffner et al. 1997b and references therein). As outlined in the following, it is the paleogeographic preconditioning, in combination with the collision processes, that led to the special appearance and exhumation of ECM. In cross-section view, they nowadays appear as basement culminations, for which the basement-cover contacts provide excellent structural markers.

The ECM build up four major groups (Figure 2.1), which are the Gastern/Aar, Aiguilles Rouges/Mont Blanc, Belledonne–Grandes Rousses–Oisans (in the following named Belledonne-Oisans massifs) and Argentera massifs (also called Argentera-Mercantour massif in the literature). Often, these major units can be further subdivided, because of the occurrence of sedimentary relics between crystalline units (e.g. Escher et al. 1993; Pfiffner et al. 1997a; Lacombe and Mouthereau 2002; Bellahsen et al. 2014; Berger et al. 2017a). The latter relate to Mesozoic (half-) graben structures, which became squeezed between basement blocks during Alpine deformation. The Aar and Mont Blanc massifs experienced the most pronounced shortening (~40–60%), highest peak metamorphic temperatures (400–450°C) and therefore also the most pronounced vertical rock uplift (Figure 2.3; see also Herwegh et al. 2020). The Belledonne-Oisans massifs are characterized by much less internal deformation/shortening and vertical uplift compared to the ECM in the northeast with an again somewhat increased shortening in the Argentera massif (Figure 2.3).

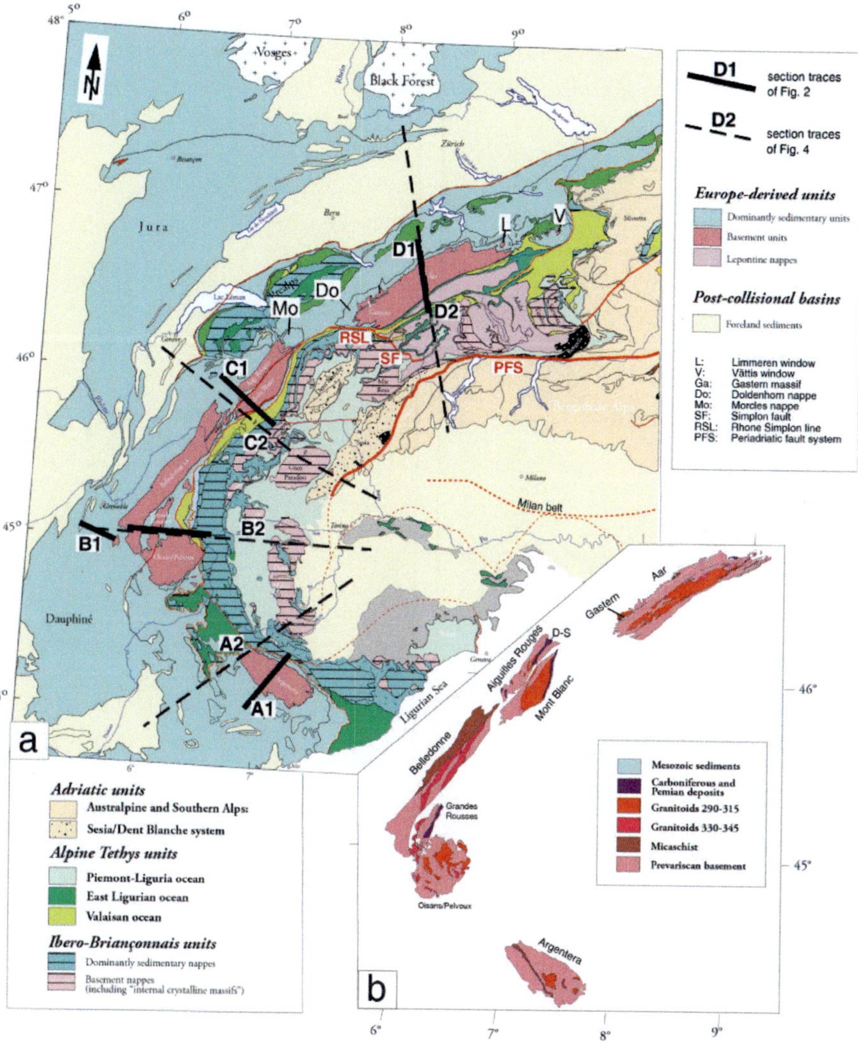

Figure 2.1. *(a) Tectonic map (after Schmid et al. 2004 and Bousquet et al. 2012) showing the Central to Western Alps and their ECM as well as the four major section traces: (A) Argentera massif, (B) Belldonne s.l. massif, (C) Mont Blanc/Aiguilles Rouges massifs, (D) Aar massif. (1) and (2) always indicate traces of massif- and lithospheric-scale sections, respectively. (b) Detailed view of major rock units of the ECM (modified after Berger et al. 2016 and Fréville et al. 2018). D-S: Permo-Carboniferous trough from Dorénaz-Salvan*

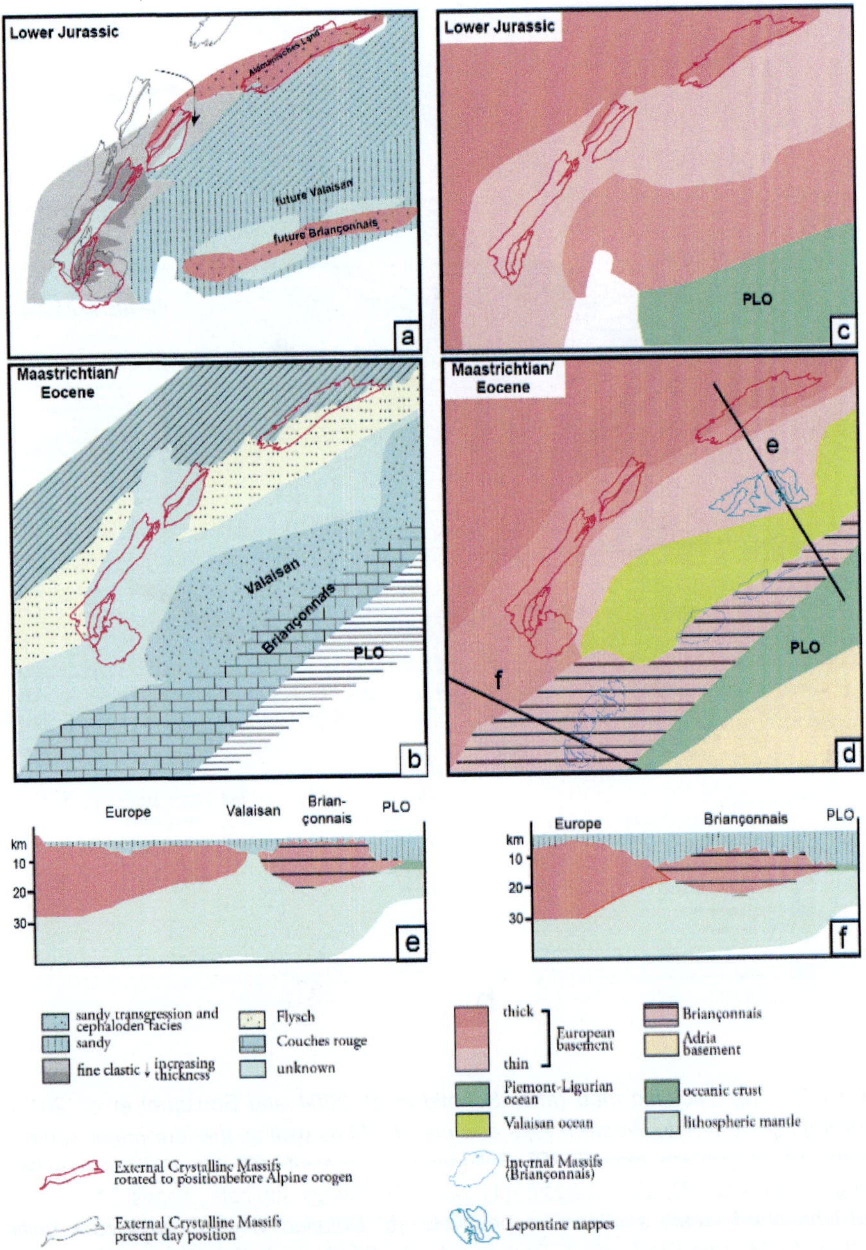

Figure 2.2. *The paleogeography of the Western/Central Alps in two steps. (a and c) Lower Jurassic; (b and d) upper Cretaceous/Eocene*

COMMENT ON FIGURE 2.2.– *The paleogeography is represented in two ways. In the left column (a and b), the sedimentary record is represented (Trümpy 1965; Gwinner 1971) and in the right column (c and d) the interpretation of the crustal basement structure is shown. For orientation, the outlines of the ECM are also displayed. To illustrate the change of the crustal structure along strike, two cross-section sketches are shown (e and f). The locations of the sketches are given in (d).*

Over the last two centuries, scientists studied the formation processes of these basement domes. Intrusion contacts of plutonic bodies within ECM were recognized early and provoked theories on emplacement by intrusion processes. Given the relics of Ordovician and Variscan ages and structures in the ECMs, it is nowadays clear that these phenomena cannot account for the young Alpine exhumation. Alternatively, concepts of classical compression tectonics (Steck 1966; Burkhard 1990, 1999; Pfiffner 1993; Pfiffner *et al.* 1997a; Burkhard and Sommaruga 1998; Lacombe and Mouthereau 2002; Krayenbuhl and Steck 2009; Egli and Mancktelow 2013; Egli *et al.* 2017), buoyancy-driven processes and slab dynamics combined with different parameters (Herwegh *et al.* 2017, 2020; Nibourel *et al.* 2018, 2021a, 2021b) have been suggested. Also in terms of deformation style different origins have been suggested: (i) basal thrusting or thrust imbrication (Burkhard 1990, 1999; Burkhard and Sommaruga 1998; Bigot-Cormier *et al.* 2006), (ii) pure ductile folding (Krayenbuhl and Steck 2009), (iii) folding and thrusting (Burkhard 1990, 1999), (iv) ductile shearing (Choukroune and Gapais 1983; Leloup *et al.* 2005; Rossi *et al.* 2005; Sanchez *et al.* 2011a), (v) sequences of initial ductile shearing with subsequent embrittlement and progressive exhumation (Bellahsen *et al.* 2012, 2014; Wehrens *et al.* 2016, 2017; Herwegh *et al.* 2020; Nibourel *et al.* 2021a).

In the context of this study, we will present a summary of the current state of knowledge on the exhumation processes, structures, kinematics and geodynamic consequences necessary to form and exhume the ECM. We first introduce the geological setting common to all ECM followed by the geological setting and tectonic evolution of each ECM individually. A next section deals with the lithosphere-scale structural-tectonic framework of the ECM. These multi-scale geometries are then discussed in the context of the evolution of the Western to Central Alpine arc including effects of inheritance of Mesozoic extensional structures and passive margin geometries, deformation structures, kinematics as well as the P-T evolution. In this context, we will use the term "vertical" and "horizontal" tectonics for cases, respectively, where the vertical and horizontal displacement vectors dominate over the other. Finally, the general ECM evolution is discussed in light of Alpine geodynamic processes.

2.2. General geological setting

The evolution of the ECM relates closely to preconditioning by (i) inheritance of pre-Alpine deformation structures, (ii) structure of the Mesozoic European passive continental margin and its southward transition into the Penninic realm as well as (iii) Cenozoic closure of the Tethys ocean. This section summarizes related aspects that are relevant for the rise of ECM.

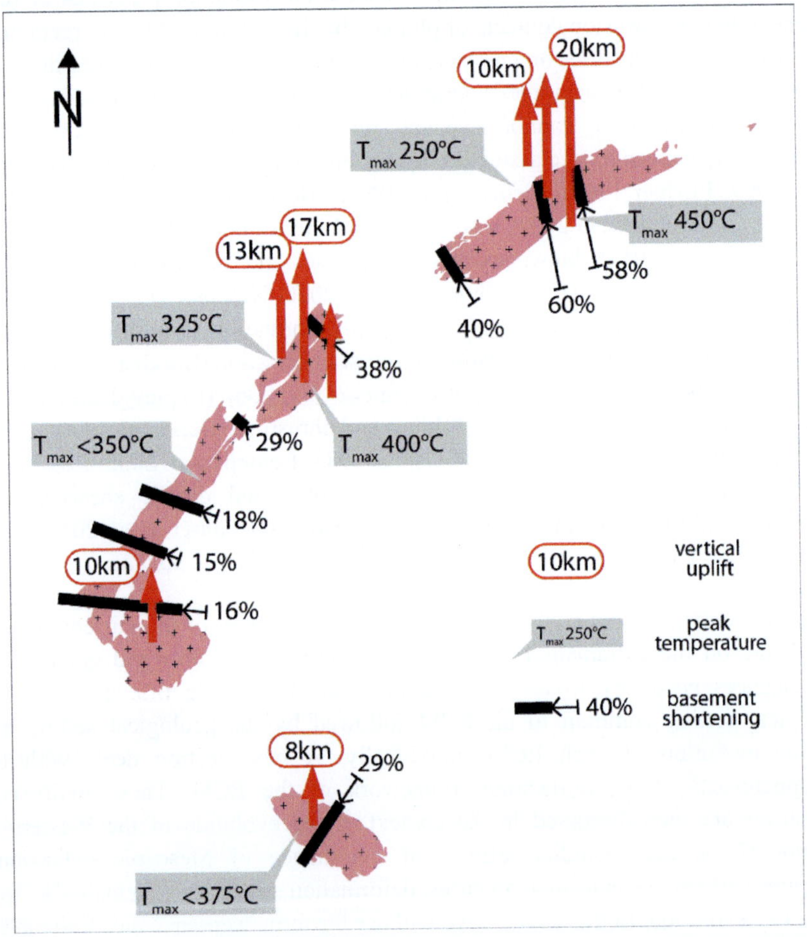

Figure 2.3. *Representation of general basement shortening, total rock uplift and peak temperature of the ECM. Data from Bellahsen et al. (2014), Bigot-Cormier et al. (2006), Boutoux et al. (2016), Challandes et al. (2008), Goncalves et al. (2012), Herwegh et al. (2017, 2019), Jourdan et al. (2014), Nibourel et al. (2021a, 2021b), Rolland et al. (2003), Rossi et al. (2005) and Sanchez et al. (2011a)*

In terms of the pre-Alpine basement, polymetamorphic gneisses are most abundant in the crystalline cores of all ECM, which in parts document and preserve their pre-Alpine geological history (Figure 2.1). Panafrican and Ordovician deformation structures are often only preserved as relics. More prominent is the preservation of Variscan to late/post-Variscan deformation and magmatic structures, although often overprinted by intense Alpine deformation. Indeed, Alpine deformation in ECM often reactivates pre-existing Variscan structures such as fault zones, lithological contacts or foliations (e.g. Bogdanoff et al. 2000; Bellahsen et al. 2014; Wehrens et al. 2017; Herwegh et al. 2020). It is this inheritance of old structures and their reactivation, which often predefines the massif-internal strike of basement units and fault structures. Pre-Alpine basement units can generally be subdivided into (1) polycyclic basement units, which are intruded by (2) Variscan and (3) post-Variscan magmatic groups (Figure 2.1; Guillot and Menot 2009; Fréville et al. 2018). In contrast to this well-established subdivision, detailed knowledge on metamorphic conditions and age constraints of the different metamorphic and magmatic pre-Alpine events differ between ECM (e.g. Paquette et al. 1989; Mercolli et al. 1994; von Raumer et al. 1999; Rubatto et al. 2001, 2010; Berger et al. 2017a; Fréville et al. 2018). The largest uncertainties relate to the Ordovician metamorphic evolution inside the polycyclic basement, including a HP/LT as well as a Barrovian-type metamorphism of sediments and older gneisses, associated with granitoid plutonism (= “old orthogneisses”, e.g. Abrecht et al. 1994; Rubatto et al. 2010; Fréville et al. 2018; Jacob et al. 2023).

These metasedimentary and metagranitoids rocks were then overprinted during the Variscan orogeny by multiple magmatic and metamorphic events (e.g. von Raumer et al. 1999; Rubatto et al. 2001; Guillot and Menot 2009; Fréville et al. 2018; Jouffray et al. 2020). During Variscan times, an early group of plutons intruded, becoming deformed and metamorphosed with ongoing orogeny (e.g. Franks 1968), preceding Late Carboniferous to Permian plutons (group 2; e.g. Debon and Lemmet 1999; Bussy et al. 2000; Guillot and Menot 2009; Berger et al. 2017a; Fréville et al. 2018). Since then, no deformation of these plutons has been reported prior to their Alpine overprint.

A post-Variscan tectonic regime with initial transpression and subsequent transtension dominated in late Carboniferous and Permian times leading to the formation of rift and strike-slip structures in the basement units. Local basins and half-grabens formed, into which volcano-sedimentary sequences were deposited. The preserved Permian-Carboniferous domains vary in lateral extent, composition and age (e.g. Decarlis et al. 2013). These intracontinental basins with their associated magmatism also occurred outside the ECM and have been recently summarized by Decarlis et al. (2013), Manzotti et al. (2014) and Ballèvre et al. (2018). During Alpine orogeny these basins became inverted and integrated into the ECM architecture (e.g. Badertscher and Burkhard 1998).

In Triassic times, the ECM basement later involved in the ECM, was progressively covered by the transgression of the Tethys ocean and associated platform sedimentation (e.g. Rohr 1926). In the lower Jurassic, rifting started first far south from the later ECM, at the border between the Piemont-Ligurian ocean and the future Briançonnais swell, which by then represented the most distal part of the European continental margin (Figure 2.2). At that time, the Aar massif was a structural high (part of the Alemannisches Land; Trümpy 1949), with minor subaquatic exceptions at its southwestern border. In contrast, the central ECM (Aiguilles Rouges/Mont Blanc, Belledonne-Oisans) formed the base of a large Lower Jurassic basin with highly variable sediment thicknesses (Gwinner 1971). In subsequent Middle-Late Jurassic and Cretaceous times, the ECM were part of the European margin with slightly variable carbonate sedimentation rates along the European shelf (Pfiffner et al. 2011). During Cretaceous times, Mont Blanc and Belledonne-Oisans massifs were in close proximity to the evolving Valaisan trough, a situation that is different for the Aar and Argentera domains (Gwinner 1971; Berger and Herwegh 2015). The latter occurs south of the L-shape bend in the margin geometry (section 2.5). In the Aar massif domain, the necking zone of the European continental crust must have been south of the Gotthard nappe. This can be inferred from reduced crustal thicknesses of the Lepontine gneiss nappes compared to those of the Aar massif and Gotthard nappe. Consequently, the hyper-extended parts of the margin must have been located in the Penninic units to the south, that is, in the realm of the Valais trough. In the Argentera section, no Valaisan was present. Hence, there is a pronounced NE-SW change in the extent and geometry of the Valaisan trough, which affected sedimentation directly south of the Gotthard nappe as well as the Mont Blanc and Belledonne-Oisans massifs (Figure 2.2).

During Upper Cretaceous times, a major switch from divergence and oceanic accretion to convergence took place (e.g. Pfiffner 2015). From this time onward, the former shelf sediments and their (ECM) basement substratum evolved into the hinterland of the subduction zone, which eventually became the Northern Alpine foreland. Underfilled flexural foreland Flysch basins with deep water turbidites developed and migrated with progressive shortening from the SE toward the NW, building irregular basin shapes (e.g. Sinclair and Allen 1992; Lihou 1995; Sinclair 1997). Low sediment supply indicated low hinterland relief pointing to flexure of the European plate by the negative buoyancy of its subducting lithospheric mantle (Ford et al. 2006). Taveyannaz Sandstones (34–32 Ma; Fischer and Villa 1990; Ruffini et al. 1995) are the youngest turbiditic units that terminate sedimentation in the underfilled Flysch basins (Kerckhove 1969; Fry 1989; Sinclair 1997; Mulder et al. 2010; Lu et al. 2018). These sediments were deposited on top or slightly south/southeast of the future Aar-, Aiguilles Rouges- and Belledonne-Oisans massifs. At the important time marker of ~32 Ma, foreland sedimentation changed from under- to overfilled resulting in the foreland Molasse basin reaching from the NW of the Belledonne-Oisans massifs to the Gastern/Aar massifs (Figure 2.1; Homewood et al. 1986; Pfiffner 1986; Schlunegger et al. 1997; Sinclair 1997). The Molasse sediment spectrum contains crucial information on the erosion of units on top of the emerging ECM. In contrast, to the west of the Belledonne–Argentera massifs, a small segmented Molasse basin can be identified (Ford et al. 2006), suggesting along strike variations, the amount of flexural accommodation of sedimentation space. Indeed, eroded units above the rising western ECM were deposited in small tectonically controlled basins only (Bellahsen et al. 2014).

2.3. Geology and tectonics of the ECM

2.3.1. *The Argentera massif*

2.3.1.1. *Geological setting*

The Argentera-Mercantour massif (abbreviated in the following as the Argentera massif) is a NW-SE striking that is a ~50 km long and ~30 km wide lentoid structure (Figure 2.1). In addition to the general pre-Alpine basement, Permian and Early Triassic silts and sandstones are frequently attached to the basement, whereas the overlying Triassic to Early Cretaceous carbonates are detached above Late Triassic evaporites (Faure-Muret 1955). Within the Variscan basement, the overall structure and particularly the schistosity and shear zones result from the superposition of Variscan and Alpine deformation histories (Faure-Muret 1955; Malaroda et al. 1970). During the Variscan orogeny, the Argentera massif experienced a polyphase deformation evolution culminating in a high-grade metamorphism and widespread partial melting in a context of dextral transcurrent deformation (Corsini and Rolland 2009). The Variscan metamorphic foliation is parallel to the massif's strike, and overall very steep, with a slight fan-like arrangement (Figure 2.4(a)). To the north, the contact with the Briançonnais along the Penninic Front is very steep as a result of transposition during the Alpine transpressive deformation.

2.3.1.2. *Tectonic evolution*

A great part of Alpine deformation and associated metamorphism in the Argentera massif is concentrated in a network of ductile shear zones that mostly reactivated late-Variscan ones with local occurrences of late Carboniferous metasediments (Sanchez et al. 2011a; Simonetti et al. 2018; Figure 2.4(a)). The Triassic rocks, close to the basement-cover contact, acted as a decoupling surface being deformed within the shear zones. Micron-scale phengite, chlorite and green biotite define an Alpine pervasive foliation, which is distributed over the entire massif (Figure 2.4(a)). The Alpine Argentera shear zone network is mainly right-lateral and forms a large-scale flower structure (Figure 2.4(a); Sanchez et al. 2011a, 2011b). Two main NW-SE trending shear zones, the Valletta shear zone and the Bersezio shear zone, crosscut the massif displaying dextral transpressive shearing (Figure 2.4(a); Corsini et al. 2004; Baietto et al. 2009; Sanchez et al. 2011a; Carosi et al. 2016; Simonetti et al. 2018). These shear zones merge into E-W trending top-to-the-south ductile thrusts. The Fremamorte shear zone crosscuts the Argentera granite and merges to the east into the transpressive Autier and Casterino shear zones (see Figure 2 in Sanchez et al. 2011). $^{40}Ar/^{39}Ar$ dating of Alpine syn-kinematic phengites within the compressional N-S Fremamorte-Autier shear zone provided an age of 33.6 ± 0.6 Ma (Sanchez et al. 2011a). To the SE, dating of the Casterino shear zone gave an age of 26.8 ± 0.7 to 26.3 ± 0.7 Ma. To the NW, the Bersezio and Valetta shear zones show deformation activity at 22.2 ± 0.3 Ma and 20.5 ± 0.3 Ma, respectively (Corsini et al. 2004; Sanchez et al. 2011a). The conditions during shear deformation, corresponding to maximal burial conditions of the NE part of the massif, were estimated at 375 ± 30°C and 4.8–7 kbar (Figures 2.5 and 2.6 and literature therein). This deformation relates to underthrusting underneath the Penninic front. Uplift of the massif occurred in two phases. A first phase of exhumation of the NE internal compartment of the massif is recorded by zircon fission track (ZFT) thermochronology, which provided similar age ranges as the $^{40}Ar/^{39}Ar$ ages. Fully reset ages of 29–20 Ma were obtained in the hanging-wall's NE part of the Bersezio-Fremamorte shear zone, while non-reset ZFT ages occur in the footwall's SW part of the shear zone (Figures 2.4(a) and 2.5; Bigot-Cormier et al. 2006). Further exhumation in the last 13 Ma is recorded by apatite U-Th/He ages (Bogdanoff et al. 2000; Bigot-Cormier et al. 2006; Sanchez et al. 2011b). Significant differences in AFT ages across the strike of the main faults have been obtained from the NE (12.9 Ma) to the SW (5.2 Ma) and are linked to components of differential vertical uplift along these right-lateral transpressive structures (Sanchez et al. 2011b). Such right-lateral movements reactivated the former mylonitic shear zones in a brittle-ductile regime within a similar kinematic framework.

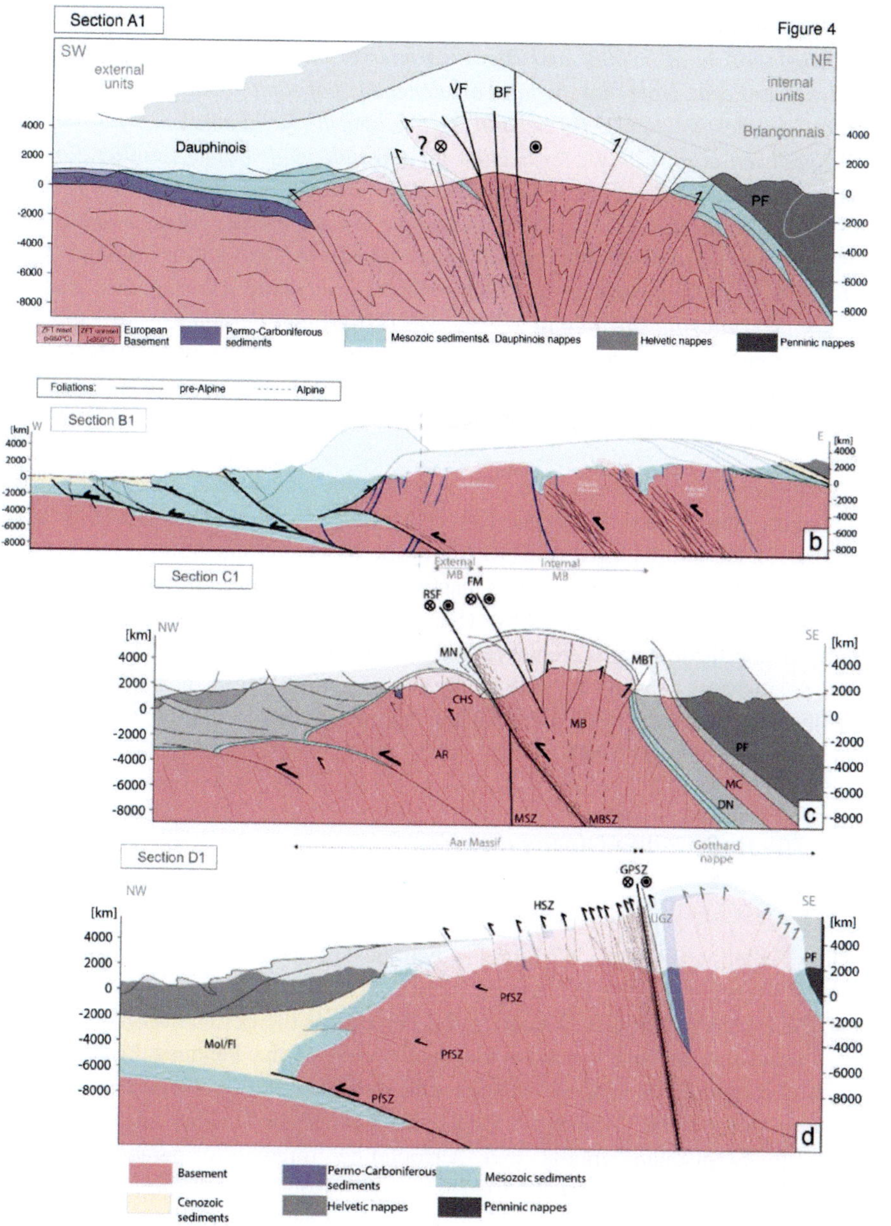

Figure 2.4. *Massif-scale tectonic cross-sections through the different ECM*

COMMENT ON FIGURE 2.4.– *(a) Section A1 through the Argentera massif. Modified after Bogdanoff et al. (2000) and Bigot-Cormier et al. (2006). VF: Valetta Fault, BF: Bersezio Fault. Note that the total displacements of some of the faults are under debate (see question mark). (b) Section B1 through the Belldonne massif. Modified after Bellahsen et al. (2014). (c) Section C1 through Mont Blanc/Aiguilles Rouges massifs. Modified after Egli and Mancktelow (2013) and Rossi et al. (2005). MB: Mont Blanc massif, RSF: Rhône-Simplon Fault, MBSZ: Mont Blanc shear zone; FM: Faille du Midi, MBT: Mont Blanc backthrust, MN: Morcles/Mont Jolie nappe, DN: Diablerets nappe, MC: Mont Chétif imbricate; AR: Aiguilles Rouges massif, MSZ: Miéville shear zone. (d) Section D1 through the Aar massif. Modified after Herwegh et al. (2017, 2020). Mol/Fl: Molasse/Flysch sediments, UGZ: Urseren-Gavera zone; HSZ: Handegg shear zones, PFZ: Pfaffenchopf shear zones, GPSZ: Grimsel Pass shear zones, PF: Penninic Front.*

A foreland Flysch basin (the Annot sandstone) developed during the first exhumation phase in the Oligocene (34–30 Ma, e.g. Jourdon et al. 2014). Similar apatite U-Th/He and FT ages of 6–10 Ma in the basement and Annot Flysch show that the basement-cover decoupling is inactive, and hence the entire massif has been passively exhumed since about 8 Ma (Sanchez et al. 2011b), but with continuing activity of the dextral faults (Sanchez et al. 2010; Figure 2.5). Similar fault-striae-based paleo-stress tensors as well as focal mechanisms from current seismicity confirm the persistence of a predominant dextral strike-slip component (Sanchez et al. 2010; Bauve et al. 2014).

2.3.2. *The Belledonne-Oisans massifs*

2.3.2.1. *Geological setting*

The Belledonne-Oisans s.l. massifs can be subdivided into the Oisans, the Grandes Rousses and the Belledonne sensu stricto massifs (Figure 2.1). In map view, the Oisans massif (also known as the Pelvoux or the Ecrins massif) is ~30 km wide, round-shaped, and located immediately west of the Penninic Front (Figure 2.1). It is composed of various Variscan units deformed during the Alpine orogeny: from east to west, the Combeynot Stephanian granite and its host micaschists, the La Meije Stephanian granite and surrounding migmatitic paragneisses and amphibolites (see Guillot et al. 2009). The Grandes Rousses massif is a 40 km long, elongated basement massif, located west of the Oisans massif (Figure 2.1). It consists of a Stephanian granite (Le Rochail), intruded into migmatites in the south of the massif. In the north, micaschists are associated with Stephanian volcanosedimentary rocks. The Belledonne Massif s.s. is the longest, yet narrowest, ECM (Figure 2.1). Along more than two-thirds of its length (total >110 km), it is separated by the "synclinal median" (composed of Mesozoic rocks, see below) into two parts: the external and internal Belledonne s.s. The external part consists mainly of micaschists, while the internal part is composed of an alternation of migmatites, lower Carboniferous granites, paragneisses, upper Devonian-lower Carboniferous magmatic rocks, and the Chamrousse ophiolite. The internal part also contains retrogressed eclogites.

Only the northern part of the Oisans massif is briefly described in the following. The southern part is more complex, probably still under-documented and would require more space for an exhaustive review (see Ford 1996; Lazarre et al. 1996; Dumont et al. 2011). Also the northern Belledonne part cannot be treated in detail, as there exist almost no studies on the Alpine evolution, except for Marquer et al. (2006).

A major former Jurassic half-graben basin (Bourg d'Oisans basin) inherited from the Piemont-Ligurian rifting is located between the Belledonne and the Grandes Rousses massifs (Barfety et al. 1979). The Jurassic rifting and associated sedimentation were controlled by the Col d'Ornon Lower Jurassic (and Middle Jurassic?) east dipping normal fault. This fault can be followed northward, where it dips toward the west, probably due to reorientation during Alpine deformation (Marquer et al. 2006). Today, the Jurassic metasediments are strongly folded (Figure 2.4(b)). The restored basin has a length of 20 km and a thickness of 3 km (Bellahsen et al. 2014). Post-rift metasediments from upper Jurassic to lower Cretaceous times represent a reduced thickness on the western Rochail massif (southern Grandes Rousses, Barféty and Gidon 1983). A similar inherited syn-rift basin is recognized near La Mure, southwest of the Oisans massif. Although debated, two additional syn-rift basins are often described: (i) west of the Belledonne massif, Butler et al. (2006) used balanced cross-sections to propose the existence of a thick Mesozoic syn-rift basin, while it is absent in other sections (Dumont et al. 2008); (ii) east of the Grandes Rousses massif a further syn-rift basin has been proposed, although its thickness is debated (compare Bellahsen et al. 2012 and Dumont et al. 2008).

In all these massifs the granites are generally only weakly, but pervasively, deformed. Some shear zones or faults dissect them but deformation is very local and heterogeneously distributed (Figure 2.4(b)). In contrast to granitic bodies, the Variscan micaschists are always intensely deformed in a pervasive manner. This deformation style suggests that at the P-T conditions of the Oisans massif (~300°C, 3 kbar), Alpine deformation mainly localized in the foliated micaschists, while the granites were probably too resistant.

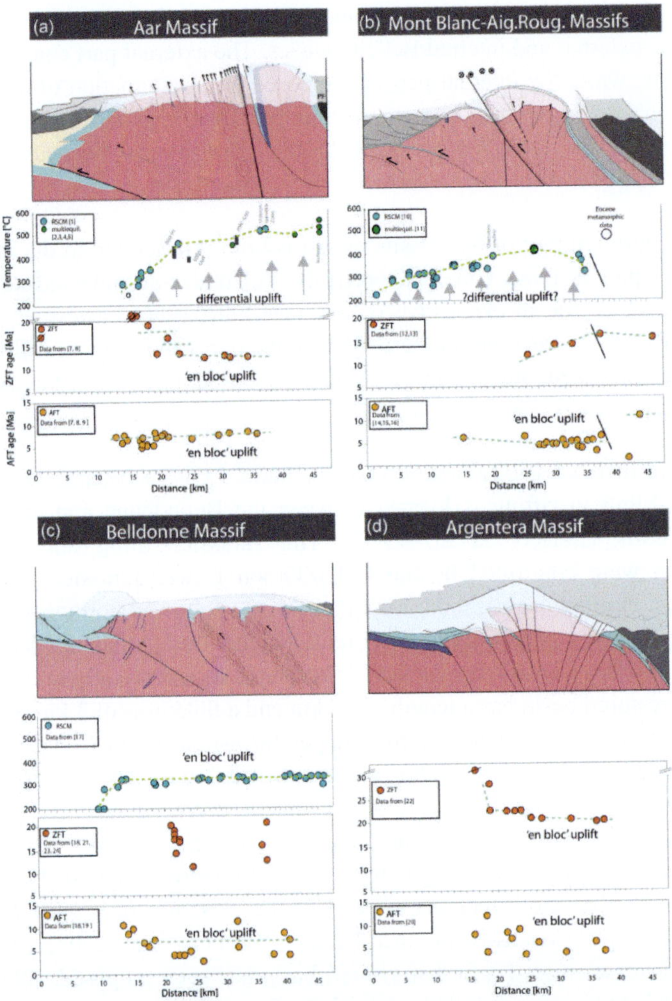

Figure 2.5. *Temperature-time relationships for the a) Argentera massif, b) Belledonne s.l. massif, c) Mont Blanc/Aiguilles Rouges massifs and d) Aar massif*

COMMENT ON FIGURE 2.5.– *Data sources: First row: simplified tectonic sections (from Figure 2.4); second row: T_{peak} versus horizontal distance; third and fourth rows: ZFT and AFT ages versus horizontal distance. [1]Nibourel et al. (2021b), [2]Herwegh et al. (2017), [3]Oliot et al. (2010), [4]Goncalves et al. (2012), [5]Todd and Engi (1997), [7]Herwegh et al. (2020), [8]Michalski and Soom (1990), [9]Valla et al. (2012), [10]Boutoux et al. (2016), [11]Rolland et al. (2003), [12]Glotzbach et al. (2008), [13]Leloup et al. (2005), [14]Soom (1990), [15]Fügenschuh and Schmid (2003),*

[16]Glotzbach et al. (2011), [17]Bellanger et al. (2015), [18]Seward et al. (1999), [19]Sabil (1995), [20]Bigot-Cormier et al. (2006), [21]Lelarge (1993), [22]Bogdanoff et al. (2000), [23]van der Beek et al. (2010), [24]Girault et al. (2022).

2.3.2.2. *Alpine tectonic evolution*

2.3.2.2.1. The cover

In the Grandes Rousses and Oisans massifs, the sedimentary cover is either detached from the basement or deformed disharmonically above deformed basement units (mapped as antiforms). In the east, below the Penninic Front, top-to-the-west shearing is intense but localized in the form of thrusted thin imbricates. Here, the sediment cover is detached from the basement and is stacked along a few thrusts, which root at the base of the cover (Figure 14 in Bellahsen et al. 2012; Ceriani et al. 2001). Toward the west, however, the sedimentary cover is disharmonically folded above basement cores, most likely because of the rheological influence of thick sediment units originating from the inherited basins and burial underneath the internal units (Bellahsen et al. 2012; Lafosse et al. 2016). There, no major detachment between basement and sedimentary cover occurred, implying a low strength contrast between basement and cover: under at least 10 km burial, the European crust became brittle-ductile and the basement strength was reduced.

The deformation sequence in the sedimentary cover has been studied through the foliation sequence that can be observed, especially between the Grandes Rousses and the Belledonne massifs and north of the Oisans massif (e.g. Gratier and Vialon 1980; Dumont et al. 2008; Bellahsen et al. 2012; Boutoux et al. 2014). The main foliation strikes N-S and formed coevally with basement shortening. It is rather steep in the western part, attesting for E-W contraction in the inherited Jurassic basins. It has a shallower dip toward the east, where inherited basins are absent or less well expressed. Moreover, E-W shortening, manifested by imbricate thrusting below the basal thrust of the Penninic Front, also resulted in shallower dip angles of the foliation within deformed cover sediments. In terms of deformation sequences, interpretations vary in details. In the Bourg d'Oisans basin, according to Dumont et al. (2008), the deformation sequence is comprised of four phases, each of which formed its own foliation. An early (Pyrenean?) S1 was followed by an early Alpine, E-W-trending, N-dipping S2 foliation. The subsequent S3 foliation trends N-S and dips toward the E. It is associated with the main top-to-west shearing. S4 has roughly the same orientation as S3, is west-dipping, but related to top-to-the-east shearing. According to Bellahsen et al. (2012) and Boutoux et al. (2014), only two distinct foliations are observed: an S1 developed early, coeval with an early shortening of the crust, locally consisting of top-to-the-east shearing (especially close to the basement-cover interface) and an S2 is the main foliation and developed later during top-to-the-west shearing.

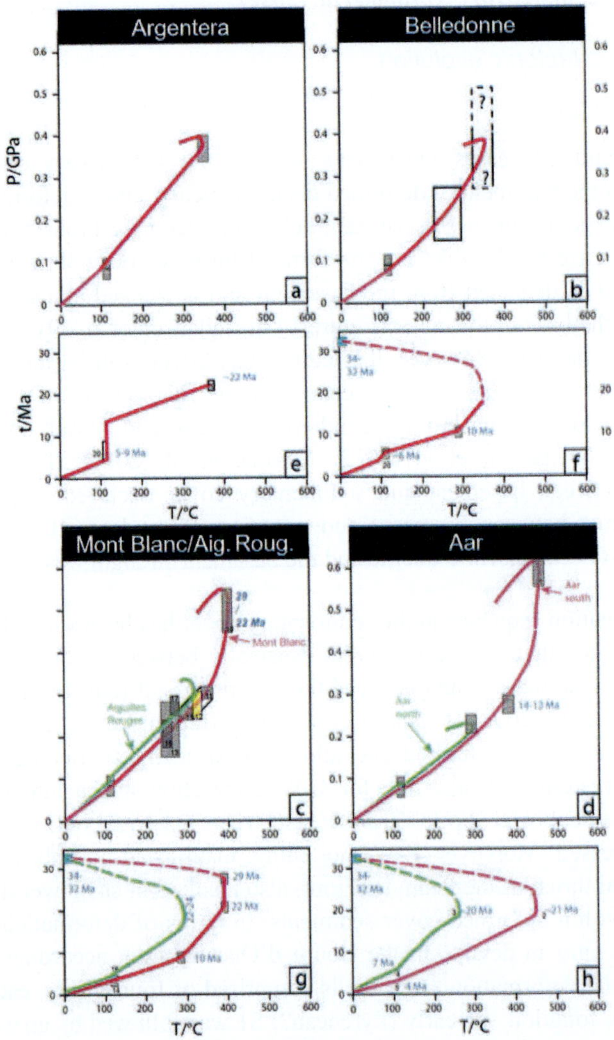

Figure 2.6. *Evolution of pressure–temperature–time in P/T (top row) and T-t (bottom row) diagrams. (a + e) Argentera massif using isotope and FT data; (b + f) Belldonne-Oisans massif with RSCM data in the sediments and isotope data in the basement. FT and isotope ages are used; (c + g) Aiguilles Rouges/Mont Blanc using RSCM data, isotope PT and age data and P-T and t data from clefts; (d + h) Aar massif using RSCM data, isotope PT and age data and P-T and t data from clefts. Data sources are the same as in Figure 2.5*

To summarize, the main E-W shortening (Oligocene) event is characterized by a N-S trending main foliation with simultaneous basement shortening and the top-to-the-west shearing in the basement. Variations in dip angles relate to variable amounts of shearing and/or effects of inherited structures from either Variscan deformation or Jurassic basin evolution.

2.3.2.2.2. The basement

$^{40}Ar/^{39}Ar$ ages on white micas indicate a first Alpine deformation of the Oisans massif during Rupelian times: strike-slip shear zones (dextral NE-SW and sinistral NW-SE) are dated at around 31–34 Ma (Simon-Labric et al. 2009) and distributed, reverse shear zones provide ages between 33 and 25 Ma (Bellanger et al. 2015). The latter shear zones are characterized by anastomosing shear bands that strike N-S and dip around 45° toward the east. This deformation occurred at temperatures of ~300°C and pressures of ~3 kbar (Simon-Labric et al. 2009; Bellanger et al. 2015). No change in temperature or pressure during their activity was documented, implying rather isothermal/isobaric deformation conditions. This suggests that neither cooling nor exhumation took place between 33 and 25 Ma. The total shortening of the External Zone accommodated by these shear zones is ~16 km (Bellahsen et al. 2014). Cooling and exhumation started around 25–27 Ma in the Oisans massif, or possibly before, as indicated by the oldest ZFT data in the Oisans massif (at La Meije; van der Beek et al. 2010; Figure 2.5).

In the Grandes Rousses massif, anastomozing shear zones in the basement have also been documented (Bellahsen et al. 2012). They strike N-S to NNE-SSW and dip 45° toward the East. This deformation occurred at temperatures of ~300°C and pressures of ~3 kbar (Crouzet et al. 2001; Bellanger et al. 2015). The shear zones have not been dated but are considered to have approximately the same age as those in the Oisans massif. The onset of cooling of the Grandes Rousses Massif is dated at ~25 Ma (or slightly before) by thermopaleomagnetism (Crouzet et al. 2001) and LT-thermochronology (Girault et al. 2022).

In the northern Belledonne massif, the Alpine deformation consists of shear zones mainly west verging (Girault et al. 2022) but also conjugate (west and east-verging) in some places (Marquer et al. 2006). These deformations are most likely Oligo-Miocene as in the other massif. However, only a few radiometric age dates that yield 15 Ma are available (Egli et al. 2017), an age which probably reflects late movements along the shear zones. In the South Belledonne massif, ZFT data indicate the onset of cooling at ~18 Ma (Figure 2.8) but this may have started earlier, around 27 Ma, like in the Oisans massif (see discussion in Girault et al. 2022).

The 18–25 Ma time period corresponds to the age of activation of the Subalpine fold-and-thrust-belt where thrusts root in a Triassic or Liassic detachment (Deville et al. 1994; Philippe et al. 1998; Girault et al. 2020). The main exhumation of the Belledonne-Oisans massifs is most likely linked to the late Oligocene (?) to early Miocene activation of such frontal crustal ramps, which caused the detachment and thrust activation in cover sediments. This suggests that localized shortening (i.e. along a thrust) started just after the cessation of the distributed shortening phase (Bellahsen et al. 2014). The frontal thrusts accommodated ~16 km of shortening (Figure 2.4(b)). The available thermochronological data (Girault et al. 2022) indicate "en bloc" cooling and exhumation of the massif from 20 to 25 Ma onward (Figure 2.5).

2.3.3. *The Mont Blanc/Aiguilles Rouges massifs*

2.3.3.1. *Geological setting*

The Mont Blanc massif represents a NE-SW striking ~80 km long and ~25 km wide dome-like structure, with culmination axes plunging toward the NE and SW (Figure 2.1). It is bound to the NW by the overturned sedimentary series of the Chamonix syncline (Figure 2.4(c); Paréjas 1922; Epard 1986; Belliere 1988) and to the SE by the massif's stratigraphically normally oriented Mesozoic cover (e.g. Grasmück 1961; Egli and Mancktelow 2013). Generally, the tectonic contact of the steeply SE-dipping Mont Blanc shear zone (MBSZ, Figure 2.4(c)), with its top-to-NW-directed shear sense indicators, represents the northwestern boundary (Eltchaninoff-Lancelot et al. 1982; Butler 1983; Gourlay and Ricou 1983; Platt 1984; Leloup et al. 2005), the tectonic character of the contact of which, however, is debated by Epard (1986). Egli and Mancktelow (2013) interpret the MBSZ as a relatively broad shear zone extending from the basement into the cover, without "discrete" tectonic contact and that the sediments are therefore strongly sheared in an overturned position but still (par)-autochthonous on the Mont Blanc basement. The MBSZ has a long lasting history initiating as a pre-Alpine tectonic structure synchronous with the intrusion of the Mont Blanc granite (Belliere 1988) that has been reactivated during Alpine deformation (Belliere 1988; Rossi et al. 2005; Rolland et al. 2008). Toward the SE, the Faille du Midi forms a second ductile shear zone separating the External Mont Blanc in the north from the southern Internal Mont Blanc (Figure 2.4(c); von Raumer 1971; Epard 1990; Leloup et al. 2005; Egli and Mancktelow 2013; Egli et al. 2017). Some steep localized shear zones, with evidence for strong Alpine metasomatic alteration with lower crustal or mantle fluid signatures, dissect the Internal Mont Blanc massif, which mainly consists of the Mont Blanc granite (Figure 2.4(c); Bertini et al. 1985; Rossi et al. 2005; Rolland et al. 2008, Rossi and Rolland 2014; Rolland and Rossi 2016). These authors define a fan-like geometry with mainly reverse faulting (Bertini et al. 1985), corresponding to dominantly top-to-NW- and top-to-SE-directed movements in the NW and SE parts, respectively (Figure 3 in Rossi et al. 2005). At the SE massif margin the Mont Blanc back-thrust occurs, which geometrically represents a continuation of this fan-structure (Figure 2.4(c)) and placed basement rocks on top of their Mesozoic cover (Antoine et al. 1975; Escher et al. 1993). This situation has been interpreted to result from back-thrusting toward the SE (e.g. Guermani and Pennacchioni 1998; Leloup et al. 2005; Egli and Mancktelow 2013) or back-folding (Escher et al. 1993; Escher and Beaumont 1997; Steck et al. 2001). The Mont Blanc massif was interpreted to represent the antiformal core of the Lower Helvetic Morcles nappe (Lugeon 1914; Ramsay 1981; Escher et al. 1993), which forms a large-scale recumbent fold (Masson et al. 1980; Ramsay 1981; Dietrich and Casey 1989; Escher et al. 1993; Pfiffner 1993; von Tscharner et al. 2016). However, in the same tectonic position, the lateral continuation toward the SW, which forms the Mont Joly nappe, shows a much less developed and therefore smaller inverted limb (Epard 1986). Controversial interpretations exist as to whether the Morcles/Mont Joly nappe (Figure 2.1) represents the cover of the Mont Blanc massif or if it is derived from the SE of it (e.g. compare Boutoux et al. 2014 and Leloup et al. 2005). While different authors agree that the Morcles/Mont Joly nappe represents an inverted limb in continuity to the Chamonix Syncline (Figure 2.4(c)), major differences in interpretation and origin of the location of the normal limb exist. Escher et al. (1993), Bauville et al. (2013) and Boutoux et al. (2014) root this limb into the Faille du Midi, whereas Leloup et al. (2005) and Egli and Mancktelow (2013) suggest a continuation corresponding to the normal sedimentary sequence on the SE rim of the Mont Blanc massif.

With the same strike, the Aiguilles Rouges massif is located in front of the Mont Blanc massif (Figure 2.1). Like the latter, it represents a dome-like structure with its long axis plunging toward the NE and SW. It has a normal stratigraphic Mesozoic cover, dipping toward the SE into the Chamonix syncline as well as to the NW into the N-Alpine foreland (Figure 2.4(c)). It contains the Permo-Carboniferous half-graben of Dorénaz-Salvan (Figure 2.1(b)), which is shortened and in parts inverted, owing to Alpine compression (Badertscher and Burkhard 1998). At the NW rim of the half-graben, the Miéville shear zone (Figure 2.4(c)) represents the major massif internal large-scale shear zone (Reinhard and Preiswerk 1927; Steck and Vocat 1973; Kerrich et al. 1980; Fitz Gerald and Stunitz 1993; Egli et al. 2017). The Miéville shear zone and the Permo-Carboniferous half-graben are both exposed in the northern Aiguilles Rouges massif and disappear underneath the MBSZ toward the south (Figure 2.4(c)). In accordance with seismic sections from other Alpine transects and based on a tectonic restoration, Burkhard and Sommaruga (1998) suggest the occurrence of a sole thrust ramp at the northern rim of the Aiguilles

Rouges massif which descends underneath the Mont Blanc massif toward the S (Figure 2.7(c); see also Lacombe and Mouthereau 2002; Egli and Mancktelow 2013; Bellahsen et al. 2014; Boutoux et al. 2016; Egli et al. 2017).

2.3.3.2. *Tectonic evolution*

The earliest deformation is related to in-sequence thrusting of Mesozoic sediments forming the Upper Helvetic nappes, which consists from top to bottom of the Wildhorn, Diablerets, Ardon and Morcles nappes (Figure 2.1; Masson et al. 1980; Ramsay 1981; Escher et al. 1993; Pfiffner 1993; Burkhard and Sommaruga 1998; Pfiffner et al. 2011; Bellahsen et al. 2014). This nappe stacking must have initiated after both the out-of-sequence thrusting of Penninic units at the Penninic front at 34–30 Ma (Fügenschuh and Schmid 2003; Simon-Labric et al. 2009) and the deposition of the youngest sediments on top of the ECM, but before the updoming of the Mont Blanc/Aiguilles Rouges massifs. The thrust of the Helvetic nappes on top of it (Figure 2.1) yield radiometric $^{40}Ar/^{39}Ar$ dating on neocrystallized white micas with an age range from 32 to 15 Ma (Figure 2.8). The main thrusting phases were: 28–23 Ma for the Wildhorn nappe (Kirschner et al. 1996, 1999), 30–28 Ma for the Diablerets nappe (Kirschner et al. 2003), 27 Ma for the Ardon (Kirschner et al. 2003) and 30–15 Ma for the Morcles nappe (Kirschner et al. 1995, 2003). Note that these ages need to be treated with care, since they document neocrystallization of white mica, which might have been long lasting and strongly depended on fluid presence. In the basement of the Mont Blanc massif, the oldest Alpine radiometric age yields 29.44±0.95 Ma for allanite, interpreted to be synkinematically grown along the steep reverse faults (Cenki-Tok et al. 2014). Together with white mica Rb-Sr ages of 30–27 Ma in the adjacent Mont Chétif imbricate (Egli et al. 2016), this age range represents fluid-assisted (Marshall et al. 1998; Rossi et al. 2005; Cenki-Tok et al. 2014) underthrusting and shortening of the Mont Blanc/Aiguilles Rouges area under the Penninic front contemporaneous with an embryonic stage of Helvetic thrusting. During this Oligocene deformation, a pervasive foliation evolved in the sedimentary cover of the Mont Blanc massif (S2 of Egli and Mancktelow 2013). Reactivation of an anastomosing network of steep reverse shear zones with steep lineations represented the next stage of basement internal deformation at conditions of 5±0.5 kbar and 400±25°C (Figures 2.5 and 2.6; Rolland et al. 2003; Rossi et al. 2005). The present 45°-dipping Mont Blanc shear zone reflects the largest structure that was already activated during the Late Oligocene as a basal thrust of the Morcles nappe. Aforementioned fan-like geometry of the steep Mont Blanc massif internal shear zones with their SE-directed sense of shears in the NW and SE parts of the massif, respectively (Figure 2.4(c); Rossi et al. 2005) related to a dextral transpressive pop-up system at about 22–18 Ma (Rolland et al. 2008). Despite this widespread fault network, however, deformation was preferentially localized along two major massif internal fault structures, the Mont Blanc and Faille du Midi shear zones (Leloup et al. 2005; Egli and Mancktelow 2013; Egli et al. 2017), as well as in the sediments of the Helvetic root zone to the SE (Figure 2.4(c); Egli and Mancktelow 2013).

The Mont Blanc massif experienced near T_{max} conditions from 29 Ma till 22 Ma (Rolland et al. 2003; Rossi et al. 2005; Bellahsen et al. 2014; Cenki-Tok et al. 2014; Boutoux et al. 2016), with temperatures as high as 400°C (Figures 2.5 to 2.7; Boutoux et al. 2016; Girault et al. 2020). Boutoux et al. (2016) therefore suggested that these Oligocene reverse shear zones did not significantly exhume the Mont Blanc massif, although this still needs to be confirmed by barometric data. Afterwards, during NW-SE shortening, the massif experienced a substantial decompression (see Herwegh et al. 2020 and references therein) requiring unloading and exhumation by reactivation of the aforementioned shear zones/reverse faults in the time interval from 22 to 18 Ma, and SE-directed reverse faulting at 16 Ma (Rolland et al. 2007) along the Mont Blanc back thrust (Figure 2.4(c)). Frequent horizontal quartz-rich veins developed simultaneously with the steep lineations (related to the reverse faulting) and contain relic monazite, whose ages are as young as 12 Ma and adularia of 10 Ma (Rossi and Rolland 2014; Bergemann et al. 2019). Note that during this period of deformation and rock uplift some kind of decoupling at the massif's base may be necessary (see sections 2.5.5 and 2.5.6).

During continued NW-SE shortening, uplift of the Lepontine area in the Central Alps was synchronous to the formation of the Simplon-Rhone extensional system from 20 Ma onwards (Campani et al. 2010), with SW-directed movements of the hanging wall above the Simplon low-angle normal fault (Figure 2.1; see also Mancktelow 1985; Steck 2008; Campani et al. 2010; Figure 15 in Egli and Mancktelow 2013). As a consequence, the NE-SW to NNE-SSW trending steep Rhone-Simplon fault formed a dextral lateral ramp disconnecting the NW from the SE block. Under this transpressive regime, the Mont Blanc massif formed a restraining bend relative to the Penninic Alps with deformation being accommodated along dextrally reactivated transcurrent faults on its NW (Mont Blanc and Faille du Midi shear zones) and SE (Helvetic root zone) rims (Egli and Mancktelow 2013). $^{40}Ar/^{39}Ar$ ages on synkinematic white mica in phyllonitic shear zones suggest a strike-slip activity till at least 9.5 Ma (Egli et al. 2017). Besides strike-slip movements in the above mentioned major shear zones, vertical veins indicate additional distributed shear movements internally within the Mont Blanc/Aiguilles Rouges massifs. These vertical veins were active until 8–7 Ma (Bergemann et al. 2019) confirming the findings of Egli et al. (2017). The strike-slip deformation was followed by an open transverse folding across the entire area (Egli and Mancktelow 2013; Egli et al. 2017).

Since pressure estimations are lacking for the Aiguilles Rouges massif, such estimates can only be made by transforming the temperatures of Boutoux et al. (2016) into pressures assuming a geothermal gradient of 25°C/km (Figure 2.5) yielding maximum burial depths of about 12–13 km. In this massif, a pervasive, steep SE dipping foliation and reverse faults represent the oldest Alpine structures (Figure 2.4(c)). They are associated with top-to-NW movements (Ayrton 1980). They probably evolved at or slightly after peak metamorphic temperatures of 320°C some 20–18 Ma ago (Burkhard and Sommaruga 1998; Leloup et al. 2005; Boutoux et al. 2016). The subvertical Miéville shear zone represents the major exposed fault structure (Reinhard and Preiswerk 1927; Steck and Vocat 1973; Kerrich et al. 1980; Fitz Gerald and Stünitz 1993). Despite its pre-Alpine origin, recent studies suggest, at least in part, an Alpine reactivation at ~15 Ma (Rolland et al. 2008; Egli et al. 2017). Based on these steep fabrics, an early stage with a strong vertical rock uplift component can also be inferred for the Aiguilles Rouges massif.

The relative temporal sequence of up-doming of the Aiguilles Rouges and Mont Blanc massifs is controversially debated. While Leloup et al. (2005) suggested an earlier up-doming of the Aiguilles Rouges and subsequent out-of-sequence thrusting of the Mont Blanc massif, Burkhard and Sommaruga (1998) favored an in-sequence deformation with rock uplift first in the Mont Blanc followed by the Aiguilles Rouges massif, with the latter model being in line with kinematic analyses of Egli and Mancktelow (2013). More recently, in light of new thermochronological data, Boutoux et al. (2016) instead inferred a contemporaneous uplift of both massifs. An early uplift along steep reverse systems in both massifs was more pronounced in the Mont Blanc massif (Figure 2.6(b)), which was affected by a much more intense transpressional component compared to the Aiguilles Rouges massif. However, both massifs experienced a younger stage (< 9 Ma) of "en bloc" exhumation, which occurred on top of a basal thrust (Figure 2.5; Glotzbach et al. 2011; Boutoux et al. 2016).

In the case of the Mont Blanc area, ZFT isoages are slightly tilted, indicating the absence of abrupt changes in relative vertical movements of the massif interior (Glotzbach et al. 2011) and of the activity of the basal thrust. Temperatures were partly too low in the Aiguilles Rouges massif to reset ZFT (Soom 1990; Vernon et al. 2008). The inferred "en bloc" exhumation persisted through time, as indicated by similar isoage trends for ZFT (12–11 Ma), ZHe (~9–6 Ma; Boutoux et al. 2016) and AFT (~4 Ma; Rahn et al. 1994; Glotzbach et al. 2008; Boutoux et al. 2016). This massif-scale "en bloc" exhumation related to thrust tectonics is consistent with a pervasive deformation along brittle structures without major offsets in the thrust's hanging wall (Egli and Mancktelow 2013; Egli et al. 2017).

Mean AHe data of the Mont Blanc and Aiguilles Rouges massifs yield ~1.5–2 Ma (Glotzbach et al. 2008; Central Mont Blanc massif) or ~4–6 Ma (Valla et al. 2012, NE rim of Aiguilles Rouges massif). This spatial age difference might be attributed to a more pronounced erosional unloading in the case of the central Mont Blanc area, which may be accelerated by glacial carving after 2–1.5 Ma (Glotzbach et al. 2008, 2011; Valla et al. 2012; Fox et al. 2015).

2.3.4. *The Aar/Gastern massifs*

2.3.4.1. *Geological setting*

The Aar massif is located in the Central Swiss Alps and is the largest of the ECM (Figure 2.1). Like the Mont Blanc/Aiguilles Rouges area, it also expresses a dome-like structure with a NE-SW trending dome axis, the latter being horizontal in the central part but plunging toward both the NE and the SW at its rims (Schmid 1975; Pfiffner et al. 1997a, 2011). Also the Gastern and Aiguilles Rouges massifs both have the same more external tectonic position compared to their big sisters, respectively, the Aar and Mont Blanc massifs. The Gastern Massif occurs only at the NW end of the Aar massif. The Aar massif's long axis spans over ~120 km, even up to ~150 km if extended to include the basement windows underneath the Helvetic sedimentary nappes to the east (e.g. Limmeren and Vättis windows; Figure 2.1). Its maximum surface width is ~25 km. The basement-cover contact in the northern Aar massif moderately dips toward the north (Figure 2.4(d)) in the E and steepens in the central part toward the Eiger-Mönch-Jungfrau region in the W (Pfiffner et al. 1997a, 2011; Mair et al. 2018; Nibourel et al. 2021b). The autochthonous to parautochthonous sediment cover was overthrusted by the Helvetic sedimentary nappes (Druesberg-Axen nappes in the E and Wildhorn nappe complex in the W; Pfiffner et al. 2011 and references therein). In the northernmost parts, Cenozoic orogenic sediments of the under- and overfilled Flysch- and Molasse basins are located in between the autochthonous sediments of the ECM and the Helvetic nappes (Figure 2.4(d)). The western end of the Aar massif is covered by a large-scale recumbent thrust-fold of the Doldenhorn nappe, with a stretched and ductiley deformed inverted limb (Burkhard 1988; Herwegh and Pfiffner 2005; Krayenbuhl and Steck 2009). Similarly to the Mont Blanc/Morcles nappe system (Escher et al. 1993; Pfiffner 1993; Steck et al. 1997), the Doldenhorn nappe also contains sheared basement parts (Aar massif) in the core of the fold (Burkhard 1988; Herwegh and Pfiffner 2005; Krayenbuhl and Steck 2009; Berger et al. 2020). Multistage-deformation during Alpine NW-SE shortening inverted a former sedimentary basin by NW-directed thick-skinned thrusting onto the Gastern Massif. Note that both the Doldenhorn nappe and Gastern Massif gradually disappear toward the central part of the Aar massif (Figure 2.1). The Aar massif's southern rim is steeply dipping to the

SE, where an abrupt tectonic transition to the sediments of the Urseren-Garvera zone and the adjacent basement units of the Gotthard nappe occurs (Figure 2.4(d)). At the southern rim of the central Aar massif, the crustal-scale dextral strike-slip Grimselpass shear zone (Figure 2.4(d)) dissects the massif in large parts in an ENE-WSW direction, being slightly oblique to the massif's strike (Figure 2.1). At the southern end, the massif terminates in the Rhone-Simplon fault (Figures 2.1 and 2.4(d); Mancktelow 1985, 1990; Steck and Hunziker 1994; Steck 2008; Campani et al. 2010, 2014). Within the massif, several NE-SW trending steeply SE-dipping large-scale faults strongly dissect the entire Aar massif from N to S (Figure 2.4(d); Steck 1968; Choukroune and Gapais 1983; Wehrens et al. 2016, 2017; Herwegh et al. 2020; Baumberger et al. 2022). Along some of these faults, narrow steep Mesozoic and/or Permo-Carboniferous sediment wedges exist, which are also intensely deformed (Figure 2.4(d); Heim and Heim 1916; Morgenthaler 1921; Rohr 1926; Muller and Arbenz 1938; Kammer 1989; Herwegh et al. 2020; Nibourel et al. 2021b). The relationship between basement-internal deformation, sediment wedges and autochthonous / parautochthonous sediments has been investigated in detail in the massif's central part as well as at its western and eastern ends (Steck 1968; Krayenbuhl and Steck 2009; Nibourel et al. 2018, 2021a, 2021b). The occurrence of syn- and anticlines in the sediments with NE and SW plunging axes, respectively, and an apparent geometric correlation of the basement-cover contact, led to the interpretation of a general deformation by folding under non-coaxial shearing (Steck 1968; Escher et al. 1993; Krayenbuhl and Steck 2009). Recent investigations of these regions revealed, however, a first stage of imbricate thrusting, with faulting in the rigid basement units and a ductile accommodation of the displaced basement blocks by folding within the mechanically weaker carbonate-shale sequences of the cover (Labhart 1966; Mair et al. 2018; Buess 2019; Herwegh et al. 2020; Nibourel et al. 2021b). A structural feature completely absent in the south is the appearance of a spaced-cleavage at the massif's NW rim, which transforms into a multitude of phyllonitic thrust planes, repeated on a decameter spacing over a vertical distance of about 1,500 m (Figure 2.4(d); Herwegh et al. 2020). These features dip weakly to moderately toward the SE and can be correlated with subsurface thrust planes found in seismic sections (Pfiffner et al. 1997a). Herwegh et al. (2017, 2020) therefore interpreted them as major crustal-shear zones.

2.3.4.2. *Tectonic evolution*

In the Central Swiss Alps, stacking of the proximal Penninic nappes terminated before the Bergell intrusion (30–32 Ma; von Blanckenburg 1992), which crosscut the thrust contacts (Rosenberg et al. 1995). Subsequent emplacement of distal Penninic nappes and shearing off of Upper Helvetic sediment nappes occurred gradually, mainly by in-sequence thrusting (except for the major out-of-sequence Glarus thrust, e.g. Schmid 1975; Burkhard 1988; Pfiffner 1993; Herwegh et al.

2008; Pfiffner et al. 2011). Deposition of the lower to mid Oligocene Taveyannaz formation in underfilled Flysch basins represents the upper time limit (34–29 Ma, Fischer and Villa 1990; Ruffini et al. 1995; Lu et al. 2018) for the onset of Helvetic nappe thrusting (Figure 2.8). This N- to NW-directed thrusting culminated between 28 and 23 Ma (Huon et al. 1994; Kirschner et al. 1995, 2003; Cardello et al. 2019). The emplacement of the Gotthard nappe must have taken place after the detachment of the Upper Helvetic nappes but prior to the onset of exhumation of the Aar massif at 22–20 Ma (Berger et al. 2011). Janots et al. (2009) defined a maximum time range for this emplacement from 30 to 18 Ma. Based on aforementioned time constraints this can now be further narrowed to 28–22 Ma. For the Helvetics, this stage presented the transition from thin- to thick-skinned tectonics (Pfiffner 2016), characterized by the shearing-off of thin sediment-basement slices (Mair et al. 2018; Nibourel et al. 2021a) and the formation of the Doldenhorn nappe some 22 Ma ago (Burkhard 1988; Herwegh and Pfiffner 2005; Krayenbuhl and Steck 2009; Berger et al. 2020). It is the emplacement of this last large-scale Helvetic nappe which terminated the first stage of horizontal tectonics of the Alpine continent-continent collision (Herwegh et al. 2020). Note that the N-S to NW-SE shortening in the Helvetic realm by thrusting and folding occurred simultaneously with top-to-SW normal faulting along the Simplon ductile shear zone (35–20 Ma; Steck 1984; Steck und Hunziker 1994) in the SE and S of the Aar massif.

The Aar massif experienced peak metamorphic conditions at 22–20 Ma (Figure 2.6; Challandes et al. 2008; Rolland et al. 2009; Berger et al. 2017b). A pervasive steep SE-dipping foliation formed, either by intensifying pre-existing foliations in pre-Alpine gneisses or by the formation of a new foliation in the case of the post-Variscan plutonites. The onset of steep reverse faulting coupled with erosion is responsible for decompression under nearly isothermal conditions (Figures 2.5 and 2.6; Challandes et al. 2008; Rolland et al. 2009; Diamond and Tarantola 2015; Berger et al. 2017; Wehrens et al. 2017; Herwegh et al. 2020). The previously mentioned widespread dissection of the Aar massif by steep reverse shear zones (Steck 1968; Wehrens et al. 2016, 2017) led to gradual differential uplift from the S (max) to N (min) and produced substantial vertical movements (Figures 2.3, 2.4(d), 2.5 and 2.6; Herwegh et al. 2017, 2020; Nibourel et al. 2021a, 2021b). Along this gradient, maximum rock exhumation rates changed from 1.55 to 0.1 km/Ma (Herwegh et al. 2020). Associated advective heat transfer and elevated temperatures resulted in localization of deformation, including an enhanced flattening component in the southern Aar massif (Choukroune and Gapais 1983; Marquer and Gapais 1985; Wehrens et al. 2017).

Simplon low-angle normal faulting in the SW of the Aar massif started at that time (~20 Ma) under transpressive conditions (Figure 2.1; Mancktelow 1985;

Campani et al. 2010, 2014; Haertel and Herwegh 2014). With progressive NW-SE shortening, this transpression also affected the Aar massif's southwestern domain, by a gradual reactivation of former reverse faults (Ziegler and Isler 2013) as dextral strike-slip to oblique-slip faults in the Lötschen valley (Berger et al. 2020). Dextral shearing also progressed in the NE direction of the Aar massif's central southern rim, forming there the crustal-scale Grimselpass shear zone (Figures 7 and 12, respectively, in Wehrens et al. 2017 and Herwegh et al. 2020). Dextral shearing initiated first under ductile deformation conditions at about 12 Ma (Rolland et al. 2009; Berger et al. 2016, 2017a; Wehrens et al. 2016). With progressive shortening and slowing vertical tectonics, the basement rock units with their pervasive fault zone network (Wehrens et al. 2017; Baumberger et al. 2022) cooled down, resulting in a transition from deformation by viscous granular flow to frictional granular flow in today's exposed ultra-fine-grained polymineralic fault rocks such as gouges, which have been dated to be active till at least 8 Ma (Pleuger et al. 2012). Embrittlement associated dilatancy combined with strike-slip shearing and the evolution of subvertical pull-apart tubes (see Figure 10 in Belgrano et al. 2016) probably established a pervasive vertical fluid flux in this domain, as suggested by Challandes et al. (2008). The lower crustal fluid signature is expressed in the stable isotope composition of calcites (Rossi and Rolland 2014). Note that the dextral strike-slip shearing and its role for uprising hydrothermal waters continues till the present day (Pfeifer et al. 1992; Hofmann et al. 2004; Belgrano et al. 2016; Diamond et al. 2018; Egli et al. 2018; Berger and Herwegh 2019). Dextral shearing terminated in the Furka region. Further to the East there exist only several small-scale low-grade ductile faults with only the Clavaniev Zone (Disentis-Oberalp Pass) as a somewhat larger brittle dextral fault with N-block up movements (Leu and Wyss 1992; Pleuger et al. 2012). In addition, moderately NW-dipping but SE-directed small back thrusts occur at the southern boundary of the eastern Aar massif (Nibourel et al. 2021a). This SW-NE change in deformation style along the Aar massif's southern domain probably relates to a lack of Simplon low-angle faulting in the eastern domain (Figure 2.1).

In the north of the Aar massif, flat to moderately SE dipping thrust zones with a NW-directed transport initiated at about 13–12 Ma (Herwegh et al. 2020). Since they were active contemporaneously to the dextral strike-slip shearing in the south, the Aar massif must have experienced a large-scale strain partitioning. Herwegh et al. (2020) showed that these decameter-spaced fault zones define a crustal-scale shear domain, which occurs at repetitive intervals with depth, as can be inferred from NRP-20 seismic surveys (Pfiffner et al. 1997b). It is this continuous NW-thrusting that led to the exhumation of the entire Aar massif "en bloc", as is evident by non-disrupted isoage lines of ZFT at 13–12 Ma and AFT at 8–5 Ma along N-S transects throughout the Central Aar massif (Figure 2.5; see also Herwegh et al.

2020). Using ZFT and AFT data and a geothermal gradient of 25°C/km, maximum rock uplift rates of 0.65 km/Ma and 0.8 km/Ma, respectively, were calculated for the Central Aar massif (Herwegh et al. 2020). These rates are in the same range as those suggested for two pulses of slightly enhanced exhumation of about ~0.7 km/Ma for the time intervals of 16–14 and 10–7 Ma estimated by fission track analyses and thermal modeling by Glotzbach et al. (2010). Recently, Nibourel et al. (2021b) showed that ZFT, ZHe and AFT ages become younger toward the east, suggesting a gradual massif parallel growth of the Aar massif. Hence, the early N-directed thrusting must have initiated at its northern rim in the case of the Central Aar massif. With progressive shortening, however, thrust-domains stepped into deeper crustal levels but additionally grew to the NE and probably also the SW. As a consequence, the Central northern rim region accommodated most strain – a fact that is impressively manifest by the Aar massif's steep north face in the Wetterhorn-Schreckhorn-Eiger-Mönch-Jungfrau region (Scabell 1926; Collet and Parejas 1931; Kammer 1989; Haenni and Pfiffner 2001; Mair et al. 2018). The previously mentioned discrete lower greenschist facies shear zones (thrusts) led to rigid basement slivers wedging into mechanically weaker sediments. The latter were deformed in a ductile manner, resulting in recumbent sequences of narrow syn- and anticlines. This ductile deformation style in the sediments led to the wrong conclusion of an isoclinally folded basement-cover contact (e.g. Ramsay et al. 1983; Giba 2006; Krayenbuhl and Steck 2009; Bellahsen et al. 2014), a structural misinterpretation that has been continuously re-used in a variety of tectonic sections since the times of Albert Heim (1921). With progressive down stepping of the thrust domains and migration to the NW from 12 to 6 Ma, deformation proceeded into the Alpine foreland. As a consequence, imbricate thrusting of the Subalpine Molasse (von Hagke et al. 2012, 2014; Mock et al. 2020) as well as thrusting of the Jura Mountains along a basal detachment within Triassic evaporites (Laubscher 1961; Philippe et al. 1996; Burkhard and Sommaruga 1998; Mosar 1999; Sommaruga 1999; Sommaruga et al. 2012) or even within the basement (Guellec et al. 1990; Jouanne et al. 1995; Phillipe et al. 1996; Mosar 1999; Becker 2000; Lacombe and Mouthereau 2002; Ustaszewski et al. 2006, 2007; Madritsch et al. 2008; Lacombe and Bellahsen 2016) occurred. Since ~4 Ma, strike-slip faults cut the Triassic detachment horizon in the central Swiss Plateau Molasse, requiring, at least since then, basement-involved deformation in this domain (Mock and Herwegh 2017). From 5 to 1 Ma, exhumation of the Central Aar massif occurred at rates of 0.3–0.5 km/Myr (Reinecker et al. 2008; Glotzbach et al. 2010; Valla et al. 2012; von Hagke et al. 2012, 2014; Fox et al. 2016) but accelerated afterwards due to glacial erosion (Haeuselmann et al. 2007). Recent seismic activity underneath the Aar massif suggests that deformation is ongoing (mainly strike-slip focal plane solutions for the SW Aar massif; Deichmann 2014) and so is the interplay between deformation and erosion, as manifest by the recent relative uplift, which increases from about 0.15 to 0.7 mm/a from N to S (Schlatter 2006).

2.4. The ECM embedded in a crustal-scale context

Over the last decades, major advances in knowledge on the lithosphere-scale structure of the Alps were provided by controlled-source seismic reflection/refraction profiles (NFP20: Pfiffner et al. 1997b; TRANSALP: Transalp Working Group 2002; ECORS-CROP: Roure et al. 1996; Schmid and Kissling 2000), 3D teleseismic tomography (Lippitisch et al. 2003) and 3D local P-wave earthquake tomography (Diehl et al. 2009). In this review, we mainly rely on the data of the latter study and associated tectonic cross-sections of Schmid et al. (2017) for the Argentera, Belledonne-Oisans and Mont Blanc/Aiguilles Rouges massifs and of Herwegh et al. (2017, 2020) in the case of the Aar massif (Figure 2.7). Note that pre-existing cross-sections of Schmid et al. (2017) were partially modified by integrating the detailed sections of the ECM from Figure 2.4. For detailed information on the data sources, spatial resolution and interpretations of the equal P-wave velocity contour lines of the Alpine crust shown in Figure 2.7, as well as on Moho depths, the reader is referred to Kissling et al. (2006), Diehl et al. (2009), Wagner et al. (2012), Spada et al. (2013) and Schmid et al. (2017). Note that the lower crust (colored in purple), and the Moho at its base, is defined by a high velocity gradient (Vp: 6.5–8 km/s normally over a depth interval of less than 2km), with Vp of 7.25 km/s as first order approximation of the Moho in local earthquake tomography imaging (see Wagner et al. 2012). In all four cross-sections (Schmid et al. 2017), the European lower crust and associated Moho dip gently in an undisturbed manner toward the Adriatic plate (Figure 2.7). This general inclination also persists for the overlying proximal parts of the European middle to upper crust ($Vp < 6.0$). This dip direction of the velocity isoline $Vp = 6.0$ km/s changes by 180° into a foreland oriented one with the structural rise of the basement-cover contacts of the ECM, which is best seen in the case of the Argentera massif (Figure 2.7(a)). Aar, Mont Blanc/Aiguilles Rouges and Belledonne massifs show a more complex shape with bulged Vp contours (Figures 2.7(b), (c) and (d)) owing to crustal thickening and more complex basement-internal deformation. Taking the undisturbed proximal European crust, we define the Vp interval between 6.0 and 6.5 km/s as a transitional domain (bright purple) between the lower and the middle crust. Despite the previously mentioned massif internal variability of Vp 6.0 km/s contours, in general, the vertical width of this transition zone increases from the European proximal to the distal part of the crust. This means that crustal material characterized by P-wave velocities that are too high compared to an undisturbed crust must be accumulated underneath the ECM. For the ECORS-CROP profile, Schmid and Kissling (2000) interpreted such enhanced velocity domains as a duplication of the lower crust. The gradual increase in P-wave velocities, however, argues against such a duplication, which should be represented by an abrupt velocity jump (Schmid et al. 2017). Following Fry et al. (2010) and Herwegh et al. (2017), we therefore interpret this volumetric increase of the lower- to mid-crustal transition as a mixing domain, in which mobilized lower crust becomes enriched within mid-crustal rocks.

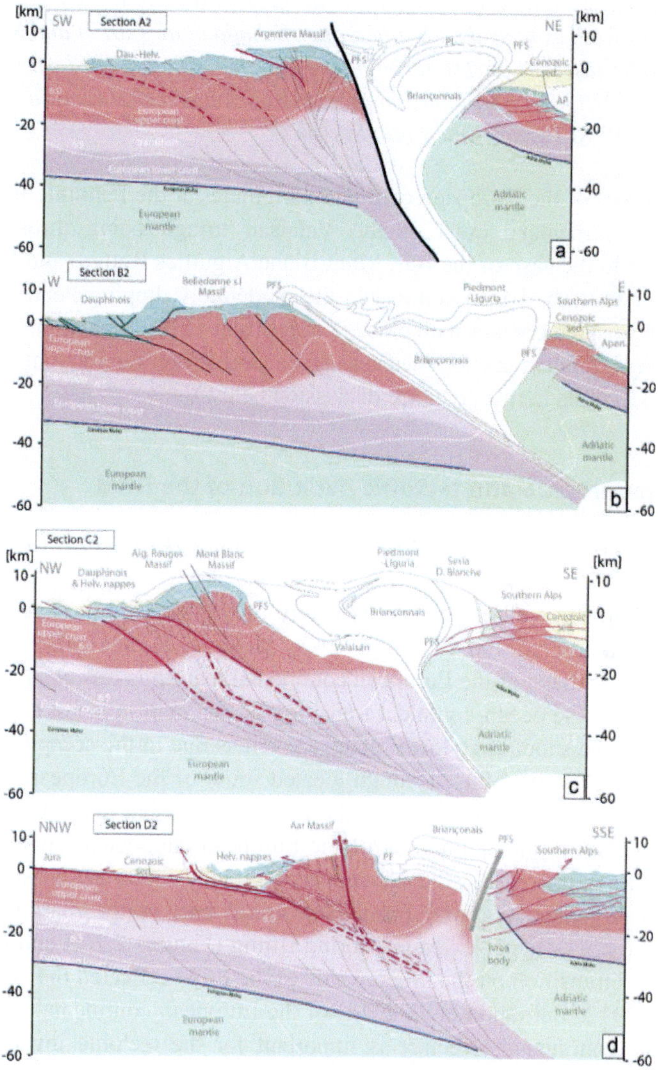

Figure 2.7. *Lithospheric-scale tectonic cross-sections through the different ECM with contours (stippled lines) of P-wave velocities from Diehl et al. (2009)*

COMMENT ON FIGURE 2.7.– *Same color coding of the European and Adriatic middle to upper crust as applied in Figure 2.4 for the ECM to visualize the inverted former passive continental margins. In contrast, the Penninic domains are kept colorless. White stippled lines represent contours of equal P-wave velocities after Diehl et al. (2009). (a) Section A1 through the Argentera massif. Modified after Schmid et al. (2017) and Leclère et al. (2014). Ap: Apennines. (b) Section B1 through the Belldonne massif. Modified after Schmid et al. (2017). (c) Section C1 through Mont Blanc/Aiguilles Rouges massifs. Modified after Schmid et al. (2017) and Schmid and Kissling (2000). (d) Section D1 through the Aar massif. Modified after Herwegh et al. (2017, 2020). Gt: Gotthard nappe, AVT: Ausserberg-Avat-Tavetsch, PF: Penninic front. PFS: Periadriatic Fault System.*

The distal part of the European continent terminates at the Penninic front, which encompasses sedimentary units of the Valaisan trough being thrust over the European plate in the case of the Aar, Mont Blanc/Aiguilles Rouges and Belledonne massifs (Figures 2.7(b–d)). In contrast to the moderately dipping Penninic front in these sections, a steep contact with close juxtaposition of continental basement units resulted in the Argentera section, because of sinistral strike-slip movements and the lack of Valaisan units in this area (see discussion below).

2.5. Paleogeographic and tectonic evolution of the ECM

2.5.1. *Influence of Mesozoic rifting on the later exhumation of the ECM*

In contrast to the well-established structure of the Adriatic passive continental margin (e.g. Manatschal and Bernoulli 1999; Mohn et al. 2010; Bellahsen et al. 2014), the ancient shape of the European continental margin is less constrained. This is mainly because the ocean-continent transition of this margin is not exposed since it is situated on the subducting lower plate, as well as due to the complex shape and evolution of the related Valaisan trough located south of the European margin. All ECM belonged to the proximal part of the European passive margin, i.e. to a part that experienced only minor thinning of the European continental crust (see above and Figure 2.2(a)). In contrast, the adjacent crust directly south-southeast-east of this area (Aar to Argentera massifs) formed the distal part of the margin, in which continental crust was hyper-extended during rifting (Figures 2.2(e) and (f)). Hence the ocean-ward transition on the internal side of the ECM reflected the necking zone (Mohn et al. 2012; Beltrando et al. 2014) of the European continental margin. This change in the continental structure is important for the tectonic inversion of the ECM. Therefore, the sedimentary records from Lower Jurassic to Cretaceous times

(Trümpy 1960; Gwinner 1971; Stampfli 2001; Pfiffner et al. 2011; Decarlis et al. 2017) as well as retro-deformed sections were used to gain insight into the margin evolution prior to Alpine inversion (Figure 2.2(a)). Specifically, the type and occurrence of syn-rift sediments allowed us to reconstruct the near-surface development of the ocean-continent transition. The latter is interpreted to be related to the evolution of the entire lithospheric structure, providing indirect information on its lateral and subsurface variation.

Two major extension-related stages preconditioned the southern European margin: (i) a Jurassic and (ii) a Cretaceous one. (i) A major change in plate tectonic configuration and lithospheric thickness occurred in Lower Jurassic times with rifting of the Piemont-Ligurian ocean (e.g. Trümpy 1975; Froitzheim and Eberli 1990; Manatschal and Bernoulli 1999; Mohn et al. 2010; Decarlis et al. 2017). A Lower Jurassic European passive margin formed (Figures 2.2(a) and (e)), located at the southern rim of the future Briançonnais domain. Between this Lower Jurassic European passive continental margin and the later ECM and their half-grabens, fine-clastic sediments witness a hemi-pelagic basin from Argentera/Belledonne to the Aar massif (Belledonne area: Ramsay 1963. Torrenthorn and Gotthard area: Trümpy 1952, 1960). The variable spatial extent of these sediments (bright blue in Figures 2.2(a) and (e)) suggests a decrease in basin width from the NE toward the SW at this time. The basin formed because of lithospheric thinning and subsidence under extensional conditions (Figure 2.2). (ii) The second and most important stage of structural ECM preconditioning related to Cretaceous rifting, which was responsible for the opening of the Valaisan trough (Trümpy 1960; Stampfli 1993). Bündnerschiefer (or schists lustrés), as well as local occurrence of ophiolitic sequences, clearly document substantial subsidence, hemi-pelagic to pelagic sedimentation (Lemoine 2003) and locally partial formation of oceanic crust (Beltrando et al. 2012). The Valaisan trough, however, showed an opposite basin polarity compared to the Jurassic basin, with a trough-closure in the SW but general trough widening/opening toward the NE (Figure 2.2(b)). Note that this trend is in line with a successive decrease in width of the Briançonnais swell from SW to NE (Stampfli 1993; Pfiffner 2015). This swell terminated to the NE, where the Valaisan trough and Piemont-Ligurian ocean merged into a single basin (Schmid et al. 2004; Handy et al. 2010). As a consequence of this paleogeographic reconstruction, the domain between Argentera and Briançonnais swell did not experience severe Cretaceous extension. The occurrence of shallow marine sediments and the lack of evidence for deeper marine fine clastic Valaisan sediments imply at that time a connection of continental crust between the western Briançonnais and the European plate (Decarlis et al. 2017; Figures 2.2(b) and (d)). As discussed in the next section, three Mesozoic rift-related geometric features strongly affected the ECM during the later Alpine convergence history: (i) The overall L-shaped margin geometry with the approximately ENE-WSW paleo-alignment between Aar and Belledonne massifs and the NW-SE alignment of Oisans and Argentera, possibly indicating a first order margin segmentation (Figure 2.2(a) and 2.9(a)). (ii) Second order segmentation of the passive margin along NW-SE trending lateral transfer zones between Belledonne-Oisans and Mont Blanc/Aiguilles Rouges massifs, as well as between the latter and the Aar massif. These transfer zones can be inferred from the lateral offsets of the basement outlines (Figure 2.2(a); see also Burkhard 1988). (iii) A trough-ward bulging of the passive margin south of the Aar massif (Lepontine gneiss nappes, Figure 2.2(a)).

2.5.2. *Eocene Alpine convergence*

With 1–2 cm/yr inferred NNW-directed movement of the Adriatic plate during the Eocene (55 – 35 Ma) the Valais trough became closed (e.g. Handy et al. 2010; Schmid et al. 2017; van Hinsbergen et al. 2020). Basal amalgamation of first the Briançonnais and then the Valaisan units below the Adriatic upper plate resulted in the Penninic nappe stack (e.g. Argand 1916; Schmid et al. 1996; Ceriani et al. 2003; Wiederkehr et al. 2008; Berger et al. 2011; stage 1 of Schmid et al. 2017; Figures 2.7, 2.8 and 2.9(b)). At this stage (Figure 2.9(b)), the previously mentioned transition between Europe-related basement (preserved as Argentera/Belledonne-Oisans ECM) and the western end of the Valais trough acted as a lateral NNW-SSE striking ramp, disconnecting the shortening in the Valaisan units from the more westerly located European margin units (Figure 2.9(b); Ford et al. 2006). This ramp has to be postulated from geometric plate tectonic reasoning but is not exposed because it was overprinted and covered by later W-directed movements of the Penninic front (Figure 2.9(c); stage 3 of Schmid et al. 2017). It is currently unclear if and how this transpression affected strike-slip faulting in the Argentera massif and the pre-Priabonian N to NW-directed deformation in the Pelvoux massif (southern part of Belledonne-Oisans; Dumont et al. 2011). All other ECM were not yet severely affected by this early convergence.

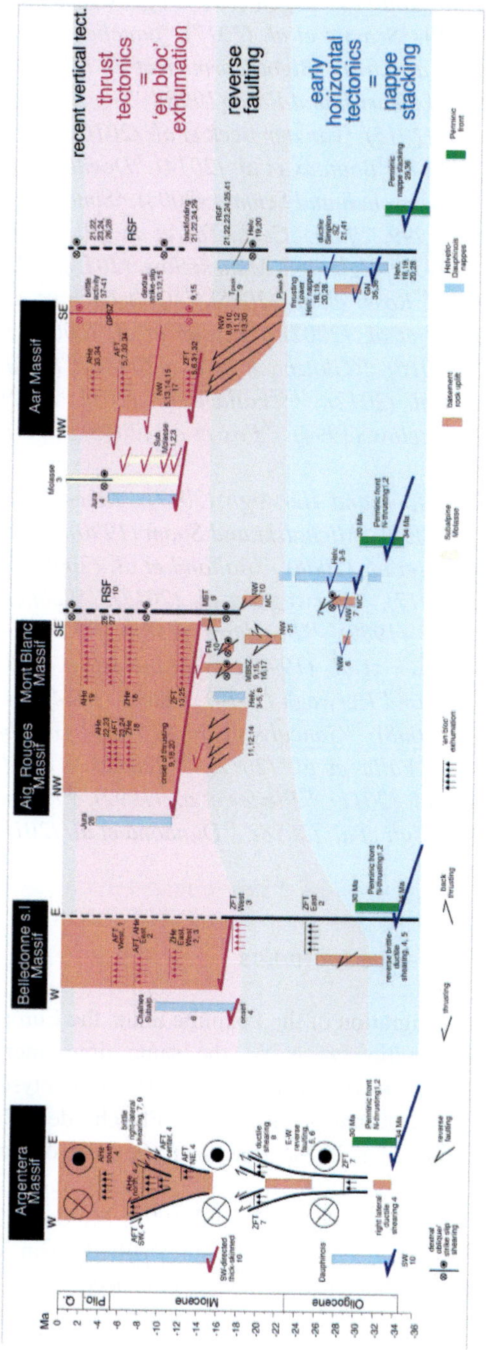

Figure 8

Figure 2.8. *Orogenic time table of the structural and kinematic evolution of ECM and their geodynamic frameworks: Argentera massif, Belldonne-Oisans massif, Mont Blanc/Aiguilles Rouges massifs and Aar massif. The latter three experienced two stages of horizontal and vertical tectonics*

COMMENT ON FIGURE 2.8.– *Data sources: Argentera:* [1]*Fügenschuh and Schmid (2003),* [2]*Simon-Labric et al. (2009),* [3]*Schmid et al. (2017),* [4]*Sanchez et al. (2011b),* [5]*Corsini et al. (2004),* [6]*Jourdan et al. 2014;* [7]*Bigot-Cormier et al. (2006),* [8]*Sanchez et al. (2011),* [9]*Bogdanoff (2000),* [10]*Lickorish and Ford (1998).*
Belledonne (s.l.): [1]*Bellanger et al. (2015),* [2]*van der Beek et al. (2010),* [3]*Girault et al. (2022),* [4]*Bellahsen et al. (2012, 2014),* [5]*Boutoux et al. (2014),* [6]*Deville et al. (1994).*
Mont Blanc/Aiguilles Rouges: [1]*Fügenschuh and Schmid (2003),* [2]*Simon-Labric et al. (2009),* [3-5]*Kirschner et al. (1996, 1999, 2003),* [6]*Cenki-Tok et al. (2014),* [7]*Egli et al. (2016),* [8]*Crespo-Blanc et al. (1995),* [9]*Egli and Mancktelow (2013),* [10]*Egli et al. (2017),* [11]*Marshall et al. (1998),* [12]*Rossi et al. (2005),* [13]*Glotzbach et al. (2011),* [14]*Boutoux et al. (2014),* [15]*Rolland et al. (2007),* [16]*Leloup et al. (2005),* [17]*Rolland et al. (2008),* [18]*Boutoux et al. (2016),* [19]*Glotzbach et al. (2008),* [20]*Burkhard and Sommaruga (1998),* [21]*Rolland et al. (2012),* [22,23]*Valla et al. (2011, 2012),* [24]*Rahn et al. (1994),* [25]*Seward and Mancktelow (1994),* [26]*Fox et al. (2015),* [27]*Vernon et al. (2008),* [28]*Becker (2000).*
Aar: [1]*von Hagke et al. (2012),* [2]*Mock and Herwegh (2017),* [3]*Mock et al. (2020),* [4]*Becker (2000),* [5]*Herwegh et al. (2020),* [6]*Michalski and Soom (1990),* [7]*Vernon et al. (2008),* [8]*Steck (1966),* [9]*Challandes et al. (2008),* [10]*Rolland et al. (2009),* [11]*Wehrens et al. (2016),* [12]*Wehrens et al. (2017),* [13]*Herwegh et al. (2017),* [14]*Labhart (1966),* [15]*Berger et al. (2017),* [17]*Kammer (1989),* [18]*Burkhard (1988),* [19]*Kirschner et al. (1996),* [20]*Huon et al. (1994),* [21]*Steck et al. (1994),* [22,23]*Mancktelow (1985, 1990),* [24]*Campani et al. (2010),* [25]*Haertel and Herwegh (2014),* [26]*Steck (1984),* [28]*Kirschner et al. (2003),* [29]*Wiederkehr et al. (2008),* [30]*Goncalves et al. (2012),* [31]*Glotzbach et al. (2010),* [32]*Vernon et al. (2009),* [33]*Valla et al. (2012),* [34]*Reinecker et al. (2008),* [35]*Janots et al. (2009),* [36]*Berger et al. (2011),* [37]*Pfeifer et al. (1992),* [38]*Hofmann et al. (2004),* [39]*Belgrano et al. (2016),* [40]*Egli et al. (2018),* [41]*Diamond et al. (2018),* [42]*Steck et al. (2008),* [43]*Lu et al. (2018).*

2.5.3. *Early Oligocene cover nappe tectonics*

Synchronous with northward migration of the Penninic units, the European plate bent southward toward the subduction zone. At the same time interval these basement units formed the substratum of the Dauphinois-Helvetic Flysch basins. The youngest Flysch sediments (Annot Sandstone, Flysch de Champsaur, Taveyannaz sandstone), dated around 34–32 Ma, document the onset of burial of the ECM (Figures 2.6 and 2.8; Kerckhove 1969; Fry 1989; Fischer and Villa 1990; Ruffini et al. 1995; Sinclair 1997; Bürgisser and Ford 1998; Mulder et al. 2010; Lu et al. 2018). After turbiditic sedimentation in the underfilled basin, the upper Helvetic nappes formed (Wildhorn, Diablerets) and were emplaced above the future

ECM (Ramsay et al. 1983; Escher et al. 1993; Pfiffner 1993; Pfiffner et al. 2011). During this stage of classical nappe tectonics (Figures 2.7 and 2.8), stretching lineations and associated kinematics indicated thrust directions diverging gradually from N-directed toward the WNW-NW directed thrusting from the Belledonne to the Aar massif (Figure 2.9(c); see Schmid et al. 2017 and references therein). This change manifested (i) European-Adriatic collision dynamics of the curved plate boundaries, (ii) the strong uplift of the ICM (Briançonnais) within a dextral transpressive regime along the Periadriatic fault system and (iii) the indentation of the strong Adriatic mantle wedge into lower-mid crustal quartz-dominated European crust (see Schmid et al. 2017 and references therein). Here (Figure 2.9(c)), the Southern Steep Belt including the dextral Periadriatic fault system at 32–22 Ma (Schmid et al. 1989; Berger et al. 1996) and the sinistral Argentera-Cuneo transcurrent fault (Schmid et al. 2017), respectively, formed the SE and S boundary of the internal Alps, the latter being also affected by the ductile Simplon shear zone (34–25 Ma, Steck et al. 2015 and references therein).

2.5.4. *Late Oligocene onset of thick-skinned tectonics*

Geometrically as well as rheologically, a major change in large-scale deformation occurred, once the former European passive continental margin became integrated into the collision. This transition appeared first in the westernmost Alps owing to earlier juxtaposition of basement units of the Briançonnais and European margin basement units (see L-shaped margin geometry in Figure 2.9(a) and (b)), but progressed toward the Central Alps (i.e. SW-NE trending margin domains like Mont Blanc–Aar massifs, Figures 2.2 and 2.9(c)) with closing of the Valaisan trough (Schmid et al. 1996; Simon-Labric et al. 2009). Basement-involved Lower Helvetic "nappes" evolved, particularly in the eastern ECM domains (Doldenhorn nappe, Morcles/Mont Joly nappe, Figures 2.4(c) and (d)). Although incorporated into the Helvetic nappe stack, these tectonic units show some major differences compared to the already emplaced Upper Helvetic cover nappes: (1) Formation of large-scale recumbent folds with often intensely stretched inverted limbs. Here, calcite dominated lithologies combined with deformation temperatures above 300°C made them prone to act as lubricant layers, hosting the basal thrusts (Ramsay et al. 1983; Pfiffner 1993; Herwegh and Pfiffner 2005; Ebert et al. 2007; Ebert et al. 2008; Herwegh et al. 2008). (2) More internally, basement units were incorporated in the cores of the fold structures (Escher et al. 1993; Herwegh and Pfiffner 2005; Krayenbuhl and Steck 2009; Berger et al. 2020). In contrast to the mechanically soft sediments with their ductile folding and shearing, deformation temperatures between 300 and 400°C were too low for ductile folding in these quartz-/feldspar-dominated basement units. Instead, pervasive spaced cleavages evolved, allowing for a dispersed basement internal shear deformation. (3) Paleogeographically, these nappes derived from the proximal passive margin, showing limited displacements in the range of around 10–15 km (Herwegh and Pfiffner 2005; Ebert et al. 2007), which is much less compared to the Upper Helvetic nappes. In spite of these differences, common aspects between Lower Helvetic and Upper Helvetic nappes also exist: (1) in-sequence stacking, (2) passive upward bulging of Upper as well as Lower Helvetic nappes during vertical uplift of the Aar and Mont Blanc/Aiguilles Rouges massifs, (3) thick sedimentary sequences from rift basins, which were inverted during collision where (4) basement units represent sheared-off basement highs (Burkhard 1988; Herwegh and Pfiffner 2005; Krayenbuhl and Steck 2009). All these points indicate that the Lower Helvetic nappes represent paleogeographically, kinematically as well mechanically a transition between the older Upper Helvetic nappe stacking and exhumation of the Eastern ECM with their strong vertical exhumation components.

During emplacement of the Lower Helvetic nappes, the ECM reached maximum burial depths and associated peak metamorphic conditions (Figure 2.6): for Argentera (~15 km, 375±30°C; Sanchez et al. 2011a), Oisans (~10 km, < 350°C; Simon-Labric et al. 2009; Bellanger et al. 2015), Aiguilles Rouges (~12 km, < 320°C; Boutoux et al. 2016), Mont Blanc (~17km, ~400°C; Rolland et al. 2003; Rossi et al. 2005) and Aar (~20 km, 450°C; Goncalves et al. 2012; Herwegh et al. 2017) massifs. Because of the L-shaped margin geometry of the European passive margin (see above), onset of peak metamorphism occurred first in the western massifs (Argentera, Belledonne-Oisans, Mont Blanc) during southward bending of the European lithosphere between 32 and 29 Ma. Generally, peak metamorphism lasted in all massifs till ~22–20 Ma (Figure 2.8). During this time interval, the ECM (especially Oisans, Grandes Rousses, Belledonne and Mont Blanc) did not experience much exhumation or much cooling. The activity of the Periadriatic fault system in its western portion is related to fast vertical movements and some dextral displacement. This can be followed till 19 Ma (fast uplift processes between T_{max} and zircon FT data in the Central Alps (Rubatto et al. 2009), illite gouge ages of Riedel shears near Gesero (Zwingmann and Mancktelow 2004)) and at ~23–20 Ma latest movements recorded by ZFT data and dated fault gouges along the Canavese Line (Berger et al. 2012; Pleuger et al. 2012).

2.5.5. *Early Miocene – early ECM exhumation*

A tectonically determined change in deformation style took place during Early Miocene times: the continental units that failed to subduct and therefore experienced collisional deformation, metamorphism and onset of exhumation (Herwegh et al.

2017, 2020). Major vertical movements took place (Figures 2.3, 2.5, 2.6 and 2.9(d)) in the Aar and Mont Blanc/Aiguilles Rouges massifs, which experienced substantial differential uplift (see Figure 2.5 for peak T) with strong nearly isothermal decompression (Figure 2.6; see also Challandes et al. 2008; Bellahsen et al. 2014; Boutoux et al. 2016). The Aar massif was dissected by steep reverse (normal) faults (see sections 2.5 and 2.6, e.g. Wehrens et al. 2017; Nibourel et al. 2018, 2021a; Herwegh et al. 2020). These structures explain the gradient observed in peak metamorphism (Figures 2.5 and 2.6). In the Mont Blanc massif, the MBSZ was still active (Leloup et al. 2005; Rolland et al. 2008; Egli and Mancktelow 2013). This activity of this structure explains a potential jump in thermal peak conditions between the Mont Blanc and Aiguilles Rouges massifs (Figure 2.5). In the Belledonne-Oisans and Argentera, ~25–20 Ma an early stage of "en bloc" rock uplift occurred during exhumation (Figure 2.5, see RSCM for Belldonne s.l. and ZFT for Argentera). These differences in amount and style of ECM rock uplift, as well as the limited shortening in the Belledonne (16–18%) and Argentera (< 29%) area, compared to Mont Blanc/Aiguilles Rouges (38%) and Aar (40–60%, Figure 2.3), result from the paleogeographic margin shape (e.g. Bellahsen et al. 2014) and oblique movement directions during collision compared to rifting-related opening directions (Figure 2.9).

2.5.6. *Early–late Miocene thrust tectonics – late ECM exhumation*

A change in major structural style occurred in Miocene times (Figures 2.7 and 2.9(e)). In all ECM moderately dipping thrust planes developed (Figure 2.7), above which the massifs were displaced toward the foreland and vertically exhumed "en bloc" as manifested by the horizontal isoage lines in the plots of the low temperature thermochronometers (Figures 2.5 and 2.8). The initiation of these crustal ramps may have been diachronous from south to north. Starting as early as late Oligocene-Early Miocene underneath the Argentera and Belldonne massifs, and then progressing to the east where they evolved during the Early Miocene below the Aiguilles Rouges and the Aar massifs including also lateral growth parallel to the latter axes. In some massifs (Argentera, Mont Blanc, Aar), synchronous activity of major dextral strike-slip faults indicates transpressive movements by strain partitioning between the foreland-directed thrusting and hinterland strike-slip faulting. This strain partitioning allowed disconnection of the ECM from deformation in the internal zone of the Central and Western Alps (Figure 2.9). The latter experienced pronounced upper-crustal, orogen-parallel extension by brittle faulting (Bertrand and Sue 2017 and references therein). With further shortening, deformation progressed from the ECM out into the foreland. As a consequence, thrust systems in the Subalpine Molasse (Pfiffner 1986; Burkhard and Sommaruga 1998; von Hagke et al.

2012, 2014; Mock and Herwegh 2017; Mock et al. 2020) and the major stage of Jura fold-and-thrust belt formation took place (e.g. Burkhard and Sommaruga 1998; Sommaruga 1999; Becker 2000). Foreland-directed thrusting gradually rotated laterally from top NW-directed to SW-directed, respectively, from the Jura mountains in the north down to the Digne thrust system in the South (Figure 2.9(e); Fry 1989; Bürgisser and Ford 1998; Lickorish and Ford 1998), being laterally accommodated by strike-slip faults.

2.5.7. *Pliocene to recent tectonic and surface processes*

With progressive exhumation, the interplay of surface processes and tectonic forcing became increasingly important for shaping today's 3D surface expression of the ECM. In this light, effects of long-term plate-scale processes (slab steepening, slab tearing and detachment, crustal indentation: Schmid and Kissling 2000; Lippitsch et al. 2003; Kissling 2008; Fry et al. 2010; Duretz et al. 2011; Zhao et al. 2015, 2016; Herwegh et al. 2017, 2020; Kissling and Schlunegger 2018; Kästle et al. 2020) and short-term local-scale processes (erosion processes, rock distribution and their erodibility, climatic changes (e.g. Kühni and Pfiffner 2001; Willet 2010; Sternai et al. 2012; Glaus et al. 2019) interfered with each other.

Over the last 10 Ma, exhumation rates in the ECM varied in space and time (e.g. Glotzbach et al. 2008, 2010, 2011; Reinecker et al. 2008; Vernon et al. 2008; Valla et al. 2011, 2012, 2016; Fox et al. 2015, 2016; Nibourel et al. 2021a, 2021b). At the Miocene–Pliocene transition, only low exhumation rates of < 0.4 km/m.y. occurred in the Argentera and Pelvoux (Fox et al. 2015), but with substantially enhanced rates of 0.6–0.8 km/m.y. in the Northern Belledonne, Mont Blanc/Aiguilles Rouges and Aar massifs. This resulted from a continuation of the N-migration of the Adriatic plate with progressive crustal NW-SE shortening in the latter massifs and strike-slip movements in the case of the Argentera massif.

For the time interval 4–2 Ma, a general background exhumation rate of < 0.4 km/m.y. was assessed from the Western to Central Alps, with somewhat enhanced rates of 0.6–0.7 km/m.y. at the SW and NE ends of the Aar massif, the transition between Belledonne-Oisans and Mont Blanc/Aiguilles Rouges massifs, and the southern end of Belledonne-Oisans as well as Argentera massifs (Fox et al. 2015). As suggested for the eastern Aar massif by Nibourel et al. (2021b), these domains could document growth stages of the ECM parallel to their axes. Alternatively, enhanced surface erosion of the cover sediments (e.g. slates) compared to less erodible basement rocks could have accounted for the increased exhumation. At the end of this period, pronounced valley-scale topography existed (Sternai et al. 2012).

A two to three times increase in erosion rate occurred within the last 2 Ma, with rates between 1 and 1.5 km/m.y. These high rates affected the entire area from the Argentera to Mont Blanc/Aiguilles Rouges massifs, but without comparable change in the Eastern Alps (Fox et al. 2015; Figure 2.10). Despite this W to E change, average topography remained constant, suggesting enhanced rock uplift in the Western Alps (Fox et al. 2015). This long-wavelength change in erosion rates along strike of the massifs must have been due to a crustal-scale process (Duretz et al. 2011; Fox et al. 2015). A gap at 120–150 km in the otherwise continuous high-seismicity-velocity region of the European slab suggests slab detachment underneath the Western Alps (Figure 2.10; Lippitsch et al. 2003; Faccenna et al. 2004; Kissling et al. 2006; Beller et al. 2018; Kissling and Schlunegger 2018; Kästle et al. 2020). Owing to the lost negatively buoyant European lithospheric mantle, remaining crustal units above the slab tear experienced positive buoyancy resulting in the pronounced vertical rock uplift (Figure 2.10). Particularly intensified glacial erosion (Haeuselmann et al. 2007; Glotzbach et al. 2008; Reinecker et al. 2008; Vernon et al. 2008; Valla et al. 2012; Sternai et al. 2019) accommodated this uplift (Fox et al. 2015), maintaining the mean surface elevation and a constant topographic load (Figure 2.10). It is this stage that mainly shaped today's ECM surfaces.

The recent geodynamic setting and associated stress fields (Figure 2.10) control the present day earthquake focal mechanisms (Delacou et al. 2004; Bauve et al. 2014; Singer et al. 2014; Diehl et al. 2017), GPS geodesy (D'Agostino et al. 2008), fault activities and surface uplift (Sue et al. 2007) as well as the tilting of the surrounding foreland (Champagnac et al. 2008), which might be related to persisting slab tear and plate convergence between Adria and Europe. With the notable exception of the eastern boundary of the Oisans massif with the onset of extensional movements at 3–4 Ma (Bilau et al. 2021), the lack of jumps in recent surface uplift patterns indicates the paucity of single fault-bound changes (e.g. Egli and Mancktelow 2013; Nibourel et al. 2018, 2021a, 2021b) but rather suggest spatially gradual changes in surface uplift. This uplift recently accommodates massif-internally, mainly along steep inherited fault structures. In contrast to their initiation as ductile faults at depth, they meanwhile transferred into brittle, in parts non-cohesive, fault zones owing to ongoing deformation during exhumation. In this sense, the ECM are back at a stage of vertical tectonics, with a strong vertical but limited horizontal movement component (Sanchez et al. 2018 and references therein). If slab tear progresses further to NE, the steep faults in the Aar massif will remain active, with the future eastward propagation of this massif most likely also then experiencing enhanced exhumation rates there (Figure 2.10).

2.6. Geodynamic forcing for the exhumation of the ECM

As shown above, there are local variations in kinematics, style of exhumation structures and overall stress field but also many common aspects between the individual ECM. Hence, the overall driving forces controlling the pop-up of ECM must have had a common geodynamic ground requiring information from the pre-ECM uplift history of the Central to Western Alps.

In contrast to classical ocean-continent subduction, with its associated slab-pull forces, an alternative geodynamic scenario has recently been proposed (see summary in Kissling and Schlunegger 2018 and references therein), because of (i) a plate tectonic stable position of Central Europe over the last 100 Ma (Torsvik et al. 2012) and the continuous NNW to NW migration of the Adriatic plate; (ii) the subducting oceanic lithosphere being attached to the European plate (e.g. Laubscher et al. 1992) and (iii) the model of the oceanic lithosphere slab-break off at about 35 Ma ago (Davis and von Blanckenburg 1995). Therefore, slab steepening and retreat of the lower plate (first oceanic, then European continental) has recently been suggested as an alternative scenario to classical subduction (Kissling and Schlunegger 2018).

In this light, rendering end-member geodynamics solely based on ECM kinematics is difficult because of the variable position of individual ECM with respect to the N to NNW moving Adriatic plate (e.g. Maffione et al. 2008). For example, the dominantly transpressional tectonics in the Argentera section mainly results from the lateral transfer movements between European and Adriatic plate and the young anticlockwise rotation, owing to the rollback processes of the Apennine slab (see above and Figure 2.9; Schmid et al. 2017 and references therein). In contrast, the convergence direction perpendicular to the NE-SW aligned Mont Blanc/Aiguilles Rouges to Aar ECM provide information relevant for orogen perpendicular and/or parallel geodynamic processes. Such processes need to be in line with plate tectonic movement vectors as well as forcing induced by the remnant of the European slab (Figure 2.10). For this reason, this section focuses on the latter part of the tectonic history, for which two end-member scenarios, namely "Adriatic-Push" and "Orogeny Rollback" have been proposed in the literature.

2.6.1. *The "Adria-Push" scenario*

Exhumation of the ECM in this scenario is attributed to a wedge evolution in response to the NNW movement of Adria (e.g. Burkhard 1990, 1999; Burkhard and Sommaruga 1998; Lacombe and Mouthereau 2002; Butler 2013; Bellahsen et al. 2014; Pfiffner 2015; Lacombe and Bellahsen 2016). In this "Adria-Push" model, the kinematics of Adria is controlled by the large-scale plate movement of Africa. Owing to convergence, the Penninic domains (Piemont-Ligurian ocean till Valaisan trough) are subducted beneath Adria by slab pull forces. After the end of the continental subduction stage (by the end of Eocene), the continent-continent collision started, with the underthrusting of Europe beneath the Adriatic wedge and subsequent shortening, collisional metamorphism, relief formation, and initiation of syn-orogenic Molasse type basins. At the lithospheric scale, the Adriatic mantle indented into the European lower crust (Schmid et al. 1996; Pfiffner et al. 1997b).

Different phases of convergence have been identified (cover nappes, thick-skinned tectonics, fold-and-thrust-belt formation). During some of these periods, steep shear zones or faults have formed (see above). Previous models therefore either suggested a shear zone formation at low angles and progressive shear zone steepening with shortening (Burkhard 1990, 1999), the formation of buckle folding (e.g. Krayenbuhl and Steck 2009) on top of a basal detachment fault, or bulk horizontal shortening and strain partitioning (Choukroune and Gapais 1983; Marquer and Gapais 1985). However, in the case of the Aar massif, arguments for an already steep initiation of the shear zones (Schneeberger et al. 2016; Nibourel et al. 2021a) as well as the N-S metamorphic gradient, speak against these previous explanations (see Herwegh et al. 2020 and orogeny slab rollback scenario below). In the case of the Adria-push scenario, such high-angle structures can be explained by a combination of the three following processes: reactivation of steep inherited structures (in Belledonne, Marquer et al. 2006), transpressive movements (Aar, Mont Blanc, Argentera, see above), and/or high angle between the orientation of the shortening and the orientation of the shear zones in ductile regime (e.g. Fossen and Cavalcante 2017). Progressive convergence between Adria and Europe finally led to the formation of the crustal thrust ramps, above which the ECM experienced "en bloc" exhumation.

2.6.2. *The "Orogeny Slab Rollback" scenario*

Within the last decades, "Orogeny Slab Rollback" was suggested as an alternative scenario for the Alpine orogeny (Froitzheim et al. 1997; Kissling 2008; Fry et al. 2010; Magni et al. 2012; Singer et al. 2014; Schlunegger and Kissling 2015; Herwegh et al. 2017, 2020; Kissling and Schlunegger 2018; Dal Zilio et al. 2020). Even in this scenario, the Penninic realm subducted underneath the Adriatic plate, but by a slab rollback process (Kissling and Schlunegger 2018), which is in line with the paleogeographic stable position of Europe (Torsvik et al. 2012). In this case, slab rollback of the European plate induced plate suction forces, which passively moved Adria toward the NNW.

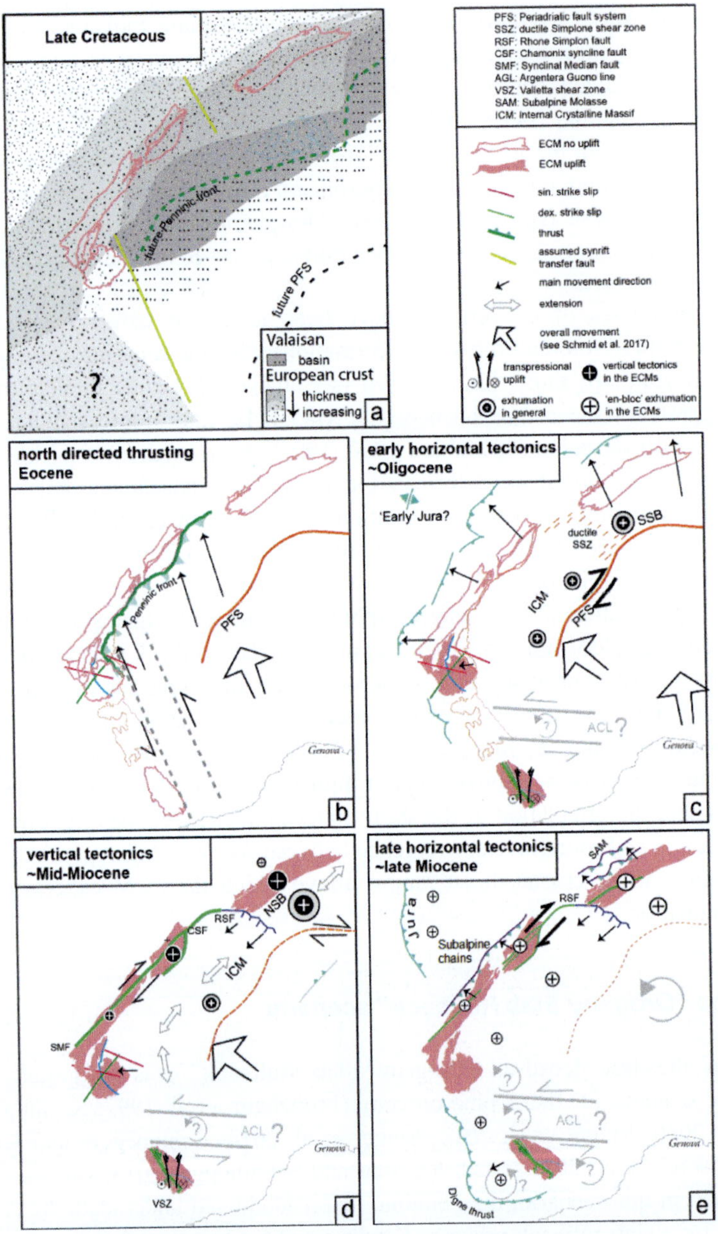

Figure 2.9. *Schematic tectonic evolution of the ECM within the Western and Central Alpine geodynamic framework*

COMMENT ON FIGURE 2.9.– *(a) Late Cretaceous arcuate-shaped geometry of European passive continental margin and adjacent Valais trough (details in Figure 2.2). (b) Eocene closure of Valais through NNW-directed thrusting of the Penninic Front and synchronous activity of a lateral sinistral transfer zone at its Western margin (stage 1 of Schmid et al. 2017). (c) Oligocene NW-directed movement of Adria, dextral movement along the Periadriatic fault system with synchronous Simplon shearing and vertical exhumation of the Internal Crystalline Massifs (ICM) transpose deformation into the ECM realm. Here, horizontal tectonics by thrusting of Dauphinois-Helvetic nappes buried the ECM, the latter reaching maximum peak metamorphic conditions. Southern Belledonne-Oisans (Pelvoux) and Argentera massifs experienced first vertical uplifts. (d) WNW-directed movement of Adria and internal Alpine uplift induced orogen-parallel stretching (see also Hubbard and Mancktelow 1992). In the ECM differential uplift and evolution of steep reverse faults manifest onset of vertical tectonics in Aar- and Belledonne-Oisans massifs. Dextral transpression in the West resulted in the formation of strike-slip faults (e.g. Synclinal Median fault, Rhone-Simplon fault). (e) Change to horizontal tectonics with a strain partitioning: (i) ECM from Belledonne-Oisans until Aar massifs experienced "en bloc" exhumation on top of N to SW-directed basal thrusts. Deformation propagated into the Alpine foreland. (ii) Dextral strike-slip faulting persisted. Counter-clockwise rotations occurred in the SW Alps. The amount of rotation is up for debate (see question marks in the figures and text).*

Particularly for the steep shear zones in the Mont Blanc/Aiguilles Rouges and Aar massifs with the steep metamorphic gradients (Figure 2.5), the "Orogeny Slab Rollback" scenario highlights a vertical uplift component, which can be explained by progressive decoupling in the lower crust and buoyancy-driven vertical motions. Still under compressive conditions, a rheological weakening at the lower to mid-crustal transition allowed decoupling and a reduction of negative buoyancy in the European middle to upper crust induced by the European lithospheric slab underneath. Consequently, positive buoyancy forces in the mid- to upper crustal levels dominated, resulting in N-directed mass movement vectors being moderately inclined toward the south (Herwegh et al. 2017). Associated inclined stress tensors were in perfect orientations to activate the steep reverse faults found in many of the EMCs – a fact that was corroborated by numerical modeling (Dal Zilio et al. 2020). These structures, in combination with the non-linear metamorphic gradients, are the main evidence for the vertical tectonic style, with differential uplift being active from 22 to 16/13 Ma (Figures 2.5 and 2.8). Spatially, vertical tectonics was dispersed and most severe in the Aar- and Mont Blanc/Aiguilles Rouges massifs and somewhat less expressed in the Belledonne s.l. massifs (Figures 2.4, 2.7 and 2.9(d)), while its effect is blurred because of the 30 million years long transpression-induced pop up of the Argentera massif (Figure 2.8). However, transpression allowed for the partial strike-slip activation of steep shear zones in the case of the SW Aar (Wehrens et al. 2017) and Mont Blanc (Egli et al. 2017) massifs.

For the Miocene stage of ECM exhumation, indentation of the Adriatic mantle into the European lower crust is kinematically in line with the "Adriatic Push" and the "Orogeny Slab Rollback" scenarios. It is this indentation, combined with rollback in the Appennines, that contributed to the oroclinal bending of the NW-Alps. Also, the "en bloc" ECM rock uplifts on top of major basal thrusts in combination with propagation of the deformation front into the foreland can be explained by this late and still ongoing convergence (Figures 2.7, 2.8 and 2.9(e)).

2.6.3. *Why the ECM do not continue further to the east?*

The Aar massif represents the eastern-most exposed ECM with a still further eastward plunging massif axis in the subsurface (e.g. Engadine window; Hitz and Pfiffner 1994; Pfiffner et al. 2011; Zerlauth et al. 2014; Nibourel et al. 2021a, 2021b). To the east, it must terminate somewhere between Eastern Switzerland and the Tauern window, which represents the most eastward end of European basement outcrop within the Alps. Owing to its different tectonic history, the metamorphic grade and, in particular, to its former distal position on the European crust, the Tauern window is not considered as an ECM. In this sense, it rather locates at a position similar to the Lepontine gneiss nappes in the Central Swiss Alps. Accepting this argumentation, no ECM exist in the Eastern Alps. Why? To address this question, a 3D view on the present-day lithospheric structure and its role during Adriatic-European continent-continent collision is required. Within the asthenosphere beneath the Alps, two distinct high P-wave velocity anomalies exist: (i) a western one underneath the Western and Central Alps is interpreted to represent the SSE dipping European slab (Lippitsch et al. 2003); (ii) an eastern anomaly located beneath the Eastern Alps (Lippitsch et al. 2003; Mitterbauer et al. 2011; Karousova et al. 2013; Handy et al. 2015), controversy interpreted in the literature regarding its lithospheric interpretation. Lippitsch et al. (2003), Karousova et al. (2013) and Hetény et al. (2018) suggest the presence of a northeastward dipping Adriatic slab, while Mitterbauer et al. (2011) propose an original European lithospheric slab that steepened and overturned. It has been interpreted by some authors that a narrow gap underneath the Western Tauern window separates the two P-wave anomalies (Lippitsch et al. 2003; Rosenberg et al. 2015). In analogy with Luth et al. (2013) and Handy et al. (2015), we consider this gap to reflect a vertical NW-SE trending decoupling zone between the Western-Central and Eastern Alps. Hence, one can speculate that the lower crustal delamination and associated mid-upper crustal buoyancy forces required for ECM uplift within the Western-Central

European lithosphere terminate along this domain, resulting in a lack of major driving forces for potential ECM uplift in the Eastern Alps.

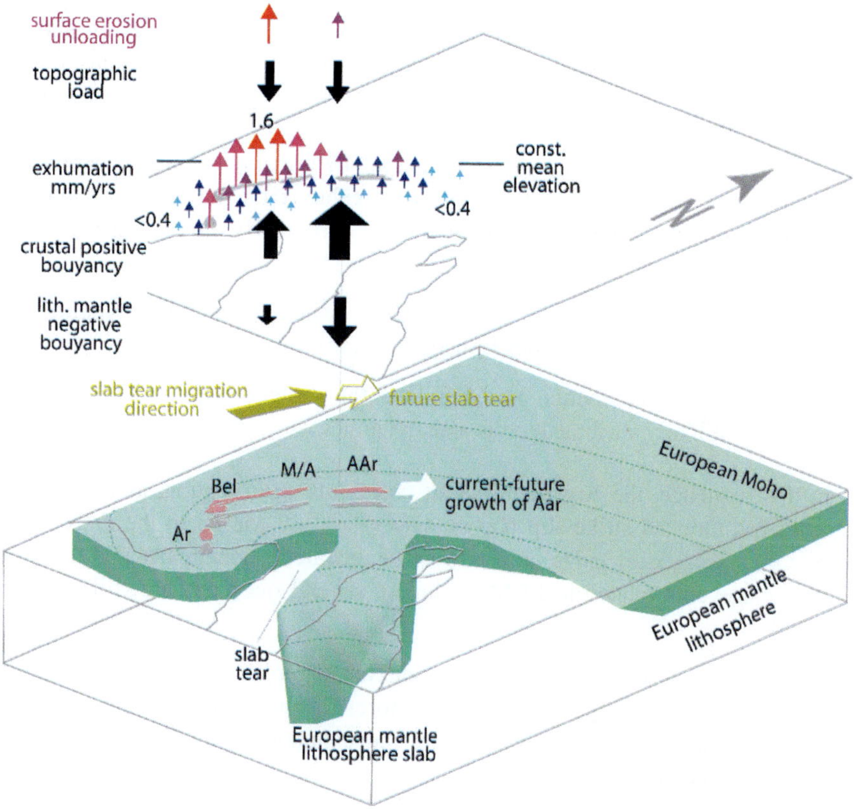

Figure 2.10. *Schematic 3D presentation of today's geometry of the European slab (bottom, after Kissling et al. 2006), its position with respect to the ECM and exhumation rates (top) during the last 2 Mys (Fox et al. 2015). Note the NE-directed migration of slab tear reducing and enhancing, respectively, negative (mantle) and positive (crust) vertical forces underneath the western ECM compared to the eastern ones. Nonetheless, mean elevation was kept constant, requiring enhanced erosion rates in the western ECM compared to their eastern counter parts*

2.7. Conclusion

The ECM represent parts of the unthinned to only weakly thinned proximal southern European continental margin. Along strike variations in the evolution of the ECM in terms of timing, type of deformation, metamorphism and exhumation are strongly controlled by the structure of this European passive continental margin in the Cretaceous and the convergence directions between the Adriatic and European plates. Thin-skinned nappe tectonics, slicing off the Helvetic/Dauphinois sedimentary cover, predates ECM exhumation. The ECM themselves show a sequence of shortening, exhumation and rock uplift, which often is fourfold:

i) early transitional thick-skinned thrust tectonics (Aar, Mont Blanc massifs);

ii) a stage of distributed basement internal deformation along moderately to steeply inclined shear zones, which helped to exhume the rocks from depth and produced considerable metamorphic gradients, particularly in the case of Mont Blanc/Aiguilles Rouges and Aar massifs;

iii) localized basal thrusting with “en-bloc” exhumation of the hanging wall (Belledonne, Mont Blanc/Aiguilles Rouges and Aar massifs);

iv) considerable exhumation gradients from the western ECM to the Aar massif, which are most likely attributed to the slab dynamics.

In this sequence, the Argentera massif represents an exception, as it undergoes transpression-induced strike-slip pop-up since about 30 Ma up to the present day. Despite increasing knowledge on the structure, geometry, kinematics and timing of deformation events related to ECM evolution, several questions are still yet to be answered:

– Large parts of the sedimentary cover and the upper basement units of the ECM have been eroded and were deposited in the foreland basins. What is the interaction between the developing orogenic wedge, the exhumation of the ECM and the evolving sedimentation sequences?

– Current P-wave tomography displays anomalies at the transition between lower to middle crust. What are the detailed structure, deformation style and rheology in these domains?

– How are the deep slab dynamics affecting the exhumation of the ECM?

– How much and when did rotations in the Western Alpine Arc occur and how severely was the Argentera massif affected by such rotations?

Field studies as well as current deep-seated research projects, such as AlpArray (e.g. Hetényi et al. 2018), may shed new light on these questions in the future.

2.8. Acknowledgments

We greatly acknowledge the thorough and constructive reviews of Neil Mancktelow and Claudio L. Rosenberg.

2.9. References

Antoine, P., Pairis, J., Pairis, B. (1975). Quelques observations nouvelles sur la structure de la couverture sédimentaire interne du massif du Mont-Blanc, entre le Col du Ferret (frontiere italo-suisse) et la Tête des Fours (Savoie, France). *Géologie Alpine*, 51, 5–23.

Argand, E. (1916). Sur l'arc des Alpes Occidentales. *Eclogae Geologicae Helvetiae*, 14, 145–191.

Ayrton, S. (1980). High fluid pressure, isothermal surfaces, and the initiation of nappe movement. *Geology*, 8, 172–174.

Badertscher, N. and Burkhard, M. (1998). Alpine inversion of the Permo-Carboniferous Salvan-Dorenaz Graben and its relation with the overlying Morcles nappe. *Eclogae Geologicae Helvetiae*, 91, 359–373.

Baietto, A., Perello, P., Cadoppi, P., Martinotti, G. (2009). Alpine tectonic evolution and thermal water circulations of the Argentera massif (South-Western Alps). *Swiss Journal of Geosciences*, 102, 223–245.

Ballevre, M., Manzotti, P., Dal Piaz, G.V. (2018). Pre-Alpine (Variscan) inheritance: A key for the location of the future Valaisan Basin (Western Alps). *Tectonics*, 37, 786–817.

Barféty, J.-C. and Gidon, M. (1983). La stratigraphie et la structure de la couverture dauphinoise au Sud de Bourg d'Oisans. Leurs relations avec les déformations synsédimentaires jurassiques. *Géologie Alpine (Grenoble)*, 59, 5–32.

Barféty, J.-C., Gidon, M., Lemoine, M., Mouterde, R. (1979). Liassic synsedimentary tectonics in the crystalline massifs, external zone of the French Western Alps Col d'Ornon fault. *Comptes rendus hebdomadaires des séances de l'academie des sciences, Série D*, 289, 1207–1210.

Barruol, G., Bonnin, M., Pedersen, H., Bokelmann, G.H., Tiberi, C. (2011). Belt-parallel mantle flow beneath a halted continental collision: The Western Alps. *Earth and Planetary Science Letters*, 302, 429–438.

Baumberger, R., Herwegh, M., Kissling, E. (2022). Remote sensing and field data based structural 3D modelling (Haslital, Switzerland) in combination with uncertainty estimation and verification by underground data. In *3D Digital Models from Terrestrial Outcrops to Surfaces*, Bistacchi, A., Massironi, M., Viseur, S. (eds). Wiley Online Library. doi: 10.1002/9781119313922.

Bauve, V., Plateaux, R., Rolland, Y., Sanchez, G., Bethoux, N., Delouis, B., Darnault, R. (2014). Long-lasting transcurrent tectonics in SW Alps evidenced by Neogene to present-day stress fields. *Tectonophysics*, 621, 85–100.

Bauville, A., Epard, J.-L., Schmalholz, S.M. (2013). A simple thermo-mechanical shear model applied to the Morcles fold nappe (Western Alps). *Tectonophysics*, 583, 76–87.

Becker, A. (2000). The Jura Mountains – An active foreland fold-and-thrust belt? *Tectonophysics*, 321, 381–406.

van der Beek, P.A., Valla, P.G., Herman, F., Braun, J., Persano, C., Dobson, K.J., Labrin, E. (2010). Inversion of thermochronological age–elevation profiles to extract independent estimates of denudation and relief history II: Application to the French Western Alps. *Earth and Planetary Science Letters*, 296, 9–22.

Belgrano, T.M., Herwegh, M., Berger, A. (2016). Inherited structural controls on fault geometry, architecture and hydrothermal activity: An example from Grimsel Pass, Switzerland. *Swiss Journal of Geosciences*, 109, 345–364.

Bellahsen, N., Jolivet, L., Lacombe, O., Bellanger, M., Boutoux, A., Garcia, S., Mouthereau, F., Le Pourhiet, L., Gumiaux, C. (2012). Mechanisms of margin inversion in the external Western Alps: Implications for crustal rheology. *Tectonophysics*, 560, 62–83.

Bellahsen, N., Mouthereau, F., Boutoux, A., Bellanger, M., Lacombe, O., Jolivet, L., Rolland, Y. (2014). Collision kinematics in the western external Alps. *Tectonics*, 33, 1055–1088.

Bellanger, M., Bellahsen, N., Jolivet, L., Baudin, T., Augier, R., Boutoux, A. (2014). Basement shear zones development and shortening kinematics in the Ecrins Massif, Western Alps. *Tectonics*, 33, 84–111.

Bellanger, M., Augier, R., Bellahsen, N., Jolivet, L., Monié, P., Baudin, T., Beyssac, O. (2015). Shortening of the European Dauphinois margin (Oisans massif, Western Alps): New insights from RSCM maximum temperature estimates and 40Ar/39Ar in situ dating. *Journal of Geodynamics*, 83, 37–64.

Beller, S., Monteiller, V., Operto, S., Nolet, G., Paul, A., Zhao, L. (2017). Lithospheric architecture of the South-Western Alps revealed by multiparameter teleseismic full-waveform inversion. *Geophysical Journal International*, 212, 1369–1388.

Belliere, J. (1988). On the age of mylonites within the Mont Blanc massif. *Geodinamica Acta*, 2, 13–16.

Beltrando, M., Frasca, G., Compagnoni, R., Vitale-Brovarone, A. (2012). The Valaisan controversy revisited: Multi-stage folding of a Mesozoic hyper-extended margin in the Petit St. Bernard pass area (Western Alps). *Tectonophysics*, 579, 17–36.

Beltrando, M., Manatschal, G., Mohn, G., Dal Piaz, G.V., Brovarone, A.V., Masini, E. (2014). Recognizing remnants of magma-poor rifted margins in high-pressure orogenic belts: The Alpine case study. *Earth-Science Reviews*, 131, 88–115.

Bergemann, C.A., Gnos, E., Whitehouse, M.J. (2019). Insights into the tectonic history of the Western Alps through dating of fissure monazite in the Mont Blanc and Aiguilles Rouges massifs. *Tectonophysics*, 750, 203–212.

Berger, A. and Herwegh, M. (2015). Abstract: Structural changes from very proximal to distal European margin (Central Alps) and their effect on Alpine collision. Alpine Workshop, Digne.

Berger, A. and Herwegh, M. (2019). Cockade structures as a paleo-earthquake proxy in upper crustal hydrothermal systems. *Scientific Reports*, 9, 9209.

Berger, A., Rosenberg, C.L., Schmid, S. (1996). Ascent, emplacement and exhumation of the Bergell pluton within the Southern Steep Belt of the Central Alps. *Schweizerische mineralogische und petrographische Mitteilungen*, 76, 357–382.

Berger, A., Schmid, S.M., Engi, M., Bousquet, R., Wiederkehr, M. (2011). Mechanisms of mass and heat transport during Barrovian metamorphism: A discussion based on field evidence from the Central Alps (Switzerland/northern Italy). *Tectonics*, 30, TC1007.

Berger, A., Mercolli, I., Kapferer, N., Fügenschuh, B. (2012). Single versus double exhumation in the internal Western Alps: A key to interpret near-surface Oligocene tectonics. *International Journal of Earth Science*, 101, 1877–1894

Berger, A., Mercolli, I., Herwegh, M., Gnos, E. (2016). Geological map of the Aar massif, Tavetsch and Gotthard nappes 1:100000. *Geological Special Map and Explanatory Notes*, 129.

Berger, A., Mercolli, I., Herwegh, M., Gnos, E. (2017a). Geological map of the Aar massif, Tavetsch and Gotthard nappes (1:100000). Explanatory notes, Landesgeologie Schweiz, Wabern.

Berger, A., Wehrens, P., Lanari, P., Zwingmann, H., Herwegh, M. (2017b). Microstructures, mineral chemistry and geochronology of white micas along a retrograde evolution: An example from the Aar massif (Central Alps, Switzerland). *Tectonophysics*, 721, 179–195.

Berger, A., Engi, M., Erne-Schmid, S., Glotzbach, C., Spiegel, C., de Goede, R., Herwegh, M. (2020). The relation between peak metamorphic temperatures and subsequent cooling during continent-continent collision (western Central Alps, Switzerland). *Swiss Journal of Geoscience*, 113, 4.

Bertini, G., Marcucci, M., Nevini, R., Passerini, P., Sguazzoni, G. (1985). Patterns of faulting in the Mont Blanc granite. *Tectonophysics*, 111, 65–106.

Bertrand, A. and Sue, C. (2017). Reconciling late faulting over the whole Alpine belt: From structural analysis to geochronological constrains. *Swiss Journal of Geosciences*, 110, 565–580.

Bigot-Cormier, F., Sosson, M., Poupeau, G., Stéphan, J.-F., Labrin, E. (2006). The denudation history of the Argentera Alpine External Crystalline Massif (Western Alps, France-Italy): An overview from the analysis of fission tracks in apatites and zircons. *Geodinamica Acta*, 19, 455–473.

Bilau, A., Rolland, Y., Schwartz, S., Godeau, N., Guihou, A., Deschamps, P., Brigaud, B., Noret, A., Dumont, T., Gautheron, C. (2021). Extensional reactivation of the Penninic frontal thrust 3 Ma ago as evidenced by U-Pb dating on calcite in fault zone cataclasite. *Solid Earth*, 12, 237–251.

von Blanckenburg, F. (1992). Combined high-precision chronometry and geochemical tracing using accessory minerals: Applied to the Central-Alpine Bergell intrusion (central Europe). *Chemical Geology*, 100, 19–40.

Bogdanoff, S., Michard, A., Mansour, M., Poupeau, G. (2000). Apatite fission track analysis in the Argentera massif: Evidence of contrasting denudation rates in the external crystalline massifs of the Western Alps. *Terra Nova*, 12, 117–125.

Bousquet, R., Oberhansli, R., Schmid, S.M., Berger, A., Wiederkeher, M., Robert, C., Möller, A., Rosenberg, C.L., Zeilinger, G., Molli, G., Koller, F. (2012). Metamorphic framework of the Alps. Commission for the Geological Map of the World.

Boutoux, A., Bellahsen, N., Lacombe, O., Verlaguet, A., Mouthereau, F. (2014). Inversion of pre-orogenic extensional basins in the external Western Alps: Structure, microstructures and restoration. *Journal of Structural Geology*, 60, 13–29.

Boutoux, A., Bellahsen, N., Nanni, U., Pik, R., Verlaguet, A., Rolland, Y., Lacombe, O. (2016). Thermal and structural evolution of the external Western Alps: Insights from (U–Th–Sm)/He thermochronology and RSCM thermometry in the Aiguilles Rouges/Mont Blanc massifs. *Tectonophysics*, 683, 109–123.

Buess, D. (2019). Geologie und Tektonik des nördlichen Randes des Aar Massivs (Urbachtal, Berner Oberland). Master's Thesis, Institut für Geologie, University of Bern.

Bürgisser, J. and Ford, M. (1998). Overthrust shear deformation of a foreland basin; structural studies south-east of the Pelvoux massif, SE France. *Journal of Structural Geology*, 20, 1455–1475.

Burkhard, M. (1988). L'Helvetique de la bordure occidentale du massif de l'Aar (évolution tectonique et métamorphique). *Eclogae Geologicae Helvetiae*, 81, 63–114.

Burkhard, M. (1990). Aspects of the large-scale Miocene deformation in the most external part of the Swiss Alps (sub-Alpine molasse to Jura fold belt). *Eclogae Geologicae Helvetiae*, 83, 559–583.

Burkhard, M. (1999). Strukturgeologie und Tektonik im Bereich AlpTransit. In *Vorerkundung und Prognose der Basistunnels am Gotthard und am Lötschberg*, Low, S. and Wyss, R. (eds). A.A. Balkema Publishers, Rotterdam.

Burkhard, M. and Sommaruga, A. (1998). Evolution of the western Swiss Molasse basin: Structural relations with the Alps and the Jura belt. *Geological Society, London, Special Publications*, 134, 279–298.

Bussy, F., Hernandez, J., von Raumer, J. (2000). Bimodal magmatism as a consequence of the post-collisional readjustment of the thickened Variscan continental lithosphere (Aiguilles Rouges-Mont Blanc massifs, Western Alps). *Earth and Environmental Science Transactions of the Royal Society of Edinburgh*, 91, 221–233.

Butler, R.W. (1983). Balanced cross-sections and their implications for the deep structure of the northwest Alps. *Journal of Structural Geology*, 5, 125–137.

Butler, R.W. (2013). Area balancing as a test of models for the deep structure of mountain belts, with specific reference to the Alps. *Journal of Structural Geology*, 52, 2–16.

Butler, R.W., Tavarnelli, E., Grasso, M. (2006). Structural inheritance in mountain belts: An Alpine–Apennine perspective. *Journal of Structural Geology*, 28, 1893–1908.

Campani, M., Herman, F., Mancktelow, N. (2010). Two- and three-dimensional thermal modeling of a low-angle detachment: Exhumation history of the Simplon Fault Zone, central Alps. *Journal of Geophysical Research: Solid Earth*, 115.

Campani, M., Mancktelow, N., Courrioux, G. (2014). The 3D interplay between folding and faulting in a syn-orogenic extensional system: The Simplon Fault Zone in the Central Alps (Switzerland and Italy). *Swiss Journal of Geosciences*, 107, 251–271.

Cardello, G.L. and Mancktelow, N. (2014). Cretaceous syn-sedimentary faulting in the Wildhorn Nappe (SW Switzerland). *Swiss Journal of Geoscience*, 107, 223–250.

Cardello, G.L., Di Vincenzo, G., Giorgetti, G., Zwingmann, H., Mancktelow, N. (2019). Initiation and development of the Pennine Basal Thrust (Swiss Alps): A structural and geochronological study of an exhumed megathrust. *Journal of Structural Geology*, 126, 338–356.

Carosi, R., D'Addario, E., Mammoliti, E., Montomoli, C., Simonetti, M. (2016). Geology of the northwestern portion of the Ferriere-Mollieres Shear Zone, Argentera massif, Italy. *Journal of Maps*, 12, 466–475.

Cenki-Tok, B., Darling, J.R., Rolland, Y., Dhuime, B., Storey, C.D. (2014). Direct dating of mid-crustal shear zones with synkinematic allanite: New in situ U-Th-Pb geochronological approaches applied to the Mont Blanc massif. *Terra Nova*, 26, 29–37.

Ceriani, S., Fügenschuh, B., Schmid, S.M. (2001). Multi-stage thrusting at the "Penninic Front" in the Western Alps between Mont Blanc and Pelvoux massifs. *International Journal of Earth Sciences*, 90, 685–702.

Ceriani, S., Fügenschuh, B., Potel, S., Schmid, S.M. (2003). Tectono-metamorphic evolution of the Frontal Penninic units of the Western Alps: Correlation between low-grade metamorphism and tectonic phases. *Schweizerische Mineralogische und Petrographische Mitteilungen*, 83, 111–131.

Challandes, N., Marquer, D., Villa, I. (2008). P-T-t modelling, fluid circulation, and 39Ar-40Ar and Rb-Sr mica ages in the Aar massif shear zones (Swiss Alps). *Swiss Journal of Geosciences*, 101, 269–288.

Champagnac, J.D., van Der Beek, P., Diraison, G., Dauphin, S. (2008). Flexural isostatic response of the Alps to increased quaternary erosion recorded by foreland basin remnants, SE France. *Terra Nova*, 20, 213-220.

Choukroune, P. and Gapais, D. (1983). Strain patterin in the Aar granite (Central Alps): Orthogneiss developed by bulk inhomogeneous flattening. *Journal of Structural Geology*, 5, 411–418.

Collet, L. and Paréjas, E. (eds) (1931). Matériaux pour la carte géologique de la Suisse. In *Géologie de la chaîne de la Jungfrau*. A. Francke, Bern.

Corsini, M. and Rolland, Y. (2009). Late evolution of the southern European Variscan belt: Exhumation of the lower crust in a context of oblique convergence. *Comptes rendus géoscience*, 341, 214–223.

Corsini, M., Ruffet, G., Caby, R. (2004). Alpine and late-hercynian geochronological constraints in the Argentera massif (Western Alps). *Eclogae Geologicae Helvetiae*, 97, 3–15.

Crespo-Blanc, A., Masson, H., Sharp, Z., Cosca, M., Hunziker, J. (1995). A stable 40Ar/39Ar isotope study of a major thrust in the Helvetic nappes (Swiss Alps): Evidence for fluid flow and constraints on nappe kinematics. *GSA Bulletin*, 107, 1129–1144.

Crouzet, C., Ménard, G., Rochette, P. (2001). Cooling history of the Dauphinoise zone (Western Alps, France) deduced from the thermopaleomagnetic record: Geodynamic implications. *Tectonophysics*, 340, 79–93.

D'Agostino, N., Avallone, A., Cheloni, D., D'Anastasio, E., Mantenuto, S., Selvaggi, G. (2008). Active tectonics of the Adriatic region from GPS and earthquake slip vectors. *Journal of Geophysical Research: Solid Earth*, 113, 1312413m. doi: 10.1029/2008JB005860.

Dal Zilio, L., Kissling, E., Gerya, T., van Dinther, Y. (2020). Slab rollback Orogeny model: A test of concept. *Geophysical Research Letters*, 47, e2020GL089917.

Davies, J.H. and von Blanckenburg, F. (1995). Slab breakoff – A model of lithosphere detachment and its test in the magmatism and deformation of collisional orogens. *Earth and Planetary Science Letters*, 129, 85–102.

Debon, F. and Lemmet, M. (1999). Evolution of Mg/Fe ratios in Late Variscan Plutonic rocks from the external crystalline massifs of the Alps (France, Italy, Switzerland). *Journal of Petrology*, 40, 1151–1185.

Decarlis, A., Dallagiovanna, G., Lualdi, A., Maino, M., Seno, S. (2013). Stratigraphic evolution in the Ligurian Alps between Variscan heritages and the Alpine Tethys opening: A review. *Earth-Science Reviews*, 125, 43–68.

Decarlis, A., Beltrando, M., Manatschal, G., Ferrando, S., Carosi, R. (2017). Architecture of the distal Piedmont-Ligurian rifted margin in NW Italy: Hints for a flip of the rift system polarity. *Tectonics*, 36, 2388–2406.

Deichmann, N. (2014). Earthquakes in Switzerland and surrounding regions 1996–2012. Version 2014.1. Report, Swiss Seismological Service – ETH Zurich.

Delacou, B., Sue, C., Champagnac, J.-D., Burkhard, M. (2004). Present-day geodynamics in the bend of the western and central Alps as constrained by earthquake analysis. *Geophysical Journal International*, 158, 753–774.

Deville, E., Blanc, E., Tardy, M., Beck, C., Cousin, M., Ménard, G. (1994). Thrust propagation and syntectonic sedimentation in the Savoy tertiary Molasse Basin (Alpine Foreland). In *Hydrocarbon and Petroleum Geology of France*, Mascle, A. (ed.). Springer Berlin Heidelberg, Berlin, Heidelberg.

Diamond, L.W. and Tarantola, A. (2015). Interpretation of fluid inclusions in quartz deformed by weak ductile shearing: Reconstruction of differential stress magnitudes and pre-deformation fluid properties. *Earth and Planetary Science Letters*, 417, 107–119.

Diamond, L.W., Wanner, C., Waber, H.N. (2018). Penetration depth of meteoric water in orogenic geothermal systems. *Geology*, 46, 1063–1066.

Diehl, T., Husen, S., Kissling, E., Deichmann, N. (2009). High-resolution 3-D P-wave model of the Alpine crust. *Geophysical Journal International*, 179, 1133–1147.

Diehl, T., Kraft, T., Kissling, E., Wiemer, S. (2017). The induced earthquake sequence related to the St. Gallen deep geothermal project (Switzerland): Fault reactivation and fluid interactions imaged by microseismicity. *Journal of Geophysical Research: Solid Earth*, 122, 7272–7290.

Dietrich, D. and Casey, M. (1989). A new tectonic model for the Helvetic nappes. *Geological Society, London, Special Publications*, 45, 47–63.

Dumont, T., Champagnac, J.-D., Crouzet, C., Rochat, P. (2008). Multistage shortening in the Dauphiné zone (French Alps): The record of Alpine collision and implications for pre-Alpine restoration. *Swiss Journal of Geosciences*, 101, 89–110.

Dumont, T., Simon-Labric, T., Authemayou, C., Heymes, T. (2011). Lateral termination of the north-directed Alpine orogeny and onset of westward escape in the Western Alpine arc: Structural and sedimentary evidence from the external zone. *Tectonics*, 30.

Dumont, T., Schwartz, S., Guillot, S., Simon-Labric, T., Tricart, P., Jourdan, S. (2012). Structural and sedimentary records of the Oligocene revolution in the Western Alpine arc. *Journal of Geodynamics*, 56–57, 18–38.

Duretz, T., Gerya, T.V., May, D.A. (2011). Numerical modelling of spontaneous slab breakoff and subsequent topographic response. *Tectonophysics*, 502, 244–256.

Ebert, A., Herwegh, M., Austin, N., Evans, B., Pfiffner, O.A. (2007). Microfabrics in carbonate mylonites along a large-scale shear zone (Helvetic Alps). *Tectonophysics*, 444, 1–26.

Ebert, A., Herwegh, M., Berger, A., Pfiffner, O.A. (2008). Grain coarsening maps for polymineralic carbonate mylonites: A calibration based on data from different Helvetic nappes (Switzerland). *Tectonophysics*, 457, 128–142.

Egli, D. and Mancktelow, N. (2013). The structural history of the Mont Blanc massif with regard to models for its recent exhumation. *Swiss Journal of Geosciences*, 106, 469–489.

Egli, D., Müller, W., Mancktelow, N. (2016). Laser-cut Rb–Sr microsampling dating of deformational events in the Mont Blanc-Aiguilles Rouges region (European Alps). *Terra Nova*, 28, 35–42.

Egli, D., Mancktelow, N., Spikings, R. (2017). Constraints from 40Ar/39Ar geochronology on the timing of Alpine shear zones in the Mont Blanc-Aiguilles Rouges region of the European Alps. *Tectonics*, 36, 730–748.

Egli, D., Herwegh, M., Berger, A., Baron, L., Baumann, R., Küng, S. (2018). Structural characteristics and evolution of an exhumed long-lived hydrothermal reservoir. *Tectonophysics*, 747–748, 239–258.

Eltchaninoff-Lancelot, C., Triboulet, S., Doudoux, B., Fudral, S., Rampnoux, J.-P., Tardy, M. (1982). Stratigraphie et tectonique des unites delphino-helvetiques comprises entre Mont-Blanc et Belledonne (Savoie-Alpes occidentales) ; implications regionales. *Bulletin de la Société géologique de France*, 7, 817–830.

Epard, J.-L. (1986). Le contact entre le socle du Mont Blanc et la zone de Chamonix, implications tectoniques. PhD Thesis, Université de Lausanne.

Epard, J.L. (1990). La nappe de Morcles au sud-ouest du Mont-Blanc. PhD Thesis, Université de Lausanne.

Escher, A. and Beaumont, C. (1997). Formation, burial and exhumation of basement nappes at crustal scale: A geometric model based on the Western Swiss-Italian Alps. *Journal of Structural Geology*, 19, 955–974.

Escher, A., Masson, H., Steck, A. (1993). Nappe geometry in the Western Swiss Alps. *Journal of Structural Geology*, 13, 501–509.

Faccenna, C., Piromallo, C., Crespo-Blanc, A., Jolivet, L., Rossetti, F. (2004). Lateral slab deformation and the origin of the western Mediterranean arcs. *Tectonics*, 23, TC1012.

Faure-Muret, A. (1955). Etudes géologiques sur le massif de l'Argentera. Les Marges continentales actuelles et fossiles autour de la Mercantour et ses enveloppes sédimentaires. In *Mémoires pour servir à l'explication de la carte géologique détaillé de la France*. Imprimerie Nationale, Paris.

Fischer, H. and Villa, I. (1990). Erste K/Ar-und 40Ar/39Ar-Hornblende-Mineralalter des Taveyannaz-Sandsteins. *Schweizerische Mineralogische und Petrographische Mitteilungen*, 70, 73–75.

Fitz Gerald, J.D. and Stunitz, H. (1993). Deformation of granitoids at low metamorphic grade 1. Reactions and grain-size reduction. *Tectonophysics*, 221, 269–297.

Ford, M. (1996). Kinematics and geometry of early Alpine, basement-involved folds, SW Pelvoux Massif, SE France. *Eclogae Geologicae Helvetiae*, 89, 269–295.

Ford, M., Duchêne, S., Gasquet, D., Vanderhaeghe, O. (2006). Two-phase orogenic convergence in the external and internal SW Alps. *Journal of the Geological Society*, 163, 815–826.

Fossen, H. and Cavalcante, G.C.G. (2017). Shear zones – A review. *Earth-Science Reviews*, 171, 434–455.

Fox, M., Herman, F., Kissling, E., Willett, S.D. (2015). Rapid exhumation in the Western Alps driven by slab detachment and glacial erosion. *Geology*, 43, 379–382.

Fox, M., Herman, F., Willett, S.D., Schmid, S.M. (2016). The exhumation history of the European Alps inferred from linear inversion of thermochronometric data. *American Journal of Science*, 316, 505–541.

Franks, G. (1968). A study of the Upper Palaeozoic sediments and volcanics in the northern part of the eastern Aar massif. Doctoral thesis, ETH Zürich.

Fréville, K., Trap, P., Faure, M., Melleton, J., Li, X.-H., Lin, W., Blein, O., Bruguier, O., Poujol, M. (2018). Structural, metamorphic and geochronological insights on the Variscan evolution of the Alpine basement in the Belledonne massif (France). *Tectonophysics*, 726, 14–42.

Froitzheim, N. and Eberli, G.P. (1990). Extensional detachment faulting in the evolution of a Tethys passive continental margin, Eastern Alps, Switzerland. *Geological Society of America Bulletin*, 102, 1297–1308.

Froitzheim, N., Conti, P.T., Van Daalen, M. (1997). Late Cretaceous, synorogenic, low-angle normal faulting along the Schlinig fault (Switzerland, Italy, Austria) and its significance for the tectonics of the Eastern Alps. *Tectonophysics*, 280, 267–293.

Fry, N. (1989). Southwestward thrusting and tectonics of the western Alps. *Geological Society, London, Special Publications*, 45, 83–109.

Fry, B., Deschamps, F., Kissling, E., Stehly, L., Giardini, D. (2010). Layered azimuthal anisotropy of Rayleigh wave phase velocities in the European Alpine lithosphere inferred from ambient noise. *Earth and Planetary Science Letters*, 297, 95–102.

Fugenschuh, B. and Schmid, S. (2003). Late stages of deformation and exhumation of an orogen constrained by fission-track data: A case study in the Western Alps. *Geological Society of America Bulletin*, 115, 1425–1440.

Giba, M.O. (2006). Strukturelle Kartierung am Nordrand des Aarmassivs, Berner Oberland. Master's thesis, Albert-Ludwigs-Universität, Freiburg im Breisgau.

Girault, J.B., Bellahsen, N., Boutoux, A., Rosenberg, C.L., Nanni, U., Verlaguet, A., Beyssac, O. (2020). 3D thermal structure of the Helvetic nappes of the European Alps: Implications for collisional processes. *Tectonics*, 39, e2018Tc005334.

Girault, J.B., Bellahsen, N., Bernet, M., Pik, R., Loget, N., Lasseur, E., Rosenberg, C.L., Balvay, M., Sonnet, M. (2022). Exhumation of the Western Alpine Collisional wedge: New thermochronological data. *Tectonophysics*, 822, 229155.

Glaus, G., Delunel, R., Stutenbecker, L., Akçar, N., Christl, M., Schlunegger, F. (2019). Differential erosion and sediment fluxes in the Landquart basin and possible relationships to lithology and tectonic controls. *Swiss Journal of Geosciences*, 112, 453–473.

Glotzbach, C., Reinecker, J., Danisik, M., Rahn, M., Frisch, W., Spiegel, C. (2008). Neogene exhumation history of the Mont Blanc massif, western Alps. *Tectonics*, 27, TC4011.

Glotzbach, C., Reinecker, J., Danisik, M., Rahn, M., Frisch, W., Spiegel, C. (2010). Thermal history of the central Gotthard and Aar massifs, European Alps: Evidence for steady state, long-term exhumation. *Journal of Geophysical Research – Earth Surface*, 115, F0J017.

Glotzbach, C., van der Beek, P.A., Spiegel, C. (2011). Episodic exhumation and relief growth in the Mont Blanc massif, Western Alps from numerical modelling of thermochronology data. *Earth and Planetary Science Letters*, 304, 417–430.

Goncalves, P., Oliot, E., Marquer, D., Connolly, J.A.D. (2012). Role of chemical processes on shear zone formation: An example from the Grimsel metagranodiorite (Aar massif, Central Alps). *Journal of Metamorphic Geology*, 30, 703–722.

Gourlay, P. and Ricou, L.E. (1983). Le jeu décrochant dextre tardif de la suture de Chamonix (Alpes françaises et suisses). *C.R. Acad. Sci.*, 296, 927–932.

Grasmück, K.P. (1961). Die helvetischen Sedimente am Nordostrand des Mont Blanc-Massivs (zwischen Sembrancher und dem Col Ferret). *Eclogae Geologicae Helvetiae*, 54, 351–450.

Gratier, J.P. and Vialon, P. (1980). Deformation pattern in a heterogeneous material: Folded and cleaved sedimentary cover immediately overlying a crystalline basement (Oisans, French Alps). *Tectonophysics*, 65(1–2), 151–179.

Guellec, S., Lajat, D., Mascle, A., Roure, F., Tardy, M. (1990). Deep seismic profiling and petroleum potential in the western Alps: Constraints with ECORS data, balanced cross-sections and hydrocarbon modeling. *The Potential of Deep Seismic Profiling for Hydrocarbon Exploration Conference*, 425–437.

Guermani, A. and Pennacchioni, G. (1998). Brittle precursors of plastic deformation in a granite: An example from the Mont Blanc massif (Helvetic, western Alps). *Journal of Structural Geology*, 20, 135–148.

Guillot, S. and Ménot, R.-P. (2009). Paleozoic evolution of the external crystalline massifs of the western Alps. *Comptes rendus géoscience*, 341, 253–265.

Guillot, S., di Paola, S., Ménot, R.-P., Ledru, P., Spalla, M.I., Gosso, G., Schwartz, S. (2009). Suture zones and importance of strike-slip faulting for Variscan geodynamic reconstructions of the external crystalline massifs of the western Alps. *Bulletin de la Société géologique de France*, 180, 483–500.

Gwinner, H.P. (1971). *Geologie der Alpen; Stratigraphie, Palaeogeographie, Tektonik*. Schweizerbart, Stuttgart.

Haenni, R. and Pfiffner, O.A. (2001). Evolution and internal structure of the Helvetic nappes in the Bernese Oberland. *Eclogae Geologicae Helvetiae*, 94, 161–171.

Haertel, M. and Herwegh, M. (2014). Microfabric memory of vein quartz for strain localization in detachment faults: A case study on the Simplon fault zone. *Journal of Structural Geology*, 68, 16–32.

von Hagke, C., Cederbom, C., Oncken, O., Stöckli, D., Rahn, M., Schlunegger, F. (2012). Linking the northern Alps with their foreland: The latest exhumation history resolved by low-temperature thermochronology. *Tectonics*, 31, TC5010.

von Hagke, C., Oncken, O., Evseev, S. (2014). Critical taper analysis reveals lithological control of variations in detachment strength: An analysis of the Alpine basal detachment (Swiss Alps). *Geochemistry, Geophysics, Geosystems*, 15, 176–191.

Handy, M.R., Schmid, S.M., Bousquet, R., Kissling, E., Bernoulli, D. (2010). Reconciling plate-tectonic reconstructions of Alpine Tethys with the geological–geophysical record of spreading and subduction in the Alps. *Earth-Science Reviews*, 102, 121–158.

Handy, M.R., Ustaszewski, K., Kissling, E. (2015). Reconstructing the Alps–Carpathians–Dinarides as a key to understanding switches in subduction polarity, slab gaps and surface motion. *International Journal of Earth Sciences*, 104, 1–26.

Haeuselmann, P., Granger, D.E., Jeannin, P.Y., Lauritzen, S.E. (2007). Abrupt glacial valley incision at 0.8 Ma dated from cave deposits in Switzerland. *Geology*, 35, 143–146.

Heim, A. (1921). *Geologie der Schweiz*. Tauchniz, Leipzig.

Heim, A. and Heim, A. (1916). Die Juramulde im Aarmassiv bei Fernigen (Uri). *Ujschu natf, Ges. Zurich*, 61, 503–530.

Herwegh, M. and Pfiffner, O.A. (2005). Tectono-metamorphic evolution of a nappe stack: A case study of the Swiss Alps. *Tectonophysics*, 404, 55–76.

Herwegh, M., Berger, A., Ebert, A., Brodhag, S. (2008). Discrimination of annealed and dynamic fabrics: Consequences for strain localization and deformation episodes of large-scale shear zones. *Earth and Planetary Science Letters*, 276, 52–61.

Herwegh, M., Berger, A., Baumberger, R., Wehrens, P., Kissling, E. (2017). Large-scale crustal-block-extrusion during late Alpine collision. *Scientific Reports*, 7, 413.

Herwegh, M., Berger, A., Glotzbach, C., Wangenheim, C., Mock, S., Wehrens, P., Baumberger, R., Egli, D., Kissling, E. (2020). Late stages of continent-continent collision: Timing, kinematic evolution, and exhumation of the northern Rim (Aar massif) of the Alps. *Earth-Science Reviews*, 102959.

Hetényi, G., Plomerová, J., Bianchi, I., Exnerová, H.K., Bokelmann, G., Handy, M.R., Babuška, V., Group, A.-E.W. (2018). From mountain summits to roots: Crustal structure of the Eastern Alps and Bohemian Massif along longitude 13.3 E. *Tectonophysics*, 744, 239–255.

van Hinsbergen, D.J., Torsvik, T.H., Schmid, S.M., Maţenco, L.C., Maffione, M., Vissers, R.L., Gürer, D., Spakman, W. (2020). Orogenic architecture of the Mediterranean region and kinematic reconstruction of its tectonic evolution since the Triassic. *Gondwana Research*, 81, 79–229.

Hitz, L. and Pfiffner, O.-A. (1994). A 3D crustal model of the Eastern external Aar massif interpreted from a network of deep seismic profiles. *Schweizerische Mineralogische und Petrographische Mitteilungen*, 74, 405–420.

Hofmann, B.A., Helfer, M., Diamond Larryn, W., Villa Igor, M., Frei, R., Eikenberg, J. (2004). Topography-driven hydrothermal breccia mineralization of Pliocene age at Grimsel Pass, Aar massif, Central Swiss Alps. *Schweizerische Mineralogische und Petrographische Mitteilungen*, 84, 271–302.

Homewood, P., Allen, P.A., Williams, G. (1986). Dynamics of the Molasse Basin of Western Switzerland. In *Foreland Basins*, Allen, P.A. and Homewood, P. (eds). Wiley-Blackwell, New York.

Hubbard, M. and Mancktelow, N.S. (1992). Lateral displacement during Neogene convergence in the western and central Alps. *Geology*, 20, 943–946.

Huon, S., Burkhard, M., Hunziker, J.-C. (1994). Mineralogical, K-Ar, stable and Sr isotope systematics of K-white micas during very low-grade metamorphism of limestones (Helvetic nappes, western Switzerland). *Chemical Geology*, 113, 347–376.

Jacob, J.-B., Guillot, S., Thiéblemont, D., Vanardois, J., Trap, P., Faure, M., Marquer, D. (2023). Paleozoic evolution and variscan inheritance in the Alps. In *Geodynamics of the Alps 2*, Rosenberg, C.L. and Bellahsen, N. (eds). ISTE Ltd, London, and John Wiley & Sons, New York.

Janots, E., Engi, M., Rubatto, D., Berger, A., Gregory, C., Rahn, M. (2009). Metamorphic rates in collisional orogeny from in situ allanite and monazite dating. *Geology*, 37, 11–14.

Jouanne, F., Ménard, G., Darmendrail, X. (1995). Present-day vertical displacements in the north-western Alps and southern Jura Mountains: Data from leveling comparisons. *Tectonics*, 14, 606–616.

Jouffray, F., Spalla, M.I., Lardeaux, J.M., Filippi, M., Rebay, G., Corsini, M., Zanoni, D., Zucali, M., Gosso, G. (2020). Variscan eclogites from the Argentera–Mercantour Massif (external crystalline massifs, SW Alps): A dismembered cryptic suture zone. *International Journal of Earth Sciences*, 109, 1273–1294.

Jourdon, A., Rolland, Y., Petit, C., Bellahsen, N. (2014). Style of Alpine tectonic deformation in the Castellane fold-and-thrust belt (SW Alps, France): Insights from balanced cross-sections. *Tectonophysics*, 633, 143–155.

Kammer, A. (1989). Alpidische Verformung des aarmassivischen Nordrandes. *Schweizerische Mineralogische und Petrographische Mitteilungen*, 69, 37–53.

Karousova, H., Plomerova, J., Babuska, V. (2013). Upper mantle structure beneath the southern Bohemian Massif and its surroundings imaged by high-resolution tomography. *Geophys. J. Int.*, 194(2), 1203–1215.

Kästle, E.D., Rosenberg, C.L., Boschi, L., Bellahsen, N., Meier, T., El-Sharkawy, A. (2020). Slab break-offs in the Alpine subduction zone. *International Journal of Earth Sciences*, 609, 587–603.

Kerckhove, C. (1969). La “zone du Flysch” dans les nappes de l’Embrunais-Ubaye (Alpes occidentales). *Géologie Alpine*, 45, 5–204.

Kerrich, R., Allison, I., Barnett, R.L., Moss, S., Starkey, J. (1980). Microstructural and chemical transformations accompanying deformation of granite in a shear zone at Mieville, Switzerland; with implications for stress corrosion cracking and superplastic flow. *Contributions to Mineralogy and Petrology*, 73, 221–242.

Kirschner, D.L., Sharp, Z.D., Masson, H. (1995). Oxygen isotope thermometry of quartz-calcite veins: Unraveling the thermal-tectonic history of the subgreenschist facies Morcles nappe (Swiss Alps). *Geol. Soc. Am. Bull.*, 107, 1145–1156.

Kirschner, D.L., Cosca, M.A., Masson, H., Hunziker, J.C. (1996). Staircase 40Ar/39Ar spectra of fine-grained white mica: Timing and duration of deformation and empirical constraints on argon diffusion. *Geology*, 24, 747–750.

Kirschner, D.L., Masson, H., Sharp, Z.D. (1999). Fluid migration through thrust faults in the Helvetic nappes (Western Swiss Alps). *Contrib. Mineral. Petrol.*, 136, 169–183.

Kirschner, D.L., Masson, H., Cosca, M.A. (2003). An 40 Ar/39 Ar, Rb/Sr, and stable isotope study of micas in low-grade fold-and-thrust belt: An example from the Swiss Helvetic Alps. *Contributions to Mineralogy and Petrology*, 145, 460–480.

Kissling, E. (2008). Deep structure and tectonics of the Valais and the rest of the Alps. *Bulletin Angewandte Géologie*, 13, 3–10.

Kissling, E. and Schlunegger, F. (2018). Rollback orogeny model for the evolution of the Swiss Alps. *Tectonics*, 37, 1097–1115.

Kissling, E., Schmid, S.M., Lippitsch, R., Ansorge, J., Fügenschuh, B. (2006). Lithosphere structure and tectonic evolution of the Alpine arc: New evidence from high-resolution teleseismic tomography. *Geological Society, London, Memoirs*, 32, 129–145.

Krayenbuhl, T. and Steck, A. (2009). Structure and kinematics of the Jungfrau syncline, Faflertal (Valais, Alps), and its regional significance. *Swiss Journal of Geosciences*, 102, 441–455.

Kühni, A. and Pfiffner, O.A. (2001). Drainage patterns and tectonic forcing; a model for the Swiss Alps. *Basin Research*, 13, 169–197.

Labhart, T.P. (1966). Mehrphasige alpine Tektonik am Nordrand des Aarmassivs: Beobachtungen im Druckstollen Trift-Speicherberg (Gadmental) der Kraftwerke Oberhasli AG. *Eclogae Geologicae Helvetiae*, 59.

Lacombe, O. and Bellahsen, N. (2016). Thick-skinned tectonics and basement-involved fold–thrust belts: Insights from selected Cenozoic orogens. *Geological Magazine*, 153, 763–810.

Lacombe, O. and Mouthereau, F. (2002). Basement-involved shortening and deep detachment tectonics in forelands of orogens: Insights from recent collision belts (Taiwan, Western Alps, Pyrenees). *Tectonics*, 21(4), 1030.

Lafosse, M., Boutoux, A., Bellahsen, N., Le Pourhiet, L. (2016). Role of tectonic burial and temperature on the inversion of inherited extensional basins during collision. *Geological Magazine*, 153, 811–826.

Laubscher, H.P. (1961). Die fernschubhypothese der jurafaltung. *Eclogae Geologicae Helvetiae*, 54, 222–282.

Laubscher, H.P., Biella, G., Cassinis, R., Gelati, R., Lozej, A., Scarascia, S., Tabacco, I. (1992). The collisional knot in Liguria. *Geologische Rundschau*, 81, 275–289.

Lazarre, J., Tricart, P., Courrioux, G., Ledru, P. (1996). Héritage téthysien et polyphasage alpin : réinterprétation tectonique du "synclinal" de l'Aiguille de Morges (Massif du Pelvoux, Alpes occidentales, France). *Comptes rendus de l'Académie des sciences. Série 2. Sciences de la terre et des planètes*, 323, 1051–1058.

Lelarge, M.L. (1993). Thermochronologie par la méthode des traces de fission d'une marge passive (dôme de Ponta Grossa, SE Brésil) et au sein d'une chaîne de collision (zone externe de l'arc alpin, France). Thesis, Université Joseph Fourier de Grenoble.

Leloup, P.-H., Arnaud, N., Sobel, E.R., Lacassin, R. (2005). Alpine thermal and structural evolution of the highest external crystalline massif: The Mont Blanc. *Tectonics*, 24, TC4002.

Lemoine, M. (2003). Schistes lustrés from Corsica to Hungary: Back to the original sediments and tentative dating of partly azoic metasediments. *Bulletin de la Société géologique de France*, 174(3), 197–209.

Leu, W. and Wyss, R. (1992). Geologische Aufnahmen und Prognoseprofil im Gebiet von Sedrun, Vorderrheintal (Alptransit/Gotthard-Basistunnel). *Bulletin der Vereinigung Schweizerischer Petroleum-Geologen und-Ingenieure*, 59, 81–92.

Lickorish, W.H. and Ford, M. (1998). Sequential restoration of the external Alpine Digne thrust system, SE France, constrained by kinematic data and synorogenic sediments. *Geological Society, London, Special Publications*, 134, 189–211.

Lihou, J.C. (1995). A new look at the Blattengrat Unit of Eastern Switzerland – Early tertiary Foreland Basin sediments from the South Helvetic Realm. *Eclogae Geologicae Helvetiae*, 88, 91–114.

Lippitsch, R., Kissling, E., Ansorge, J. (2003). Upper mantle structure beneath the Alpine orogen from high-resolution teleseismic tomography. *Journal of Geophysical Research-Solid Earth*, 108, 138, 2376.

Lu, G., Winkler, W., Rahn, M., Von Quadt, A., Willett, S.D. (2018). Evaluating igneous sources of the Taveyannaz formation in the Central Alps by detrital zircon U–Pb age dating and geochemistry. *Swiss Journal of Geosciences*, 111, 399–416.

Lugeon, M. (1914). Sur l'ampleur de la nappe de Morcles. *CR Acad. Sc.*, 153, 2029–2031.

Luth, S., Willingshofer, E., Sokoutis, D., Cloetingh, S. (2013). Does subduction polarity changes below the Alps? Inferences from analogue modelling. *Tectonophysics*, 582, 140–161.

Madritsch, H., Schmid, S.M., Fabbri, O. (2008). Interactions between thin- and thick-skinned tectonics at the northwestern front of the Jura fold-and-thrust belt (eastern France). *Tectonics*, 27, TC5005.

Maffione, M., Speranza, F., Faccenna, C., Cascella, A., Vignaroli, G., Sagnotti, L. (2008). A synchronous Alpine and Corsica-Sardinia rotation. *Journal of Geophysical Research*, 113, B03104.

Magni, V., van Hunen, J., Funiciello, F., Faccenna, C. (2012). Numerical models of slab migration in continental collision zones. *Solid Earth*, 3, 293–306.

Mair, D., Lechmann, A., Herwegh, M., Nibourel, L., Schlunegger, F. (2018). Linking Alpine deformation in the Aar massif basement and its cover units – The case of the Jungfrau–Eiger mountains (Central Alps, Switzerland). *Solid Earth*, 9, 1099–1122.

Malaroda, R., Carraro, F., Dal Piaz, G., Franceschetti, B., Sturani, C., Zanella, E. (1970). Carta geologica del Massiccio dell'Argentera alla scala 1:50.000. *Men. Soc. Ital.*, 9, P557 A663.

Manatschal, G. and Bernoulli, D. (1999). Architecture and tectonic evolution of nonvolcanic margins: Present-day Galicia and ancient Adria. *Tectonics*, 18, 1099–1119.

Mancktelow, N. (1985). The Simplon Line: A major displacement zone in the western Lepontine Alps. *Eclogae Geologicae Helvetiae*, 78, 73–96.

Mancktelow, N. (1990). The Simplon fault zone. *Beitraege zur Geologischen Karte der Schweiz = Materiaux pour la Carte Geologique de la Suisse = Materiali per la Carta Geologica della Svizzera*, 163, 5–74.

Manzotti, P., Ballèvre, M., Zucali, M., Robyr, M., Engi, M. (2014). The tectonometamorphic evolution of the Sesia–Dent Blanche nappes (internal Western Alps): Review and synthesis. *Swiss Journal of Geosciences*, 107, 309–336.

Marquer, D. and Gapais, D. (1985). Les massifs cristallins externes sur une transversale Guttanen-Val Bedretto (alpes centrales) : structures et histoire cinématique. *Comptes rendus de l'Académie des sciences. Série 2, Mécanique, Physique, Chimie, Sciences de l'univers, Sciences de la Terre*, 301, 543–546.

Marquer, D., Calcagno, P., Barfety, J.-C., Baudin, T. (2006). 3D modeling and kinematics of the external zone of the French Western Alps (Belledonne and Grand Châtelard Massifs, Maurienne Valley, Savoie). *Eclogae Geologicae Helvetiae*, 99, 211–222.

Marshall, D., Pfeifer, H.-R., Hunziker, J.C., Kirschner, D. (1998). A pressure-temperature-time path for the NE Mont-Blanc massif; fluid-inclusion, isotopic and thermobarometric evidence. *European Journal of Mineralogy*, 10, 1227–1240.

Masson, H., Herb, R., Steck, A. (1980). Helvetic Alps of Switzerland. In *26th International Geological Congress*, Schweiz. Geol. Konm. (eds). Wepf & Co. Publishers, New York.

Mercolli, I., Biino, G., Abrecht, J. (1994). The lithostratigraphy of the pre-Mesozoic basement of the Gotthard massif: A review: The pre-Alpine crustal evolution of the Aar-, Gotthard- and Tavetsch massifs. *Schweizerische Mineralogische und Petrographische Mitteilungen*, 74, 29–40.

Michalski, I. and Soom, M. (1990). The Alpine thermo-tectonic evolution of the Aar and Gotthard massifs, Central Switzerland: Fission track ages on zircon and apatite and K–Ar mica ages. *Schweizerische Mineralogische Petrographische Mitteilungen*, 70, 373–387.

Milnes, A.G. (1974). Structure of the Pennine Zone: A new working hypothesis. *Geol. Soc. Am. Bull.*, 85, 1727–1732.

Mitterbauer, U., Behm, M., Brückl, E., Lippitsch, R., Guterch, A., Keller, G.R., Koslovskaya, E., Rumpfhuber, E.-M., Šumanovac, F. (2011). Shape and origin of the East-Alpine slab constrained by the ALPASS teleseismic model. *Tectonophysics*, 510, 195–206.

Mock, S. and Herwegh, M. (2017). Tectonics of the central Swiss Molasse Basin: Post-Miocene transition to incipient thick-skinned tectonics? *Tectonics*, 36, 1699–1723.

Mock, S., von Hagke, C., Schlunegger, F., Dunkl, I., Herwegh, M. (2020). Long-wavelength late-Miocene thrusting in the north Alpine foreland: Implications for late orogenic processes. *Solid Earth*, 11, 1823–1847.

Mohn, G., Manatschal, G., Müntener, O., Beltrando, M., Masini, E. (2010). Unravelling the interaction between tectonic and sedimentary processes during lithospheric thinning in the Alpine Tethys margins. *International Journal of Earth Sciences*, 99, 75–101.

Mohn, G., Manatschal, G., Beltrando, M., Masini, E., Kusznir, N. (2012). Necking of continental crust in magma-poor rifted margins: Evidence from the fossil Alpine Tethys margins. *Tectonics*, 31, TC 1012.

Morgenthaler, H. (1921). *Petrographisch-tektonische Untersuchungen am Nordrand des Aarmassivs*. E. Birkhäuser & Cie, Basel.

Mosar, J. (1999). Present-day and future tectonic underplating in the western Swiss Alps: Reconciliation of basement/wrench-faulting and décollement folding of the Jura and Molasse basin in the Alpine foreland. *Earth and Planetary Science Letters*, 173, 143–155.

Mulder, T., Callec, Y., Parize, O., Joseph, P., Schneider, J.-L., Robin, C., Dujoncquoy, E., Salles, T., Allard, J., Bonnel, C. (2010). High-resolution analysis of submarine lobes deposits: Seismic-scale outcrops of the Lauzanier area (SE Alps, France). *Sedimentary Geology*, 229, 160–191.

Müller, F. and Arbenz, P. (1938). *Geologie der Engelhörner, der Aareschlucht und der Kalkkeile bei Inntertkirchen: Berner Oberland: mit 11 Textfiguren und VIII Tafeln*. Francke.

Nibourel, L., Berger, A., Egli, D., Luensdorf, N.K., Herwegh, M. (2018). Large vertical displacements of a crystalline massif recorded by Raman thermometry. *Geology*, 46, 879–882.

Nibourel, L., Berger, A., Egli, D., Herwegh, M. (2021a). Structural and thermal evolution of an external crystalline massif: Insights from structural field work and RSCM thermometry in the eastern Aar massif. *Swiss Journal of Geosciences*, 114, 1–43.

Nibourel, L., Rahn, M., Dunkl, I., Berger, A., Herman, F., Diehl, T., Herwegh, M. (2021b). Orogen-parallel migration of exhumation in the eastern Aar massif revealed by low-T thermochronology. *Journal of Geophysical Research: Solid Earth*, 126, e2020JB020799.

Oliot, E., Goncalves, P., Marquer, D. (2010). Role of plagioclase and reaction softening in a metagranite shear zone at mid-crustal conditions (Gotthard Massif, Swiss Central Alps). *Journal of Metamorphic Geology*, 28, 849–871.

Paquette, J.-L., Menot, R.-P., Peucat, J.-J. (1989). REE, SmNd and UPb zircon study of eclogites from the Alpine External Massifs (Western Alps): Evidence for crustal contamination. *Earth and Planetary Science Letters*, 96, 181–198.

Paréjas, E. (1922). Géologie de la zone de Chamonix, comprise entre le Mont-Blanc et les Aiguilles Rouges. *Mémoires de la Société de physique et d'histoire naturelle de Genève*, 39(7), 373–442.

Pfeifer, H.-R., Sanchez, A., Degueldre, C. (1992). Thermal springs in granitic rocks from the Grimsel Pass (Swiss Alps): The late stage of a hydrothermal system related to Alpine Orogeny. In *Water-Rock Interaction*, Kharaja, Y.K. and Maest, A.S. (eds). Balkema, Rotterdam.

Pfiffner, O.A. (1986). Evolution of the north Alpine foreland basin in the Central Alps. *Foreland Basins*, 219–228.

Pfiffner, O.A. (1993). The structure of the Helvetic nappes and its relation to the mechanical stratigraphy. *Journal of Structural Geology*, 15, 511–521.

Pfiffner, O.A. (2015). *Geologie der Alpen*. Haupt Verlag, Bern, UTB-Band 8416.

Pfiffner, O.A. (2016). Basement-involved thin-skinned and thick-skinned tectonics in the Alps. *Geological Magazine*, 153, 1085–1109.

Pfiffner, O.A., Sahli, S., Stäuble, M. (1997a). Compression and uplift of the external massifs in the Helvetic zone. *Deep Structure of the Swiss Alps: Results of NRP*, 20, 139–153.

Pfiffner, O.A., Lehner, P., Heitzmann, P., Mueller, S., Steck, A. (1997b). *Deep Structure of the Swiss Alps*. Birkhäuser, Basel.

Pfiffner, O.A., Burkhard, M., Haenni, R., Kammer, A., Kligfield, R., Mancktelow, N., Menkveld, J.-W., Ramsay, J., Schmid, S.M., Zurbriggen, R. (2011). Structural map of the Helvetic zone of the Swiss Alps. In *Geological Special Map 128 1:100'000*, Survey, S.G. (ed.). Swisstopo, Bern.

Philippe, Y., Colletta, B., Deville, E., Mascle, A. (1996). The Jura fold-and-thrust belt: A kinematic model based on map-balancing. *Mémoires du Muséum national d'histoire naturelle*, 170, 235–261.

Philippe, Y., Deville, E., Mascle, A. (1998). Thin-skinned inversion tectonics at oblique basin margins: Example of the western Vercors and Chartreuse Subalpine massifs (SE France). *Geological Society London Special Publications*, 134(1), 239–262.

Platt, J. (1984). Balanced cross-sections and their implications for the deep structure of the northwest Alps: Discussion. *Journal of Structural Geology*, 16, 603–606.

Pleuger, J., Mancktelow, N., Zwingmann, H., Manser, M. (2012). K–Ar dating of synkinematic clay gouges from Neoalpine faults of the Central, Western and Eastern Alps. *Tectonophysics*, 550, 1–16.

Rahn, M., Mullis, J., Erdelbrock, K., Frey, M. (1994). Very low-grade metamorphism of the Taveyanne Greywacke, Glarus Alps, Switzerland. *Journal of Metamorphic Geology*, 12, 625–641.

Ramsay, J.G. (1963). Stratigraphy, structure and metamorphism in the Western Alps. *Proceedings of the Geologists' Association*, 74, 357–312.

Ramsay, J.G. (1981). Tectonics of the Helvetic nappes. *Geological Society London Special Publications*, 9, 293–309.

Ramsay, J.G., Casey, M., Kligfield, R. (1983). Role of shear in development of the Helvetic fold-thrust belt of Switzerland. *Geology*, 11, 439–442.

von Raumer, J. (1971). Das Mont-Blanc-Massiv: Altkristallin im Bereich schwacher alpiner Metamorphose. *Schweizerische Mineralogische und Petrographische Mitteilungen*, 51, 193–225.

von Raumer, J., Abrecht, J., Bussy, F., Lombardo, B., Menot, R.-P., Schaltegger, U. (1999). The Paleozoic metamorphic evolution of the Alpine external massifs. *Schweizerische Mineralogische und Petrographische Mitteilungen*, 79, 5–22.

Reinecker, J., Danisik, M., Schmid, C., Glotzbach, C., Rahn, M., Frisch, W., Spiegel, C. (2008). Tectonic control on the late stage exhumation of the Aar massif (Switzerland): Constraints from apatite fission track and (U-Th)/He data. *Tectonics*, 27, TC6009.

Reinhard, M. and Preiswerk, H. (1927). *Über Granitmylonite im Aiguilles-Rouges-Massiv (Westliches Wallis)*. Birkhäuser, Basel.

Rohr, K. (1926). *Stratigraphische und tektonische Untersuchung der Zwischenbildungen am Nordrande des Aarmassivs (zwischen Wendenjoch und Wetterhorn)*. Francke.

Rolland, Y. and Rossi, M. (2016). Two-stage fluid flow and element transfers in shear zones during collision burial-exhumation cycle: Insights from the Mont Blanc Crystalline Massif (Western Alps). *Journal of Geodynamics*, 101, 88–108.

Rolland, Y., Cox, S., Boullier, A.-M., Pennacchioni, G., Mancktelow, N. (2003). Rare earth and trace element mobility in mid-crustal shear zones: Insights from the Mont Blanc massif (Western Alps). *Earth and Planetary Science Letters*, 214, 203–219.

Rolland, Y., Corsini, M., Rossi, M., Cox, S., Pennacchioni, G., Mancktelow, N., Boullier, A.-M. (2007). Comment on "Alpine thermal and structural evolution of the highest external crystalline massif: The Mont Blanc" by PH Leloup, N. Arnaud, E.R. Sobel, and R. Lacassin. *Tectonics*, 26, TC2015.

Rolland, Y., Rossi, M., Cox, S., Corsini, M., Mancktelow, N., Pennacchioni, G., Fornari, M., Boullier, A.-M. (2008). 40Ar/39Ar dating of synkinematic white mica: Insights from fluid-rock reaction in low-grade shear zones (Mont Blanc massif) and constraints on timing of deformation in the NW external Alps. *Geological Society London Special Publications*, 299, 293–315.

Rolland, Y., Cox, S.F., Corsini, M. (2009). Constraining deformation stages in brittle–ductile shear zones from combined field mapping and 40Ar/39Ar dating: The structural evolution of the Grimsel Pass area (Aar massif, Swiss Alps). *Journal of Structural Geology*, 31, 1377–1394.

Rosenberg, C.L., Berger, A., Schmid, S.M. (1995). Observations from the floor of a granitoid pluton: Inferences on the driving force of final emplacement. *Geology*, 23, 443–446.

Rosenberg, C.L., Berger, A., Bellahsen, N., Bousquet, R. (2015). Relating orogen width to shortening, erosion, and exhumation during Alpine collision. *Tectonics*, 34, 1306–1328.

Rossi, M. and Rolland, Y. (2014). Stable isotope and Ar/Ar evidence of prolonged multiscale fluid flow during exhumation of orogenic crust: Example from the Mont Blanc and Aar massifs (NW Alps). *Tectonics*, 33(9), 1681–1709.

Rossi, M., Rolland, Y., Vidal, O., Cox, S. (2005). Geochemical variations and element transfer during shear-zone development and related episyenites at middle crust depths: Insights from the Mont Blanc granite (French–Italian Alps). *Geological Society London Special Publications*, 245, 373–396.

Roure, F., Bergerat, F., Damotte, B., Mugnier, J.-L. Polino, R. (1996). The ECORS-CROP Alpine seismic traverse. *Mémoire Société géologique de France*, 170, 113.

Rubatto, D., Schaltegger, U., Lombardo, D., Colombo, F., Compagnoni, R. (2001). Complex Paleozoic magmatic and metamorphic evolution in the Argentera massif (Western Alps), resolved with U-Pb dating. *Schweizerische Mineralogische und Petrographische Mitteilungen*, 81, 213–228.

Rubatto, D., Hermann, J., Berger, A., Engi, M. (2009). Protracted fluid-induced melting during Barrovian metamorphism in the Central Alps. *Contributions of Mineralogy and Petrology*, 158, 703–722.

Rubatto, D., Ferrando, S., Compagnoni, R., Lombardo, B. (2010). Carboniferous high-pressure metamorphism of Ordovician protoliths in the Argentera massif (Italy), Southern European Variscan belt. *Lithos*, 116, 65–76.

Ruffini, R., Cosca, M., s'Atri, A., Hunziker, J. (1995). The vulcanic supply of the Taveyanne turbidites (Savoie, France): A riddle for Tertiary Alpine vulcanism. *Atti delle Accademia Science Roma*, 14, 359–376.

Sabil, N. (1995). La datation par traces de fission : aspects méthodologiques et applications thermochronologiques en contexte alpin et de marge continentale. PhD Thesis, Université Joseph-Fourier-Grenoble I.

Sanchez, G., Rolland, Y., Corsini, M., Braucher, R., Bourlès, D., Arnold, M., Aumaître, G. (2010). Relationships between tectonics, slope instability and climate change: Cosmic ray exposure dating of active faults, landslides and glacial surfaces in the SW Alps. *Geomorphology*, 117, 1–13.

Sanchez, G., Rolland, Y., Schneider, J., Corsini, M., Oliot, E., Goncalves, P., Verati, C., Lardeaux, J.-M., Marquer, D. (2011a). Dating low-temperature deformation by 40Ar/39Ar on white mica, insights from the Argentera-Mercantour massif (SW Alps). *Lithos*, 125, 521–536.

Sanchez, G., Rolland, Y., Jolivet, M., Brichau, S., Corsini, M., Carter, A. (2011b). Exhumation controlled by transcurrent tectonics: The Argentera–Mercantour massif (SW Alps). *Terra Nova*, 23, 116–126.

Sánchez, L., Völksen, C., Sokolov, A., Arenz, H., Seitz, F. (2018). Present-day surface deformation of the Alpine region inferred from geodetic techniques. *Earth System Science Data*, 10, 1503–1526.

Scabell, W. (1926). *Beiträge zur Geologie der Wetterhorn-Schreckhorn-Gruppe: Berner Oberland*. A. Franke A.G., Bern.

Schlatter, A. (2006). *Das neue Landeshöhennetz der Schweiz LHN95*. Geodätisch-geophysikalische Arbeiten in der Schweiz, Band 72.

Schlunegger, F. and Kissling, E. (2015). Slab rollback orogeny in the Alps and evolution of the Swiss Molasse basin. *Nature Communications*, 6, 8605.

Schlunegger, F., Jordan, T.E., Klaper, E.M. (1997). Controls of erosional denudation in the orogen on foreland basin evolution: The Oligocene central Swiss Molasse Basin as an example. *Tectonics*, 16, 823–840.

Schmid, S.M. (1975). The Glarus overthrust; field evidence and mechanical model. *Eclogae Geologicae Helvetiae*, 68, 247–280.

Schmid, S.M. and Kissling, E. (2000). The arc of western Alps in the light of geophysical data on deep crustal structures. *Tectonics*, 19, 62–85.

Schmid, S.M., Aebli, H., Heller, F., Zingg, A. (1989). The role of the Periadriatic Line in the tectonic evolution of the Alps. *Geological Society, London, Special Publications*, 45, 153–171.

Schmid, S.M., Pfiffner, O.A., Froitzheim, N., Schonborn, G., Kissling, E. (1996). Geophysical-geological transect and tectonic evolution of the Swiss-Italian Alps. *Tectonics*, 15, 1036–1064.

Schmid, S.M., Fügenschuh, B., Kissling, E., Schuster, R. (2004). Tectonic map and overall architecture of the Alpine orogen. *Eclogae Geologicae Helvetiae*, 97, 93–117.

Schmid, S.M., Kissling, E., Diehl, T., van Hinsbergen, D.J., Molli, G. (2017). Ivrea mantle wedge, arc of the Western Alps, and kinematic evolution of the Alps–Apennines orogenic system. *Swiss Journal of Geosciences*, 110, 581–612.

Schneeberger, R., Berger, A., Herwegh, M., Eugster, A., Kober, F., Spillmann, T., Blechschmidt, I. (2016). GTS phase VI-LASMO: Geology and structures of the GTS and Grimsel region. In *Arbeitsbericht National Cooperative for the Disposal of Radioactive Waste (NAB)*. LASMO, Wettingen.

Seward, D. and Mancktelow, N.S. (1994). Neogene kinematics of the central and western Alps: Evidence from fission-track dating. *Geology*, 22, 803–806.

Seward, D., Ford, M., Bürgisser, J., Lickorish, H., Williams, E.A., Meckel, L.D. (1999). Preliminary results of fission-track analyses in the southern Pelvoux area, SE France. *Memorie di Scienze Geologiche Padova*, 51, 25–31

Simon-Labric, T., Rolland, Y., Dumont, T., Heymes, T., Authemayou, C., Corsini, M., Fornari, M. (2009). 40Ar/39Ar dating of Penninic Front tectonic displacement (W Alps) during the Lower Oligocene (31–34 Ma). *Terra Nova*, 21, 127–136.

Simonetti, M., Carosi, R., Montomoli, C., Langone, A., D'addario, E., Mammoliti, E. (2018). Kinematic and geochronological constraints on shear deformation in the Ferriere–Mollières shear zone (Argentera–Mercantour massif, Western Alps): Implications for the evolution of the Southern European Variscan Belt. *International Journal of Earth Sciences*, 107(6), 2163-2189.

Sinclair, H. (1997). Flysch to molasse transition in peripheral foreland basins: The role of the passive margin versus slab breakoff. *Geology*, 25, 1123–1126.

Sinclair, H. and Allen, P. (1992). Vertical versus horizontal motions in the Alpine orogenic wedge: Stratigraphic response in the foreland basin. *Basin Research*, 4, 215–232.

Singer, J., Diehl, T., Husen, S., Kissling, E., Duretz, T. (2014). Alpine lithosphere slab rollback causing lower crustal seismicity in northern foreland. *Earth and Planetary Science Letters*, 397, 42–56.

Sommaruga, A. (1999). Décollement tectonics in the Jura forelandfold-and-thrust belt. *Marine and Petroleum Geology*, 16, 111–134.

Sommaruga, A., Eichenberger, U., Marillier, F.J.Y. (2012). Seismic Atlas of the Swiss Molasse Basin. *Swiss Geophysical Commission, Contribution to the Geology of Switzerland – Geophysics*, 44, 1–82.

Soom, M.A. (1990). Abkühlungs-und Hebungsgeschichte der Externmassive und der penninischen Decken beidseits der Simplon-Rhone-Linie seit dem Oligozän: Spaltspurdatierungen an Apatit/Zirkon und K-Ar-Datierungen an Biotit/Muskowit (Westliche Zentralalpen). PhD Thesis, University of Bern.

Spada, M., Bianchi, I., Kissling, E., Agostinetti, N.P., Wiemer, S. (2013). Combining controlled-source seismology and receiver function information to derive 3-D Moho topography for Italy. *Geophysical Journal International*, 194, 1050–1068.

Spillmann, P. (2011). Geologie des Kantons Uri. *Naturforschende Gesellschaft Uri, Altdorf.* 24, 224.

Stampfli, G.M. (1993). Le Briançonnais : terrain exotique dans les Alpes ? *Eclogae Geologicae Helvetiae*, 86, 1–45.

Stampfli, G.M. (2001). Permo-Mesozoic evolution of the western Tethys realm: The Neotethys East Mediterranean basin connection. *Peri-Tethyan Rift/Wrench Basins and Passive Margins*, 2001, 51–108.

Steck, A. (1966). Petrographische und tektonische Untersuchungen am Zentralen Aaregranit und seinen altkristallinen Hüllgesteinen im westlichen Aarmassiv im Gebiet Belalp-Grisighorn. *Beitr. Geol. Karte Schweiz, N.F.*, 130.

Steck, A. (1968). Die alpidischen Strukturen in den zentralen Aaregraniten des westlichen Aarmassivs. *Eclogae Geologicae Helvetiae*, 61, 19–48.

Steck, A. (1984). Tertiary deformation structures in the Central Alps. *Eclogae Geologicae Helvetiae*, 77, 55–100.

Steck, A. (2008). Tectonics of the Simplon massif and Lepontine gneiss dome: Deformation structures due to collision between the underthrusting European plate and the Adriatic indenter. *Swiss Journal of Geosciences*, 101, 515–546.

Steck, A. and Hunziker, J. (1994). The Tertiary structural and thermal evolution of the Central Alps – Compressional and extensional structures in an orogenic belt. *Tectonophysics*, 238, 229–254.

Steck, A. and Vocat, D. (1973). Zur Mineralogie der Granitmylonite von Miéville, Aiguilles Rouges Massiv. *Schweizerische Mineralogische und Petrographische Mitteilungen*, 53, 474–477.

Steck, A., Epard, J.-L., Escher, A., Lehner, P., Marchant, R., Masson, H. (1997). Geological interpretation of the seismic profiles through Western Switzerland: Rawil (W1), Val d'Anniviers (W2), Mattertal (W3), Zmutt-Zermatt-Findelen (W4) and Val de Bagnes (W5). *Deep Structure of the Swiss Alps – Results from NRP*, 20, 123–137.

Steck, A., Epard, J., Escher, A., Gouffon, Y., Masson, H. (2001). Notice explicative pour la carte tectonique des Alpes de Suisse occidentale. Explanatory note, Swisstopo, Wabern.

Steck, A., Masson, H., Robyr, M. (2015). Tectonics of the Monte Rosa and surrounding nappes (Switzerland and Italy): Tertiary phases of subduction, thrusting and folding in the Pennine Alps. *Swiss Journal of Geosciences*, 108, 3–34.

Sternai, P., Herman, F., Champagnac, J.-D., Fox, M., Salcher, B., Willett, S.D. (2012). Pre-glacial topography of the European Alps. *Geology*, 40, 1067–1070.

Sternai, P., Sue, C., Husson, L., Serpelloni, E., Becker, T.W., Willett, S.D., Faccenna, C., Di Giulio, A., Spada, G., Jolivet, L. (2019). Present-day uplift of the European Alps: Evaluating mechanisms and models of their relative contributions. *Earth-Science Reviews*, 190, 589–604.

Sue, C., Delacou, B., Champagnac, J.D., Allanic, C., Burkhard, M. (2007). Aseismic deformation in the Alps: GPS vs. seismic strain quantification. *Terra Nova*, 19, 182–188.

Todd, C. and Engi, M. (1997). Metamorphic field gradients in the Central Alps. *Journal of Metamorphic Geology*, 15, 513–530.

Torsvik, T.H., Van der Voo, R., Preeden, U., Mac Niocaill, C., Steinberger, B., Doubrovine, P.V., van Hinsbergen, D.J., Domeier, M., Gaina, C., Tohver, E. (2012). Phanerozoic polar wander, palaeogeography and dynamics. *Earth-Science Reviews*, 114, 325–368.

Trümpy, R. (1949). Der Lias der Glarner Alpen. Dissertation, ETH Zürich.

Trümpy, R. (1952). Présence probable du Jurassique supérieur et du Crétacé inférieur au SW du massif du Mont-Blanc. *Compte rendu sommaire de la Société Géologique de France*, 9, 149–150.

Trümpy, R. (1960). Paleotectonic evolution of the Central and Western Alps. *Geological Society of America Bulletin*, 71, 843–907.

Trümpy, R. (1975). Penninic Austroalpine boundary in the Swiss Alps: A presumed former continental margin and its problems. *American Journal of Science*, 275-A, 209–238.

von Tscharner, M., Schmalholz, S., Epard, J.-L. (2016). 3-D numerical models of viscous flow applied to fold nappes and the Rawil depression in the Helvetic nappe system (western Switzerland). *Journal of Structural Geology*, 86, 32–46.

Ustaszewski, K. and Schmid, S.M. (2006). Control of preexisting faults on geometry and kinematics in the northernmost part of the Jura fold-and-thrust belt. *Tectonics*, 25, TC5003.

Ustaszewski, K. and Schmid, S.M. (2007). Latest Pliocene to recent thick-skinned tectonics at the Upper Rhine Graben–Jura Mountains junction. *Swiss Journal of Geosciences*, 100, 293–312.

Valla, P.G., Shuster, D.L., van der Beek, P.A. (2011). Significant increase in relief of the European Alps during mid-Pleistocene glaciations. *Nature Geoscience*, 4, 688.

Valla, P.G., Beek, P.A., Shuster, D.L., Braun, J., Herman, F., Tassan-Got, L., Gautheron, C. (2012). Late Neogene exhumation and relief development of the Aar and Aiguilles Rouges massifs (Swiss Alps) from low-temperature thermochronology modeling and 4He/3He thermochronometry. *Journal of Geophysical Research: Earth Surface*, 117, F01004.

Valla, P.G., Rahn, M., Shuster, D.L., van der Beek, P.A. (2016). Multi-phase late-Neogene exhumation history of the Aar massif, Swiss central Alps. *Terra Nova*, 28, 383–393.

Vernon, A.J., van der Beek, P.A., Sinclair, H.D., Rahn, M.K. (2008). Increase in late Neogene denudation of the European Alps confirmed by analysis of a fission-track thermochronology database. *Earth and Planetary Science Letters*, 270, 316–329.

Vernon, A.J., van der Beek, P.A., Sinclair, H.D. (2009). Spatial correlation between long-term exhumation rates and present-day forcing parameters in the western European Alps. *Geology*, 37, 859–862.

Wagner, M., Kissling, E., Husen, S. (2012). Combining controlled-source seismology and local earthquake tomography to derive a 3-D crustal model of the western Alpine region. *Geophysical Journal International*, 191, 789–802.

Wehrens, P.C., Berger, A., Peters, M., Spillmann, T., Herwegh, M. (2016). Deformation at the frictional-viscous transition: Evidence for cycles of fluid-assisted embrittlement and ductile deformation in the granitoid crust. *Tectonphysics*, 693, 66–84.

Wehrens, P.C., Baumberger, R., Berger, A., Herwegh, M. (2017). How is strain localized in a meta-granitoid, mid-crustal basement section? Spatial distribution of deformation in the central Aar massif (Switzerland). *Journal of Structural Geology*, 94, 47–67.

Wiederkehr, M., Bousquet, R., Schmid, S.M., Berger, A. (2008). From subduction to collision: Thermal overprint of HP/LT meta-sediments in the north-eastern Lepontine Dome (Swiss Alps) and consequences regarding the tectono-metamorphic evolution of the Alpine orogenic wedge. *Swiss Journal of Geosciences*, 101, 127–155.

Willett, S.D. (2010). Late Neogene erosion of the Alps: A climate driver? *Annual Review of Earth and Planetary Sciences*, 38, 411–437.

Zerlauth, M., Ortner, H., Pomella, H., Pfiffner, O.A., Fügenschuh, B. (2014). Inherited tectonic structures controlling the deformation style: An example from the Helvetic nappes of the Eastern Alps. *Swiss Journal of Geosciences*, 107, 157–175.

Zhao, L., Paul, A., Guillot, S., Solarino, S., Malusà, M.G., Zheng, T., Aubert, C., Salimbeni, S., Dumont, T., Schwartz, S. (2015). First seismic evidence for continental subduction beneath the Western Alps. *Geology*, 43, 815–818.

Zhao, L., Paul, A., Malusà, M.G., Xu, X., Zheng, T., Solarino, S., Guillot, S., Schwartz, S., Dumont, T., Salimbeni, S. (2016). Continuity of the Alpine slab unraveled by high-resolution P wave tomography. *Journal of Geophysical Research: Solid Earth*, 121, 8720–8737.

Ziegler, H.-J. and Isler, A. (2013). *Lötschberg-Basistunnel: zusammenfassender geologischer Schlussbericht*. Swisstopo, Wabern.

Zwingmann, H. and Mancktelow, N. (2004). Timing of Alpine fault gouges. *Earth and Planetary Science Letters*, 223, 415–425.

3

A Review of How Our Understanding on the Relationships between the Alps and the North Alpine Foreland Basin has Evolved

Fritz SCHLUNEGGER[1] and Edi KISSLING[2]

[1]Institute of Geological Sciences, University of Bern, Switzerland

[2]Department of Earth Sciences, Institute of Geophysics, ETH Zürich, Switzerland

The North Alpine foreland basin or the Molasse basin that is situated on the northern side of the Alps has experienced a long history of research on its stratigraphic architecture, chronological framework, facies relationships, paleontological records and the provenance of its clastic material. Among these efforts, the most emphasis has been placed upon analyzing the Molasse basin as part of the Alpine orogenesis. Accordingly, the progress of Molasse research has been framed by advances in our knowledge, from a conceptual point of view, of how foreland basins evolve in response to crustal loading and erosional unloading in the adjacent mountain belt. Our understanding that has resulted from these conceptual investigations has changed over time, and so has our view on the development of the Alps/Molasse basin ensemble. This contribution recasts the history of research of the

Geodynamics of the Alps 3,
coordinated by Claudio L. ROSENBERG and Nicolas BELLAHSEN.

Molasse foreland basin, and it explains why particular mechanisms and the collection of related datasets have enjoyed more attention than others. It then ends with a call for a more careful exploration of how slab scale processes could have driven not only the development of the Alps, but also that of the Molasse basin, mainly because shifts in subduction and thus slab dynamics possibly explain the first order changes recorded in the evolution of this basin. These include (i) the transition from Flysch to Molasse sedimentation, which occurred in the Swiss part of the basin at 30 Ma and in the Austrian sector 10 Ma later, (ii) the ca. 18 Ma old reversal of the drainage direction in the basin axis, and (iii) the large-scale uplift and rebound of the entire basin since Pliocene times. The AlpArray initiative focuses on offering a bridge between surface dynamics, deep lithosphere structure and subduction mechanisms in the Alps. We postulate that the stratigraphic archive of the Molasse basin offers one of the cornerstones to link the processes on the surface with those in the mantle beneath the Alps.

3.1. Introduction

A foreland basin is generally defined as an elongated trough that forms between a contractional orogenic belt and the stable craton, mainly in response to flexural subsidence that is driven by tectonic loading in the adjacent mountain belt (DeCelles and Gilles 1996). Because of this implicit link between orogensis and foreland basin formation, the sedimentary archives preserved in these troughs offer very useful information on the collisional and erosional history of the adjacent mountain belt (Flemings and Jordan 1990; DeCelles et al. 1998; Garcia-Castellanos et al. 2002; Naylor and Sinclair 2008; Sinclair and Naylor 2012; Amidon et al. 2017; Horton 2018; Stutenbecker et al. 2019). This is particularly also the case for the European Alps and its North Alpine Foreland basin or the Molasse basin (Figure 3.1) situated on the northern side of this orogen, where numerous sedimentologic (Homewood and Allen 1981; Diem 1986; Keller 1989; Tessier and Gigog 1989; Homewood et al. 1986; De Ruig and Hubbard 2006; Hubbard et al. 2009; Garefalakis and Schlunegger 2019; Hülscher et al. 2019), chronologic (Schlunegger et al. 1996; Kempf et al. 1999; Strunck and Matter 2002; Reichenbacher et al. 2013; Grunert et al. 2015; Sant et al. 2017; van der Boon et al. 2018; Hofmayer et al. 2019) and structural work (Hinsch 2013; Ortner et al. 2015; Mock and Herwegh 2017) yielded a detailed picture of how this basin evolved through time (Sissingh 1997, 1998; Kuhlemann and Kempf 2002; Ford and Lickorish 2004; Berger et al. 2005). These investigations also disclosed at a high resolution how the stratigraphic records within the Molasse basin can be used to reconstruct the development of the adjacent Alps (Kuhlemann and Kempf 2002; Ford and Lickorish 2004; Schlunegger and Kissling 2015). Among the aforementioned authors, Pfiffner (1986) was the first to reconstruct the direct and causal link between orogenic processes in the Alps and the evolution of the Molasse foreland basin for a section across eastern Switzerland. In his study, Pfiffner (1986) documented that the subsidence of the foreland plate and thus the accumulation of Molasse deposits was closely linked with the tectonic thickening and the basinward propagation of the orogenic wedge. Pfiffner's work was guided by the inference that the formation of accommodation space in the basin primarily occurred through the construction of topographic loads, where a larger topography in the mountain belt was commonly associated with a greater deflection of the plate underlying the foreland basin (Beaumont 1981; Jordan 1981). Others used the same principle, but applied theoretical models instead, with an effort to identify the basin's response to episodic loading at or close to the Alpine thrust front (Sinclair et al. 1991; Sinclair and Allen 1992; Sinclair 1996; Schlunegger et al. 1997a, 1997b; Zweigel et al. 1998; Genser et al. 2007). Subsequent studies, building on this view, were conducted with the aim to quantify the subsidence history of the Molasse basin. This was achieved through dating of selected sections using a combination of magneto-polarity stratigraphy, bio- and chemostratigraphy (Burbank et al. 1992; Schlunegger et al. 1996, 1997a; Kempf et al. 1999; Strunck and Matter 2002; Sant et al. 2017; van der Boon et al. 2018; Hofmayer et al. 2019; Hülscher et al. 2019). The resulting chronologies then built the basis to identify periods of rapid sediment accumulation, which in turn were used to identify distinct phases of thrusting and loading at the Alpine thrust front (Schlunegger et al. 1997a; Kempf et al. 1999; Strunck and Matter 2002; Ford and Lickerish 2004). While these approaches were generally accepted until the end of the 1990s, research conducted in the following decades started to disclose a different picture where subduction loads at deeper crustal levels rather than surface loads exerted the largest control on the subsidence history of the Molasse basin (Pfiffner et al. 2002; Schlunegger and Kissling 2015, 2022). Our interpretation of the relationships between Alpine orogenesis and the development of the Molasse basin has thus changed during the past few decades.

The Alps themselves are bordered on both sides by a foreland basin that is referred to as the Molasse basin on its northern margin, and the Po basin on the southern side (Figure 3.1). Because the Alps are a bivergent orogen (section 3.2), we would anticipate that stratigraphic data should be collected from the sedimentary basins on either side of the Alps, that is, from the Molasse and the Po basin, to unravel the Alpine evolution. However, as will be outlined in this chapter, the deflection of a foreland plate and changes thereof offer the most important information for inferring the history of thrusting and loading in the adjacent orogen. Because the history of Alpine loading has mainly been controlled by the subduction of the European plate and not of the Adriatic plate (Holliger and Kissling 1992; Schmid et al. 1996; Rosenberg et al. 2015; Kissling and Schlunegger 2018), it is not surprising that the Molasse basin and particularly its subsidence history has been the preferred focus of research in the past few years for reconstructing the chronology of loading in the Alps. Admittedly, if we consider that the Adriatic plate has been subducted northward since 20 Ma (Handy et al. 2015), then the Po basin should also be included in the discussion because it would have operated as a pro-wedge foreland basin since then. However, we refrain from extending our discussion to the Po basin on the southern side of the Alps as this is beyond the scope of this contribution, and rather focus our review on the Molasse basin.

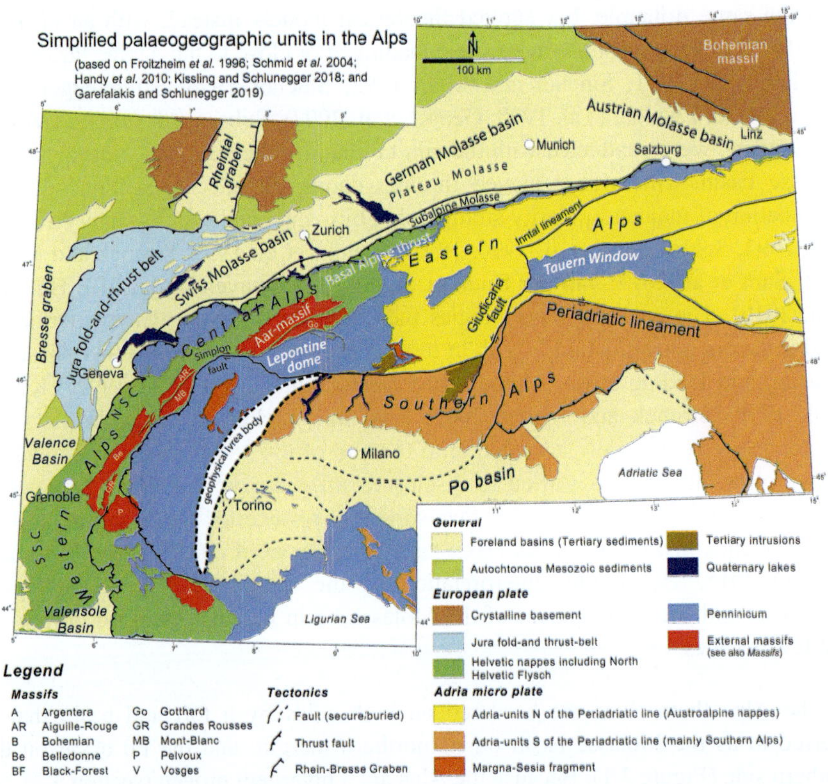

Figure 3.1. *Map illustrating the major litho-tectonic units of the Alps and the adjacent sedimentary basins. We have only labeled the units that are considered to be important for understanding the development of the Molasse basin*

Here, we present an overview of how research on the relationships between Alpine orogeny and the sedimentary processes in the Molasse basin has evolved during the past few decades. We proceed by reviewing and synthesizing data on the sedimentary record preserved in the Molasse basin, together with information on the geological architecture of the Alps. We particularly focus our work on the central, or alternatively the Swiss part of the basin, where we have mainly been working during the past few years. Where adequate, we also integrate data and information from the French and the German/Austrian parts of the basin. We will use these data to identify the major challenges upon reconstructing the causal relationships between Alpine tectonics and Molasse basin development. We then conclude this contribution with the presentation of evidence for why predominantly subduction loads need to be considered to improve our understanding of the large-scale history of the North Alpine foreland basin.

3.2. Tectonic overview of the European Alps

The Alps have an arcuate shape and stretch from the Mediterranean to the Pannonian basin over nearly 1,000 km (Figure 3.1). This doubly vergent orogen (e.g. Willett et al. 1993; Grool et al. 2018) is the result of the late Cretaceous to present continent–continent collision between the European and Adriatic plates (Schmid et al. 1996; Handy et al. 2010), which started in Late Cretacous times following a period of spreading (Lihou and Allen 1996; Schmid et al. 1996; Stampfli and Borel 2002). In this section, we present an overview on the architecture of the European Alps, thereby outlining those lithotectonic and structural features particularly situated in the West-Central and Eastern Alps (Figure 3.1), which were considered to have influenced the development of the Molasse basin.

3.2.1. *The Central and Western Alps*

In its central part, which is commonly referred to as the Central Alps, the Alpine edifice exposes a crystalline core in its center, which has been referred to as the Lepontine dome, and the external massifs adjacent to the north (e.g. Aar-massif; Spicher 1980). On its western lateral margin, the Lepontine dome (Figure 3.1) is bordered by the Simplon detachment fault, where normal faulting resulted in rapid tectonic exhumation of the dome between late Oligocene and early Miocene times, with a peak recorded by thermo-chronometric data at ca. 20 Ma (Mancktelow 1985; Hurford 1986; Mancktelow and Grasemann 1997; Schlunegger and Willett 1999; Boston et al. 2017). This was also the time when the Aar-massif, situated to the north of the Lepontine dome (Figure 3.1), experienced a period of rapid vertical extrusion (Herwegh et al. 2017, 2020). Both the area surrounding the Lepontine region and the external massifs were considered to represent important sediment sources for the Molasse basin. Therefore, the results of research on the tectonic (e.g. Mancktelow and Grasemann 1997) and topographic development of these units and the surrounding regions (e.g. Kühni and Pfiffner 2001a, 2001b) have major implications on our understanding of the petrographic signals recorded in the basin (e.g. von Eynatten et al. 1999; Spiegel et al. 2000, 2001, 2002, 2004; Stutenbecker et al. 2019; Anfinson et al. 2020), and the patterns of sediment supply to this basin (section 3.5.3).

At deeper crustal levels, seismic imaging (e.g. Kissling et al. (2006) and references therein; Fry et al. (2010)) in combination with the interpretation of negative Bouguer gravity anomalies surrounding the center of the Alps (Lyon-Caen and Molnar 1989; Holliger and Kissling 1992) suggest that the Lepontine dome is underlain by a thick crustal root made up of a stack of lower crustal material derived from the European continental plate (Fry et al. 2010). This stack of buoyant lower crustal material has been considered to maintain the topography of the Central Alps at their current elevation where the highest peaks reach ca. 4,500 m. Also beneath the Central Alps, seismic tomography data suggest that the ca. 160 km long (Lippitsch et al. 2003) lithospheric slab of the European continental plate, made up of mantle material with a relatively high density, downwarps the European foreland plate toward the south-east (Kissling and Schlunegger 2018). As outlined in section 3.7, we consider that the subduction of the European lithospheric mantle beneath the Alps largely controlled the subsidence history of the Molasse basin (Schlunegger and Kissling 2015, 2022).

The highest lithotectonic unit comprises the Austroalpine nappes that structurally overlie the (meta)sedimentary successions of the Penninic and the underlying Helvetic thrust nappes. The tectonic front of the latter units has been referred to as the Basal Alpine thrust (Figure 3.1). The Austroalpine cover nappes and the underlying Penninic sedimentary thrust sheets were the most important sediment sources of the Molasse basin until ca. 22–20 Ma. After that time, slip along the Simplon detachment fault resulted in the widespread exhumation of the crystalline core surrounding the Lepontine dome, and crystalline material became more abundant in the Molasse basin (Schlunegger et al. 1997a; von Eynatten et al. 1999; Spiegel et al. 2002; von Eynatten and Wijbrans 2003; Anfinson et al. 2020). The Jura fold-and-thrust belt, which delineates the Swiss Molasse basin on its northern margin, is the northernmost lithotectonic unit of the Central Alps. This mountain range formed on top of thick Triassic evaporates and stretches from Geneva to the north of Zurich, where the easternmost anticline (Lägern) dips beneath the deposits of the Molasse basin (Buxtorf 1916; Laubscher 1992; Dèzes et al. 2004). The total amount of shortening of the Jura Mountains was estimated to a minimum of 26 km northeast of Geneva, and to ca. 5 km in the area surrounding the Lägern north of Zurich (Burkhard 1990; Burkhard and Sommaruga 1998). The timing of folding of the Jura Mountains is poorly constrained, but most recent work, mainly based on chronological investigations of folded Molasse sediments within this mountain range, considered that the major phase of folding of the Jura Mountains occurred ca. at 10 Ma or slightly before, and was coeval with deposition of the youngest Molasse deposits (Burkhard 1990; Laubscher 1992; Philippe et al. 1996; Kälin 1997; Affolter and Gratier 2004; Cederbom et al. 2011; Willett and Schlunegger 2011). On its northern tip, the occurrence of folded quaternary meanders of the Doubs river indicates that active folding is ongoing (Madritsch et al. 2010). From the perspective of the Molasse basin, research has focused toward exploring how the folding of the Jura Mountains influenced the sedimentation and erosion of the Molasse basin (section 3.4.4).

The Western Alps are mainly characterized by their arcuate shape and their widespread exposure of Penninic sedimentary units, while Austro-Alpine cover nappes are missing. The Western Alps comprise (i) the Northern and Southern Subalpine Chains (NSC and SSC; Figure 3.1) that are made up of Mesozoic sequences deposited on the European continental plate, and (ii) exhumed European basement blocks referred to as the external massifs (e.g. Argentera, Belledonne, Mont-Blanc, Figure 3.1). These units form the external zone of the Western Alps. The internal zone comprises (i) metasedimentary rocks that were deposited in the Valais ocean separating the European margin from the Briançonnais swell, an extension of the Iberian microcontinent, (ii) shallow-marine sediments that were deposited on the Iberia microcontinent itself (Briançonnais) and (iii) the Schistes Lustrés and ophiolites that were formed in the Ligurian ocean (or Tethys) (Ford et al. 2006). Along its inner arc, the Western Alps expose the Ivrea body (e.g. Boriani and Sacchi 1973; Fountain 1976; Zingg et al. 1990; Schmid et al. 2017) that is made up of mantle rocks (Figure 3.1). Because of the high density of these rocks, Pfiffner et al. (2002) considered the Ivrea body to have influenced the deflection of the Molasse basin between the Late Oligocene and the Miocene. These authors also believed that the SW-directed displacement of this unit controlled the 20–18 Ma old reversal of the drainage direction in the Molasse basin. Seismo-tomography investigations suggest that beneath the Western Alps, the subducted European mantle slab is delaminated from the European plate (Lippitsch et al. 2003), which possibly occurred at ca. 5 Ma before present (Fox et al. 2014). According to Kissling and Schlunegger (2018), this delamination was associated with the formation of a tear, whose tip is currently situated beneath the eastern margin of the Western Alps. As discussed in section 3.7.3, delamination of the subducted mantle lithosphere at ca. 5 Ma could have resulted in a rebound of the European plate, and in an acceleration of erosion in the area surrounding the external massifs of the Western Alps (Fox et al. 2014), and possibly in the inversion of the Molasse foreland basin (Baran et al. 2014).

The southern side of the Alps is made up of the Southalpine thrust sheets, or the Southern Alps, that consist of crystalline basement rocks and sedimentary units of Adriatic/African origin (Barbieri et al. 2004; Toscani et al. 2014). This fold-and-thrust belt is bordered to the south by the Po Basin (Figure 3.1). The north-vergent Alpine thrusts are separated from the South Alpine nappes by the north-dipping Periadriatic line that accommodated most of the shortening during the Oligocene and the early Miocene by backthrusting and right-lateral slip (Schmid et al. 1996).

3.2.2. *The Eastern and Southern Alps*

Similar to the Central Alps, the tectonic architecture of the Eastern and Southern Alps is the result of the continent–continent collision between the European and Adriatic continental plates (Handy et al. 2010). The eastern part of the Alpine orogen is characterized by the large and widespread exposure of the Austroalpine cover nappes (Figure 3.1). However, also in the Eastern Alps, the large north-vergent thrust nappes that were detached from the European continental plate and dominate the Central Alps are largely absent or occur as thin slivers at the northern margin of the Alpine thrust front and within the Eastern Alps. These rocks are exposed in a tectonic window, referred to as the Tauern Window and Engandin Window (e.g. Lammerer et al. (2008); Rosenberg et al. (2018) and references therein), which is bordered on the western and eastern margins by low-angle detachment faults. Stacking and exhumation of these crustal blocks occurred simultaneously with slip along these detachment faults, starting in the Late Oligocene/Early Miocene and reaching the highest rates around 20 Ma (Fügenschuh et al. 1997, 2006; Wolff et al. 2020, 2021).

Similar to the central part, the Eastern Alps are separated from the Southern Alpine nappes by the Periadriatic fault (or lineament). This fault itself is cut and offset by the left-lateral Giudicaria fault (Figure 3.1). Ustaszewski et al. (2008) considered that slip along this fault was kinematically and temporally linked with the shortening within the Southern Alps, which commenced at ca. 20 Ma as cross-cutting relationships between sediments in the Po basin and structures in the Southern Alps imply (Schönborn 1992). In contrast to the Central Alps, however, a thick crustal root made up of lower crustal material derived from the European continental plate is largely absent beneath the core of the Eastern Alps, as cross-sections of Schmid et al. (2013) and Rosenberg et al. (2015) have shown. This is also mirrored in the lower elevation of the topography in the Eastern Alps (peak altitudes of ca. 3,500 m a.s.l.) in comparison to the Central Alps (peak altitudes ca. 4,500 m.s.l.). Seismic tomography investigations have revealed an at least 210 km long lithospheric mantle slab (Lippitsch et al. 2003) and possibly significantly longer (Dando et al. 2011; Mitterbauer et al. 2011) beneath the Eastern Alps. Its dip (toward NE: Lippitsch et al. 2003; Piromallo and Morelli 2003; Karousova et al. 2013; subvertical: Dando et al. 2011; Mitterbauer et al. 2011) and its connection to the Adriatic or the European plate is controversly discussed (i.e. Handy et al. 2015, 2021; Kaestle et al. 2020). However, we consider the northeastward dip of the high-velocity anomaly and its connection with the Adriatic plate (Lippitsch et al. 2003; Karousova et al. 2013) as reliable features because the seismic tomography procedure of Lippitsch et al. (2003) and Karousova et al. (2013) is unique as these authors incorporated a high resolution a priori 3D crustal model of the Alps in their analyses. Following these models, the Adriatic slab dips toward the NE beneath the Eastern Alps (Lippitsch et al. 2003; Schmid et al. 2004; Kissling et al. 2006), and the European slab is missing and was possibly delaminated at ca. 20 Ma (Ustaszewski et al. 2008; Handy et al. 2015). The change in subduction polarity from a SE orientation underneath the Central Alps (subduction of European mantle lithosphere) to a NNE orientation underneath the Eastern Alps (subduction of Adriatic mantle lithosphere) occurs in the vicinity of the Giudicaria fault. Accordingly, from the perspective of the subduction geometry and the fault pattern of currently exposed rocks, the boundary between the Eastern and Central Alps could be placed in the region of the Giudicaria fault. Accordingly, if the interpretation of a change of the subduction polarity at 20 Ma is valid (Ustaszewski et al. 2008) and the Lippitsch et al. (2003) interpretation of a northward dipping Adriatic slab is appropriate, then the Molasse basin and particularly the eastern German-Austrian Molasse should record a signal in the stratigraphic record that could be related to these geodynamic changes. We will re-address this question and particularly the roles of subduction loads in section 3.7, and we will use the stratigraphic development in the Molasse basin to propose that the Lippitsch et al. (2003) interpretation of a reversed subduction polarity is likely to be correct, or should at least be further considered.

3.3. Overview of the sedimentary basins surrounding the Alps

The Po and Molasse basins represent foreland sedimentary troughs on the southern and northern sides of the Alps, respectively (Figure 3.1). The Southern Alps and the Mesozoic substratum of the Southern Alpine Molasse and Flysch deposits beneath the Pliocene-Quarternary Po sediments exhibit a horst-graben-platform structure inherited from the times when the northern edge of the Adriatic microplate was a passive continental margin dominated by extensional tectonics (Bertotti et al. 1998; Scrocca et al. 2003; Berra and Carminati 2009). The thickness of the Permian-Mesozoic sediments varies between 1 and 6 km and their geometries greatly influenced the local structure of the Flysch and Molasse-type southern

Alpine foredeep deposits that show a great variability in their thickness and facies from west to east along the southern margin of the Alps (Pieri and Groppi (1981); Carminati and Doglioni (2012) and references therein). The Cretaceous-Miocene Southern Alpine foredeep basin ("early Po basin") remained generally shallow and was separated by extensive shallow marine platforms (Carminati and Doglioni 2012). Since the Late Eocene and until the Early Miocene, it played the role of a retro-foreland basin. In addition, by resting on the overriding Adriatic plate, it was largely independent of the Alpine subduction–collision dynamics on a large scale. The current ultradeep (>11 km) asymmetric Po foreland basin developed during late Miocene-Pliocene times (e.g. Argnani and Ricci Lucchi 2001) as a result of the downbending of the Adriatic lithosphere beneath the northern Apennines resulting from slab rollback (e.g. Laubscher 1988; Speranza et al. 2002; Argnani 2012; Carminati et al. 2012; Amadori et al. 2019). This was also the time when the basin took the role of a pro-foreland basin of the Apennines. This double role is illustrated by the opposite-verging thrust fronts beneath the Po Plain (Toscani et al. 2014). Note that while this downbending is directed toward the SSW (away from the Alps!), the majority of the Pliocene to present sedimentary infill in the Po basin is of Alpine origin. In conclusion, the mid-Tertiary sedimentary basin on the southern side of the Alps situated on the overriding Adriatic plate was distinctively different from the Molasse basin in the north, and the Pliocene-to-recent Po basin is a foreland basin that steeply dips toward the northern Apennines. Accordingly, while the shape of the Po basin is to a large extent controlled by the rollback subduction of the Adriatic plate, the basin itself is filled in good part by Alpine erosional products.

The Molasse basin accumulated material that was mainly derived from the Alps and partly also from the crystalline massifs on its northern margin (Füchtbauer 1964; Sissingh 1997). The Molasse basin itself has a N-S direction south of Geneva where isolated remnants of Molasse depocenters are mainly found in the Southern Subalpine Chaines (SSC). In the region surrounding Geneva, the basin adapts a SW-NE orientation and currently widens from ca. 20 to >150 km to the east of Munich (Figure 3.1). Farther east, the basin then narrows again to a few kilometers. Between the Lake of Geneva and Salzburg, the proximal basin deposits were sliced into several imbricated thrust sheets referred to as the Subalpine Molasse, while farther east, evidence for major thrusting and shortening is absent (Hinsch 2013; Ortner et al. 2015). The Bresse, Rhône and Rhine graben structures form depocenters (Figure 3.1) that have accumulated some of the Alpine material and interfered with the development of the Molasse basin at different times, mainly on its distal margin (Sissingh 1997, 1998; Berger et al. 2005).

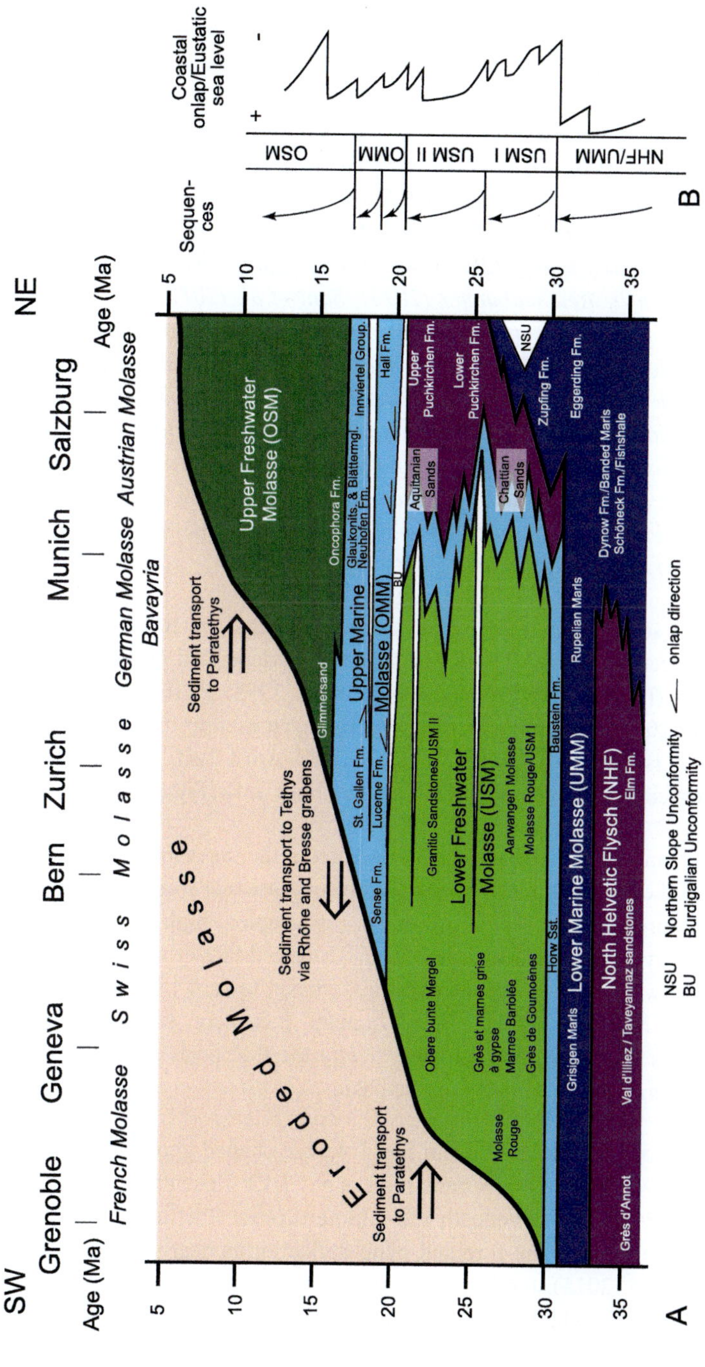

Figure 3.2. *Chronologic (Wheeler) diagram for Molasse deposits, and sequence stratigraphic concept and eustatic sea level changes*

COMMENT ON FIGURE 3.2.– *a) Chronological (Wheeler) diagram for the Molasse deposits along its topographic axis. Data for the French Molasse basin are taken from the review articles by Ford and Lickorish (2004) and Ford et al. (2006). Strunck and Matter (2002) presented a stratigraphic synthesis of the western Swiss Molasse basin. We thus adapted their lithostratigraphic scheme for this work. For the eastern Swiss Molasse basin, we employed the stratigraphic scheme by Diem (1986), Keller (1989), Schlunegger et al. (1996), Kempf et al. (1999) and Jost et al. (2016). Data for the German and Austrian Molasse basin are taken from the synthesis of Kuhlemann and Kempf (2002), which was updated based on the results of subsequent research by Reichenbacher (2013), Sant et al. (2017), van der Boon et al. (2018), Hofmayer et al. (2019) and Hülscher et al. (2019). The chart shows that the facies relationships along the topographic axis were strongly heterochronous until 20 Ma. After that time period, the stratigraphic development was quite similar along strike. Note that major erosional unconformities can be found within the Molasse suites (white fields in diagram). (b) Sequence stratigraphic concept and eustatic sea level changes, which suggest that some of the erosional unconformities could have been controlled by eustatic sea level variations (modified form Jin et al. (1995) and Zweigel et al. (1998)).*

At its proximal southern margin, the basin is delineated by the Alpine thrust front with thrust sheets made up of sediments that were deposited on the European continental margin (Pfiffner 1986; Schlunegger et al. 1993; Pfiffner et al. 1997; Hänni and Pfiffner 2001; Ortner et al. 2015). On its distal margin, the Swiss part of the Molasse basin is bordered by the Jura fold-and-thrust belt, where Molasse deposits are preserved as thin remnants in a few synclines only (Figure 3.1).

At the eastern limits of the Molasse basin in Austria, the northern basin border is defined by exhumed crystalline rocks of the European plate that form the Bohemian massif. The exposure of this latter unit as a topographic high is most likely conditioned by Hercynian faults that strike NW-SE (e.g. Malz et al. 2019), which also control the eastward narrowing of the Molasse basin toward Linz (Figure 3.1). Between the Jura Mountains and the Bohemian basement block, Mesozoic limestones and marls of the Svabian Jura that overly the European continental plate demarcate the northern margin of the Molasse basin.

In the cross-section, the Molasse basin has a wedge-shaped geometry and is up to 5 km thick at the Alpine border (Sommaruga et al. 2012; Rosenberg et al. 2015), where the Molasse deposits dip beneath the Alpine orogen (Pfiffner 1986; Ortner et al. 2015). The amplitude of the foreland plate deflection is nearly constant along strike (Rosenberg et al. 2015), but the cross-sectional width of the basin decreases from the SW to the NE, which results in an increase in the angle of dip of the foreland plate. This pattern could reflect the response of a larger flexural rigidity of the foreland plate in the SW compared to the NE (Tesauro et al. 2009; Mey et al. 2016). Alternatively, the deflection could also be influenced by the effect of NE-SW striking Variscian faults (e.g. Malz et al. 2019; Schlunegger and Kissling 2022). Simple flexural modeling has indeed shown that structural weaknesses of a foreland plate (e.g. faults) have the potential to localize the deformation and increase the deflection at that particular site (Waschbusch and Royden 1992). The oldest Eocene foreland basin deposits (Flysch) are outcropping within the Helvetic thrust sheets, while Oligocene to Miocene strata (Molasse) occur within the thrusted Subalpine and the flat-lying Plateau Molasse (Ford and Lickerish 2004).

3.4. Stratigraphic development of the Molasse basin

The stratigraphic architecture and evolution of the basin was remarkably similar along strike prior to 32/30 Ma, at least west of Munich (Figure 3.2), and then again after 20/18 Ma. However, between 30 and 20 Ma, the large-scale stratigraphic architecture of the Molasse basin discloses a different style west and east of Munich and thus in the northward prolongation of the Giudicaria fault. This along-strike difference in the sedimentation history has offered major challenges to reconcile the stratigraphic architectures along the basin axis (Figure 3.2). Similar complications for reconstructions of the facies relationships also concern the correlation of Molasse sediments in France, mainly because the region to the west of the Pelvoux massif (labeled with "P" on Figure 3.1) marked the southern termination of the Molasse basin, and no further regional accommodation space was apparently formed farther south by lithospheric flexure. Instead, sediment accumulated locally in small tectonically controlled basins, as outlined in the review article by Ford et al. (2006).

3.4.1. *Prior to 30 Ma*

The history of the Molasse basin starts with deposition of the North Helvetic Flysch, that was deposited between ca. 35 and 32 Ma in a deep marine trough by turbidity currents (Figure 3.3(a)). Currently, these deposits can be found as erosional remnants within the Helvetic thrust nappes and have been referred to as Grès d'Annot and Grès de Champsaur in France along the southern border of the Pelvoux massif (Elliott et al. 1985; Waibel 1990; Ford and Lickorish 2004), the Taveyannaz and Val d'Iliez Sandstones in western Switzerland (Homewood and Caron 1982; Lateltin 1998), and the Taveyannaz Sanstone and Elm Formation in eastern Switzerland (Sinclair 1992; Lu et al. 2018). As mentioned by Ford and Lickorish

(2004) and illustrated by Sinclair (1997), the facies relationships, ages and depositional processes are quite consistent and constant along the Alpine arc. In Switzerland and Bavaria (Figure 3.2), the deposition of the North Helvetic Flysch was followed by the regressive sequence of the 32–30 Ma old Lower Marine Molasse group (UMM) that records a gradual change from a deep marine environment toward continental conditions. The top of the Lower Marine Molasse group comprises a several meter thick succession of sandstone beds that were deposited under shallow marine, deltaic conditions and thus mark the termination of the Flysch stage of foreland basin evolution (Sinclair and Allen 1992). This unit, referred to as the Ralligen Formation and the Horw Sandstone in Central Switzerland and the Baustein Formation in Bavaria, was mapped from the Lac Léman to the Munich region; it can thus be considered as a marker bed (Diem 1986). To the south of the Pelvoux massif situated in the Western Alps (Figure 3.1), sedimentation within a flexural foredeep came to an end at ca. 30 Ma (Ford et al. 2006).

Although the western and central basin accumulated Flysch type of material during the Late Eocene, the North Helvertic Flysch turbidites did not reach the eastern part of the basin. There, the basin fill records a suite that transitions from organic-rich marls and shales (Schöneck Formation) to an alternation of micritic limestones and marls (Dynow Formation), and finally to laminated pelites with sandstone interbeds (Eggerding Formation) (Sachsenhofer et al. 2010). These sediments were deposited on the northern slope and in the foredeep of the Austrian basin (Figure 3.3(a)). Sachsenhofer et al. (2010) additionally mentioned that the northern slope was also the source of submarine slumps and slides that finally accumulated in the foredeep, where mudstone deposition in an offshore environment prevailed (Kuhlemann and Kempf 2002). Also, in the Austrian Molasse and in contrast to the Swiss and German Molasse, sedimentation on the southern margin occurred to a large extent on top of the frontal Alpine nappes (i.e. on top of the Late Triassic Dachsteinkalk) and thus in a wedge-top position (Figure 3.3(a)). These wedge-top sediments, which are referred to as the Augenstein Formation, comprise fluvio-deltaic conglomerates and sandstones with sources in the adjacent Eastern Alps. Deposition of the Augenstein unit started sometime between 35 and 30 Ma (Frisch et al. 2001) and lasted until the Early Miocene (Frisch et al. 2001), resulting in a >1.3 km thick coarse-grained suite within a wedge-top position. It thus appears that most of the erosional detritus of the Eastern Alps was stored on this wedge-top basin, which was possibly decoupled from the foredeep. We use such a configuration to explain why the adjacent foredeep was almost void of coarse-grained clastic material.

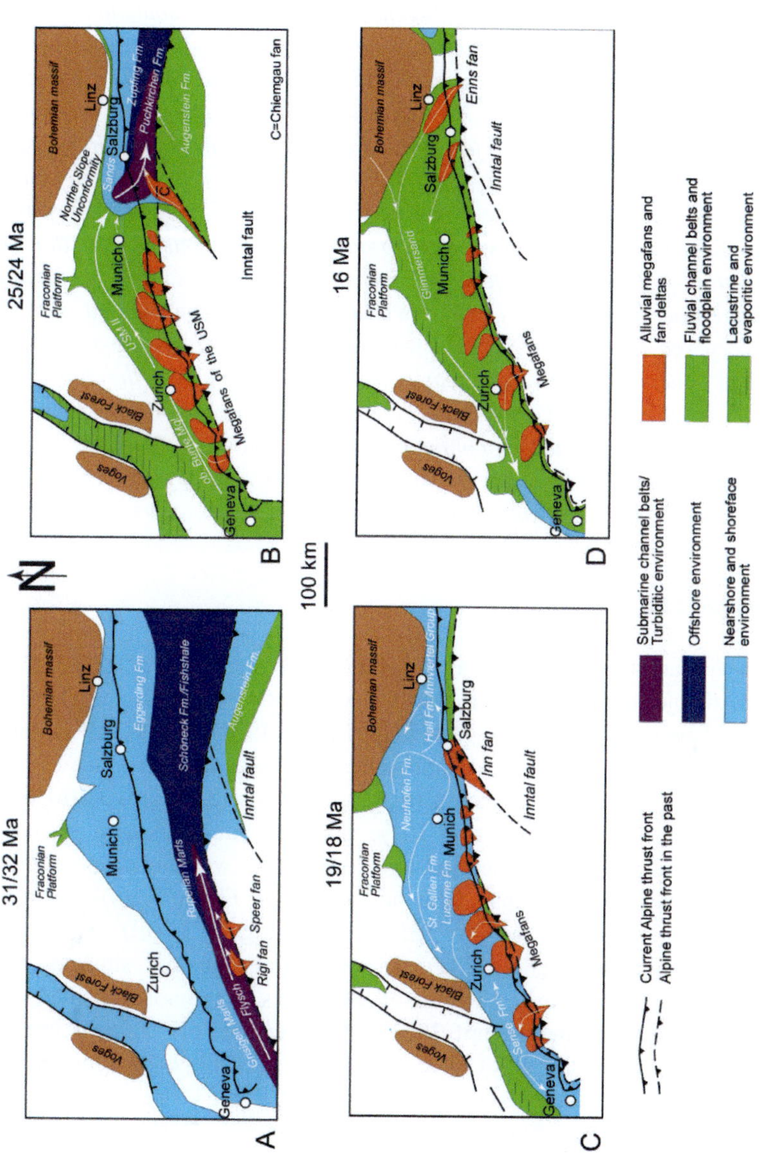

Figure 3.3. *Evolution of the Molasse basin*

COMMENT ON FIGURE 3.3.– *(a) Situation at ca. 32–31 Ma, when flysch sedimentation prevailed, particularly in the Swiss and the German Molasse basin. In the flysch basin, sediment transport was oriented toward the NE. In the Austrian Molasse, incipient deposition of the Augenstein Fm. started sometime prior to 30 Ma. (b) Situation at ca. 25–24 Ma, when alluvial megafans were constructed at the proximal basin border of the Swiss and the German Molasse basin. The distal basin was occupied by an axial, NE-directed fluvial system. East of Munich, however, the fluvial depositional systems gave way to a submarine trough, where the material was transported farther to the NE in the submarine Puchkirchen channel network. In the Austrian Molasse basin, accumulation of material (Augenstein Formation) mainly occurred in a wedged-top basin, with the consequence that only a small fraction of the clastic material received the foredeep. (c) Paleogeographic sketch for the time between ca. 19 and 18 Ma. From this time onward, sediment dispersal occurred toward the SW and was mainly accomplished by strong tidal currents that entered the Molasse basin from the Tethys and the Paratethys. (d) During deposition of the Upper Freshwater Molasse, here shown for the time at ca. 16 Ma, sediment transport continued to occur toward the SE. All four paleogeographic sketch figures are modified from Kuhlemann and Kempf (2002) and updated from Schlunegger and Kissling (2022).*

3.4.2. *Between 30 and 20 Ma*

In the Swiss and German basin east of Munich, sedimentation of the UMM was followed by the accumulation of the Lower Freshwater Molasse group (USM) that was deposited between 30 and 20 Ma (Kempf et al. 1999). The USM comprises kilometer thick coarsening-upward conglomerate sequences that accumulated at the Alpine thrust front on alluvial megafans (Sissingh 1997) (e.g. Figures 3.3(b), 3.4(a) and 3.4(b) for the Swiss Molasse). These conglomerate packages grade into kilometer thick suites of sandstone-mudstone alternations with limestone interbeds (e.g. Grés et marnes grises à gypse, Granitic Sandstones; Figure 3.2(a)) that record the occurrence of an axial, east-directed drainage (Figure 3.3(b)) and partially shallow lakes (Platt and Keller 1992). The axial drainage was situated in the topographic axis of the basin a few tens of kilometers to the north of the Alpine thrust front (Kuhlemann and Kempf 2002). Measurements of sediment transport directions (Schlunegger et al. 1997a; Kempf et al. 1999; Garefalakis and Schlunegger 2019) and heavy mineral suites (Füchtbauer 1964) disclosed a material transport toward the east by braided to meandering streams. During USM times, a coastline established near Munich (Bachmann and Müller 1992). There, coastal sands (Chattian and Aquitanian sands, Figures 3.2(a) and 3.3(b)) separated a terrestrial environment in the west from a ca. 1,000–1,500 m deep marine trough farther east (Rögl et al. 1979; Zweigel et al. 1998). Here, mass transport deposits and turbidity currents with material sources in the coastal region near Munich and in the Eastern Alps were funneled toward the east through an axial submarine channel belt (Figure 3.3(b)). The submarine channels were situated in the basin center and ended as depositional lobes ca. 20 km east of Salzburg (Hülscher et al. 2019). This resulted in the deposition of the Zupfing Formation and the Puchkirchen Formation (Figure 3.2) (de Ruig and Hubbard 2006; Hubbard et al. 2009; Bernhardt et al. 2012). While the Zupfing unit comprises the fine-grained clastic material that was deposited on the submarine floodplain and the foreland basin slope to the north, the Puchkirchen Formation represents the submarine channel belt and assembles the coarse-grained material that was deposited in the submarine topographic axis and adjacent to the Alpine front (Sachsenhofer et al. 2010).

Among the various sediment supply systems, the paleo-Inn with sources in the western part of the Eastern Alps (or alternatively the eastern margin of the Central Alps if the Giudicaria fault is considered as a boundary, see above), including the area surrounding the Bergell (Brügel et al. 2003), was one of the largest Alpine streams at that time. The course of this stream was possibly controlled by the strike-direction of the Inntal lineament (Ortner and Stingl 2001) (Figure 3.3(b)). Upon entrance into the foreland basin, the paleo-Inn resulted in the accumulation of the submarine Chiemgau fan situated west of Salzburg (Ornter and Stigl 2001; Figure 3.3(b)), which interfingered with the Puchkirchen submarine channel belt system toward the basin center (Frisch et al. 2001).

Farther east of Salzburg, deposition of the fluvio-deltaic Augenstein Formation continued in a wedge-top position (Frisch et al. 2001), and the foredeep of the Austrian Molasse started to subside rapidly (Hülscher et al. 2019). As a result, the submarine fan system of the Puchkirchen Formation prograded into the Austrian foredeep and started to seal the Eggerding Formation. During the same time, the northern slope of the Austrian Molasse was uplifted, which resulted in the formation of the Northern Slope Unconformity (NSU, Figures 3.2 and 3.3(b)). Because the Austrian foreland plate experienced a larger downward bending at that time, as inferred from the increase in subsidence rates (Hülscher et al. 2019), we consider this uplift in the region of the northern slope, and thus in the plate's forebulge area, as a flexural response to the larger bending of the plate beneath the foredeep. Indeed, flexural models (e.g. Sinclair et al. 1991) predict that a larger downward deflection of the foreland plate beneath an orogen should also be associated with a larger upward doming of the plate in the forebulge area.

For the French part of the Molasse basin that is situated in the NSC (Figure 3.1), the USM deposits (Molasse Rouge, Figure 3.2(a)) are characterized by gypsiferous marls, freshwater carbonates and paleosoils embedded within fluvial sandstones (Ford and Lickorish 2004), suggesting a sedimentary environment that is characteristic for a basin margin. According to Ford and Lickorish (2004), the USM deposits thicken along strike from the SW to the NE, which was used as evidence to propose that the basin was closed to the SW from 30 Ma onward (Ford and Lickorish 2004). Accordingly, while the facies patterns were similar along the entire Western Alps during deposition of the North Helvetic Flysch (Figure 3.2(a)), the post 30 Ma stage was characterized by a closure of the basin in the region of Grenoble.

3.4.3. *After 20 Ma*

The boundary between the pre and post 20 Ma deposits is characterized by a major unconformity that has been referred to as the Burdigalian unconformity in the literature (e.g. Sinclair et al. 1991). The temporal duration of the hiatus that is associated with this unconformity is poorly constrained, and age estimates range from ca. 0.6 Ma for the Austrian Molasse (Hülscher et al. 2019) to ca. 1 Ma for the German Molasse (e.g. Grunert et al. 2013, 2015). Farther to the SW near Zurich, the Burdigalian unconformity spans only ca. 0.2 Ma. West of Bern, however, the Sense Formation conformably covers the Obere Bunte Mergel (Figure 3.2(a)) with no evidence for unconformity (Garefalakis and Schlunegger 2019).

In the eastern Molasse basin, the post 20 Ma sedimentation started with the accumulation of the Hall Formation deposits that display N-directed clinoforms (Hülscher et al. 2019). These structures mark the incipient filling of the deep marine Puchkirchen trough, thereby giving way to shallow marine conditions at ca. 18 Ma (Innviertel group; Figures 3.2 and 3.3(c)) (Grunert et al. 2013). This transition toward shallow marine conditions was accompanied by incipient uplift and erosion of the Augenstein Formation situated on top of the orogen front (Figure 3.3(c)). During the same time, the Swiss Molasse basin became flooded by a peripheral shallow-marine sea, resulting in the deposition of the Upper Marine Molasse group (OMM; Lucerne and St. Gallen Formations) between 20 and 17 Ma (Matter et al. 1980; Homewood and Allen 1981; Allen 1984; Allen et al. 1985; Keller 1989; Jost et al. 2016) (Figures 3.2 and 3.3(c)). In Switzerland, the 20 Ma old transgression of the peripheral sea progressed from the East toward the West within a few Ka, which implies an eastward-directed tilt of the basin axis (Strunck and Matter 2002; Garefalakis and Schlunegger 2019). Also, in the Swiss part of the Molasse basin, the

OMM unit is up to 1,000 m thick close to the Alpine border. From there it thins to a few tens of meters toward distal sites. It consists of parallel laminated and cross-bedded sandstones recording the predominant activity of waves (Lucerne Formation) or tides (St. Gallen and Sense Formations) with a tidal range of 2–4 m (Allen et al. 1985; Homewood and Allen 1986; Keller 1989; Schaad et al. 1992). Large streams with sources in the Alps shed their coarse-grained material into the shallow seaway, which resulted in the construction of thick conglomerate packages adjacent to the Alpine front (Keller 1989; Kuhlemann and Kempf 2002; Figure 3.3). The most prominent ones are the Napf and Hörnli conglomerates situated in Switzerland (Figure 3.4(c)), the Pfänder conglomerates near lake Constance (Schaad et al. 1992), and the coarse-grained deposits of the Hochgrat, Nesselburg and Inn fans in the Bavarian Molasse (Kuhlemann and Kempf 2002). Palaeocurrent measurements and heavy mineral analyses revealed that the Swiss Molasse basin was a closed sedimentary sink between 20 and 18 Ma (Figure 3.4(c)), when waves and tides that entered the basin from the Tethys in the SW and the Paratethys in the NE redistributed the material supplied from the Alps within closed cells (Allen et al. 1985; Garefalakis and Schlunegger 2019). The situation then changed at 18 Ma, when the marine transgression of the St. Gallen Formation proceeded from the west to the east (Figure 3.2(a)), which implies a westward tilt of the topographic axis. Also at 18 Ma, Molasse material started to be transported toward the west (Figure 3.3(c)). A further progradation of the fan deltas with diameters >10 km finally resulted in a second regression in the Molasse basin, leading to the deposition of the Upper Freshwater Molasse group (OSM) in the entire basin. From then onwards, sediment transport occurred toward the west (Figures 3.2(a), 3.3(d) and 3.4(d)). During OSM times, major sediment sources were also situated in the Bohemian massifs, where axial streams transported large volumes of sediment to the Swiss Molasse basin, as recorded by the "Glimmersand" deposits (Hofmann 1956).

The Burdigalian transgression not only proceeded from the Paratethys to the Molasse basin, but also from the Tethys and thus from the current Mediterranean Sea. As the Molasse basin was closed in the region of Grenoble, the marine transgression from the southwest occurred via the Rhône and Bresse grabens (Allen and Bass 1993), where rifting and subsidence resulted in the establishment of a seaway linking the Molasse basin with the Tethys. During OSM times, the graben systems then served as pathways for the Alpine streams.

The material transport changed again at ca. 10 Ma when uplift of the Jura Mountains was considered to have closed the corridor to the Bresse and Rhône grabens. Sediment then re-routed toward the NE (Kuhlemann and Kempf 2002).

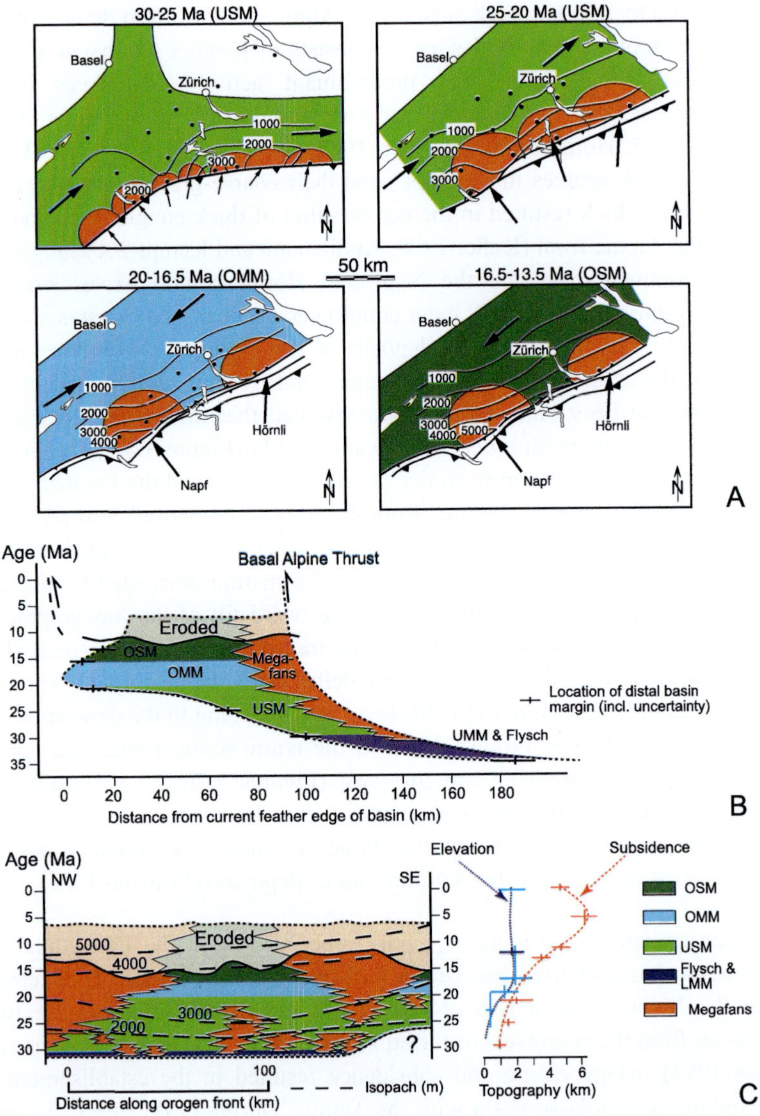

Figure 3.4. *Stratigraphic data from the Swiss Molasse basin*

COMMENT ON FIGURE 3.4.– *(a) Depositional systems, depth of the base of the Molasse sediments and material transport directions (black arrows) for different times. The black dots indicate locations where stratigraphic data are available. The contour lines indicate the depth of the base of the Molasse deposits. The locations of cities and lakes are indicated for orientation. The maps show that the amplitude of the basin deflection increased through time. This was associated by a decrease in the number of alluvial megafans (orange color) (modified from Pfiffner et al. (2002)). (b) Chronologic (Wheeler) diagram of the Swiss Molasse deposits for a section across the basin. The undulated line at ca. 12 Ma indicates the youngest ages of preserved material in the Swiss part of the Molasse basin; the overlying suite, marked as "eroded", illustrates the sediment which continued to accumulate, but eroded during the past 5–10 Ma (see Cederbom et al. (2014), for further details on the post 12 Ma accumulation and erosion history of the basin). The figure also illustrates the northward shift of the orogen front and the northward migration of the basin axis (modified from Schlunegger and Kissling (2015)). Note that we cannot explain why the basin narrowed after 19 Ma. (c) Chronologic (Wheeler) diagram of the Swiss Molasse deposits for a section along the proximal basin border. The dashed horizontal lines indicate the cumulative amount of deposited sediments for the corresponding time. The number of megafans decreased through time, while the spacing between them became larger. Because after 25–20 Ma, the same sediment flux was shared by two fluvial systems only (instead of six and more for the preceding times), the cross-sectional widths of the fans increased. The figure on the right illustrates the subsidence history at the proximal border of the Swiss Molasse basin and the development of the Alpine topography. The error bars illustrate the ranges of values that depend on the selection of the morphometric variables for the paleoelevation calculation (please see Schlunegger and Kissling (2015) for further details). Interestingly, the paleo-elevation of the Alps remained stationary, while the basin continued to subside. The paleo-elevation estimate for 12 Ma (dark blue) is taken from Campani et al. (2012). Modified from Schlunegger and Kissling (2015).*

3.4.4. *Since 10/5 Ma*

Sedimentation in the Molasse basin continued until ca. 10-5 Ma, when a large-scale phase of uplift inverted the basin (Mazurek et al. 2006; Genser et al. 2007; Cederbom et al 2011; Gustenhuber et al. 2012). Note that Cederbom et al. (2004) and Genser et al. (2007) were more specific on the assignment of an age for the uplift. In particular, Cederbom et al. (2004) used apatite fission track ages of detrital material encountered in deep wells close to the Alpine border of Central Switzerland to propose an age of 5 Ma for the onset of uplift. Genser et al. (2007) performed a backstripping of sediments at 16 drill sites in the Austrian Molasse. These authors used the age model of Jin et al. (1995) for the Molasse sediments in these drillings, and they considered an Airy-Heiskanen isostacy model for the sedimentary loads. Based on these constraints and the backstripping results, Genser et al. (2007)

suggested that the uplift started at 6 Ma. Obviously, the amount of erosion was greatest in the Swiss Molasse basin, where low temperature thermochronometric data established in deep wells (Cederbom et al. 2004, 2011) indicate that ca. 1,000 m of previously deposited Molasse sediments have been eroded in the basin axis. The uncertainties on these estimates vary significantly, and erosion amounts of more than 2,000–3,000 m have also been proposed, particularly for sites close to the Alpine thrust front (Cederbom et al. 2004). The amount of erosion then decreases to between 1,000 and 500 m in the center of the German and Austrian Molasse basin (Lemcke 1974; Gusterhuber et al. 2012; Baran et al. 2014). Gusterhuber et al. (2012) additionally identified an anomaly in this erosion pattern where the amount of erosion was possibly >1,000 m along the narrow eastern zone near Linz (Figure 3.1). The reason for this uplift and erosion anomaly is not clear, but Gusterhuber et al. (2012) proposed that differential uplift of crustal rocks surrounding the Bohemian Spur could be invoked to explain this pattern. The consequence of this differential erosion is also mirrored in the pattern of the youngest exposed rocks in the Molasse basin (Figure 3.2). In particular, while the youngest sediments have been dated to ca. 13 Ma in the Molasse basin adjacent to the Central Alps (Kempf and Matter 1999), the youngest preserved sediments in the Austrian Molasse basin are ca. 9 Ma old. This latter age is inferred from Gliridae mammal teeth (MN10) encountered in the Hauruck gravels (Daxner-Höck and Höck 2009).

The amount of postdepositional erosion in the Swiss Molasse was also estimated by Schegg et al. (1997) based on vitrinite reflectance data. These authors proposed that the amount of erosion in the Swiss Molasse basin exceeds 2 km. They additionally identified a strong eastward-directed lateral gradient of more than 1.5 km of post-depositional erosion, which was related to the folding of the Jura fold-and-thrust belt by Schegg and Leu (1998). Because of this folding, the Molasse basin should have been in a piggy-back situation, where uplift and erosion would be greatest beneath the décollement (Pfiffner et al. 1997). Kuhlemann and Kempf (2002) used these arguments to propose that the uplift of the Jura fold-and-thrust belt reversed the drainage direction toward the SE (Figure 3.2(a)) and additionally initiated a wave of erosion that progressed from the Swiss to the German, and finally to the Austrian part of the basin. These interpretations have been contested by Cederbom et al. (2011) who noted that the uplift and erosion of the Molasse basin postdates the major phase of folding of the Jura Mountains and occurred simultaneously across the entire basin. Therefore, Cederbom et al. (2011) suggested that uplift and erosion of the basin was not explicitly linked with the shortening and deformation of the Jura Mountains. Building on these results and using a numerical model, Willett and Schlunegger (2011) proposed that ongoing sedimentation in the Molasse basin resulted in large loads on the décollement underneath the Molasse, thereby shifting the sedimentary wedge to a supercritical state. Willett and Schlunegger (2011) thus used arguments offered by critical taper wedge theory (Chapple 1978; Davis et al. 1983; Dahlen 1984, 1990) to propose that the combination of (i) supercritical conditions of the Molasse sedimentary wedge and (ii) the occurrence of thick Triassic evaporates underneath the northern margin of the Molasse basin, finally allowed the release of Alpine stress in the area of the Jura fold-and-thrust belt (section 3.6.2).

The most recent interpretation for the cause of this differential erosion was proposed by Baran et al. (2014) and Winterberg and Willett (2019). Baran et al. contoured the amount of post-depositional erosion of the entire Molasse basin together with apatite fission track cooling ages in the adjacent Alps, which they inverted into a pattern of post 5 Ma exhumation rates. This compilation showed that the region surrounding the external massifs (Aar, Mt. Blanc, Ecrins; Figure 3.1) and the Swiss part of the Molasse basin experienced the highest exhumation rates during the past millions of years. Therefore, Baran et al. (2014) proposed a scenario where lithospheric unloading of the Eurasian slab underneath the Western Alps along the Eurasian-Adriatic plate boundary best explains the regional-scale uplift pattern (see also section 3.7.3). In contrast to these authors, Winterberg and Willett (2019) based their interpretation on the geometry of the drainage network surrounding the Alps, and how this has changed over time. In particular, while streams from the northern margin of the Central Alps ended in the Paratethys (and later on in the Black Sea) between ca. 10 and 5 Ma, the same streams discharged to the Mediterranean after 5 Ma via the Bresse graben, and then to the North Sea via the Rhine graben after 2 Ma (Ziegler and Fräfel 2009). In this context, it is possible that the Messinian salinity crisis initiated the re-routing of the Alpine streams via the Bresse Graben (Loget et al. (2005, Figure 1), and Amadori et al. (2018), for the impact on the Po Basin). Most important, however, the changes at 5 Ma resulted in a shortening of the streams' pathways by >1,000 km. As a consequence, the curvatures of the longitudinal profiles of these streams increased, which was associated by an elevation drop of >200 m (Schlunegger and Mosar 2011) where the Alpine streams left the Molasse basin (Ziegler and Fräfel 2009). Winterberg and Willett (2019), but also Schlunegger and Mosar (2011) considered that these mechanisms could have contributed to some of the erosional recycling of the Molasse deposits.

3.5. First order controls on the stratigraphy of the Molasse basin

There is a general consensus among stratigraphers that the ratio between accommodation space formation and sediment flux exerts the first order controls on the stratigraphic architecture of a sedimentary basin (e.g. Sinclair and Allen

1992; Bridge and Leeder 1997). Accommodation space is formed or destroyed through a change in the subsidence pattern of a sedimentary basin, and/or a shift of the basin's axis relative to the orogen's front. If these changes occur contemporaneously with modifications in supply rates of sediment to the basin, then distinct stratigraphic signals will be recorded in a sedimentary record (Allen et al. 2013; Martinius et al. 2014). This was also the reason why substantial efforts have been undertaken to quantify these variables from the stratigraphic archives in the Molasse basin. Eustatic variations in sea level represent a further parameter that results in the creation (rising sea level) or destruction (falling sea level) of accommodation space. However, because sediment flux, basin subsidence and shifts in the basin axis are ultimately controlled by tectonic processes (Flemings and Jordan 1990; Jordan and Flemings 1991), quantitative information on these variables are of primary importance if the scope lies on reconstructing the history of loading within a mountain belt. The next sections summarize the progress that has been made upon extracting these data from the Molasse basin. We also outline how eustatic variations in sea level are recorded by the Molasse deposits, and how they complicate the analysis of sedimentary records mainly because the resulting signals are likely to interfere with stratigraphic features that result from tectonic processes.

3.5.1. *Migration of the basin axis in relation to the Alpine front*

The basinward propagation of an orogen front is commonly associated with a corresponding shift of the basin axis (Sinclair and Allen 1992). If both processes occur at the same pace, then the width of accommodated space remains constant. If the propagation velocity of an orogen front slows down, but the basin axis continues to migrate at the same velocity toward more distal sites, then the basin will widen, and cross-sectional topographic gradients within a basin will decrease with consequences on the spatial distribution of the depositional systems (Flemings and Jordan 1990; Allen et al. 2013). Similar observations have also been made for the Molasse basin and the Alpine front (Sinclair and Allen 1992), both of which shifted toward the NW during the Alpine orogeneis. Reconstructions for the Swiss part of the basin (Figure 3.4(b)) show that the northwestward migration of the basin axis and the orogen front occurred at high velocities during the Flysch stage, when estimated propagation rates ranged between 3–4 mm/yr and >10 mm/yr (Homwood et al. 1986; Pfiffner 1986; Sinclair and Allen 1992; Burkhard and Summaruga 1998; Schlunegger and Kissling 2015). Kempf and Pfiffner (2004) even proposed a rate of up to 20 mm/yr. Despite some major disagreements on these velocities, most studies converge on the notion that prior to 30 Ma, the northwestward propagation of both the orogen front and the basin axis occurred at nearly the same, relatively high velocities that varied in space and time. At 30 Ma, however, the propagation rate of the Alpine front slowed down to 1–2 mm/yr, at least in the Central Alps when the Swiss Molasse basin became overfilled, while the distal margin of the basin together with the basin axis continued to migrate away from the orogen front at a rate of nearly 10 mm/yr (Schlunegger and Kissling 2015). The consequence was a widening of the basin between ca. 30 and 19 Ma (Garefalakis and Schlunegger 2019), while the cross-sectional width of the basin was nearly constant prior to 30 Ma (Schlunegger and Kissling 2015) (Figure 3.4(b)). This suggests that the width of the basin was greater during the Molasse stage of basin evolution than during the precursory Flysch period. The reason for these differences has not been explored in detail yet and awaits the results of further research. However, we tentatively consider that the difference in the subduction mechanisms between the Flysch and Molasse stages offers a possible explanation. In particular, the Flysch stage was characterized by the subduction of the relatively young (Mesozoic) oceanic lithosphere and the stretched continental margin of the European plate with a relatively low flexural rigidity, resulting in a narrow and deep flexural basin. During the overfilled Molasse stage, the old (Palaeozoic and older) and stiffer continental lithosphere of the European plate became part of the subduction process and started to become deflected. Because of the greater flexural strength of the continental plate, the foreland basin became shallower, but also much wider.

3.5.2. *Basin subsidence*

The subsidence history of the European plate underlying the Molasse basin has been determined for various scales, and related research was much stimulated by the progress in the dating of the Molasse deposits. In this context, high-resolution magneto-polarity stratigraphies (Schlunegger et al. 1996, 1997a; Kempf et al. 1999; Strunck and Matter 2002; Kirscher et al. 2016; van der Boon et al. 2018; Hofmayer et al. 2019), in combination with micromammal biostratigrapic data (e.g. Engesser 1987; Kälin and Kempf 2009; Jost et al. 2016; Engesser and Kälin 2017; Mojon et al. 2018) built the framework for a temporal correlation of the admittedly relatively poor outcrops across the basin. The new chronologies also built the basis for the calculations of subsidence rates at specific sites, which was used to identify a potential tectonic driving force for the stratigraphic development at these locations (Schlunegger et al. 1997a; Kempf et al. 1999). The related interpretations were drawn from a cross-sectional and thus a local perspective (e.g. Pfiffner 1986), and were based on the inference that possible controls on an increase (or alternatively a decrease) in the subsidence rates had to be sought in the mountain range bordering the stratigraphic section of interest (see also section 3.6.1). At a much larger scale, however, Bigi et al. (1990), Pfiffner et al. (2002) and also Rosenberg et al. (2015) were among the first who reconstructed, at a relatively high resolution, the deflection pattern of the entire foreland plate underlying the Molasse basin, at least between Geneva and Munich.

For the German Molasse east of Munich, Zweigel et al. (1998) used well and seismic data to conduct a subsidence analysis within a sequence-stratigraphic framework. These authors proposed that the first order cycles in the German Molasse, which include the transition from the Lower Marine to the Lower Freshwater Molasse, the Burdigalian transgression of the Upper Marine Molasse, and the change to the terrestrial conditions of the Upper Freshwater Molasse, coincide with changes in the eustatic sea level (Haq et al. 1987) (Figure 3.2(b)). Therefore, these authors concluded that the stratigraphic subdivision was mainly controlled by eustatic sea-level fluctuations (see also Jin et al. (1995, Figure 8)). Also, for the German Molasse basin, the results of the subsidence analysis revealed two phases of continued subsidence (Zweigel et al. 1998). The first one lasted between the Late Eocene to the Early Miocene and was characterized by an increasing subsidence rate through time. This phase of continued subsidence was interrupted by a short period of erosion in the North at ca. 20 Ma, leading to the formation of the Burdigalian unconformity (Figure 3.2(a)). From 20 Ma onward, the basin continued to subside (second phase) until 6 Ma, when the basin experienced a large-scale phase of uplift and erosion. In contrast to the first period, the post 20 Ma-subsidence was characterized by constant, or decreasing rates. Zweigel et al. (1998) related the long-term subsidence history to a tectonic driving force, which experienced major changes at ca. 20 Ma, and then again sometime between 10–5 Ma. Similarly, for the Austrian Molasse basin, Genser et al. (2007) used the results of backstripping to reconstruct two periods of subsidence. During the first period, which lasted until ca. 20 Ma, the basin experienced what they considered a phase of tectonic-driven subsidence, which was fastest at ca. 30 Ma and which then decreased continuously until 20 Ma. After that time, Genser et al. (2007) consider that subsidence appears to have been driven mainly by sediment loading. The basin finally experienced a phase of uplift and erosion from 10-5 Ma onward (please see Genser et al. (2007) for further details).

For the Swiss part of the Molasse basin, Pfiffner et al. (2002), but also Schlunegger and Kissling (2015), realized that the deflection of the foreland plate increased continuously through time and was neither associated with an advance of the Alpine thrust front (Sinclair and Allen 1992), nor with an increase in the Alpine topography (Campani et al. 2012) (Figure 3.4(c)). Because of the mismatch between topographic loading and foreland plate deflection (3.4(c)), Schlunegger and Kissling

(2015) proposed that slab loads, or at least deep crustal loads (Pfiffner et al. 2002), rather than the commonly inferred topographic loads (e.g. Sinclair et al. 1991; Schlunegger et al. 1997c; Zweigel et al. 1998; Genser et al. 2007) have to be invoked to explain the formation of accommodation space in the Molasse basin. If this inference is valid, then our views on the causal links between Alpine tectonics and Molasse basin evolution need a considerable change. Such a view also needs to consider the two phases of basin subsidence in the German and the Austrian Molasse basin (section 3.7).

3.5.3. *Estimates of sediment supply to the Molasse basin*

Sediment supply to a basin in relation to the formation rate of accommodation space is the second first order variable that controls the establishment of distinct sedimentary environments within a sedimentary basin (see also above). In addition, variations of sediment flux bear information on changes of the erosional processes in the adjacent mountain belt, which includes shifts in provenance, changes in relief and alterations of exhumation mechanisms and rates (Kühni and Pfiffner 2001a, 2001b). Therefore much effort has been undertaken to quantify the volumetric supply of material to the Molasse basin.

Among the various authors, Hay et al. (1992) were the first to determine the volumes of material deposited within circum-Alpine basins and derive an erosional budget for the Alps. This was followed by sediment budgets by Sinclair (1997), Schlunegger (1999), Kuhlemann (2000), Schlunegger et al. (2001) and Kuhlemann et al. (2001, 2002). Sediment budgets for the Alps were also presented by Willett (2010) and Baran et al. (2014), but these estimates are modified versions of the original Kuhlemann (2000) and Kuhlemann et al. (2001, 2002) dataset. Large uncertainties are commonly associated with sediment budgeting including (i) uncertainties in age assignments to the deposits, (ii) unknown volumes of sediment within the Subalpine Molasse, (iii) unknown volumes of bypassed material (e.g. Hülscher et al. 2019) and (iv) poorly constrained contribution of chemical dissolution of bedrock to the net erosion of a mountain range (Hinderer et al. 2013). Despite such uncertainties, the results of Hay et al. (1992), Sinclair (1997), Schlunegger (1999) and Kuhlemann (2000) were quite consistent regarding the temporal evolution of the erosional flux. In particular, most authors concluded on the notion that the change from the underfilled Flysch to the overfilled Molasse stage, which occurred in the Swiss and German Molasse basin at 32/30 Ma, was most likely controlled by an increase in sediment supply from originally <10,000 km^3/My to nearly 20,000 km^3/My (Figure 3.5). The increase in sediment yields was largest in the Central and Western Alps. The yields also increased in the

Eastern Alps during the same time span, but the changes were much less. After 30 Ma, sediment flux to the Swiss Molasse basin increased until 21 Ma (3.5(a)), when this variable reached a maximum value of ca. 30,000 km^3/My. Sediment flux dropped by 30% to 20,000 km^3/My between 21 and 18 Ma, and then increased again to ca. 30,000 km^3/My at 17 Ma, after which sediment flux continuously decreased until ca. 5 Ma to less than 20,000 km^3/My. At that time, sediment discharge rose again to a maximum value of more than 30,000 km^3/My as registered by quaternary sediments. Note that the 20 Ma sediment supply drop (Kuhlemann 2000) is not visible in the Baran et al. (2014) compilation because these authors averaged the sediment flux over 5 My long time intervals, which was too large to depict the inferred sediment supply variations at approximately 20 Ma (Kuhlemann 2000).

In the Eastern Alps, changes in sediment flux were much less (Figure 3.5(b)), and supply rates of material to the German and Austrian Molasse basin continuously increased from <5,000 km^3/My prior to 30 Ma to 10,000 km^3/My at 5 Ma. Apparently, a shift to lower sediment supply rates also occurred between ca. 20 and 17 Ma. Similar to the Western and Swiss Alps, the past 5 My have been characterized by a doubling of the sediment flux from originally 10,000 km^3/My prior to 5 Ma to nearly 20,000 km^3/My during the quaternary.

The patterns of sediment flux (Figure 3.5) allowed stratigraphers to solve some pending questions about the controls on the evolution of the Molasse basin. First, it is now generally accepted that the change from basin underfill to overfill, which occurred in the Molasse basin west of Munich sometime between 32 and 30 Ma (section 3.4.1, and Figure 3.2(a)), was primarily driven by the large increase in sediment supply rates to the basin. A lowering of the eustatic sea level (Figure 3.2(b)) amplified the regression trend (Diem 1986; Jin et al. 1995; Zweigel et al. 1998), but was probably not the primary driving force. This interpretation was confirmed independently by Sinclair (1997) and then later by Schlunegger and Castelltort (2016) and Garefalakis and Schlunegger (2018). Second, the Burdigalian transgression and the deposition of the OMM in Switzerland (Figure 3.2(a)) were most likely controlled by the contemporaneous reduction of sediment supply rates in combination with an increase in the eustatic sea level and a tectonically controlled widening of the basin (see section 3.5.4 and Garefalakis and Schlunegger (2019)). In this sense, the subsequent progradation of the terrestrial facies in the entire Molasse basin between at least Linz and Geneva (deposition of the OSM; Figure 3.2(a)) could also be related to the contemporaneous augmentation of sediment flux between ca. 18 and 17 Ma.

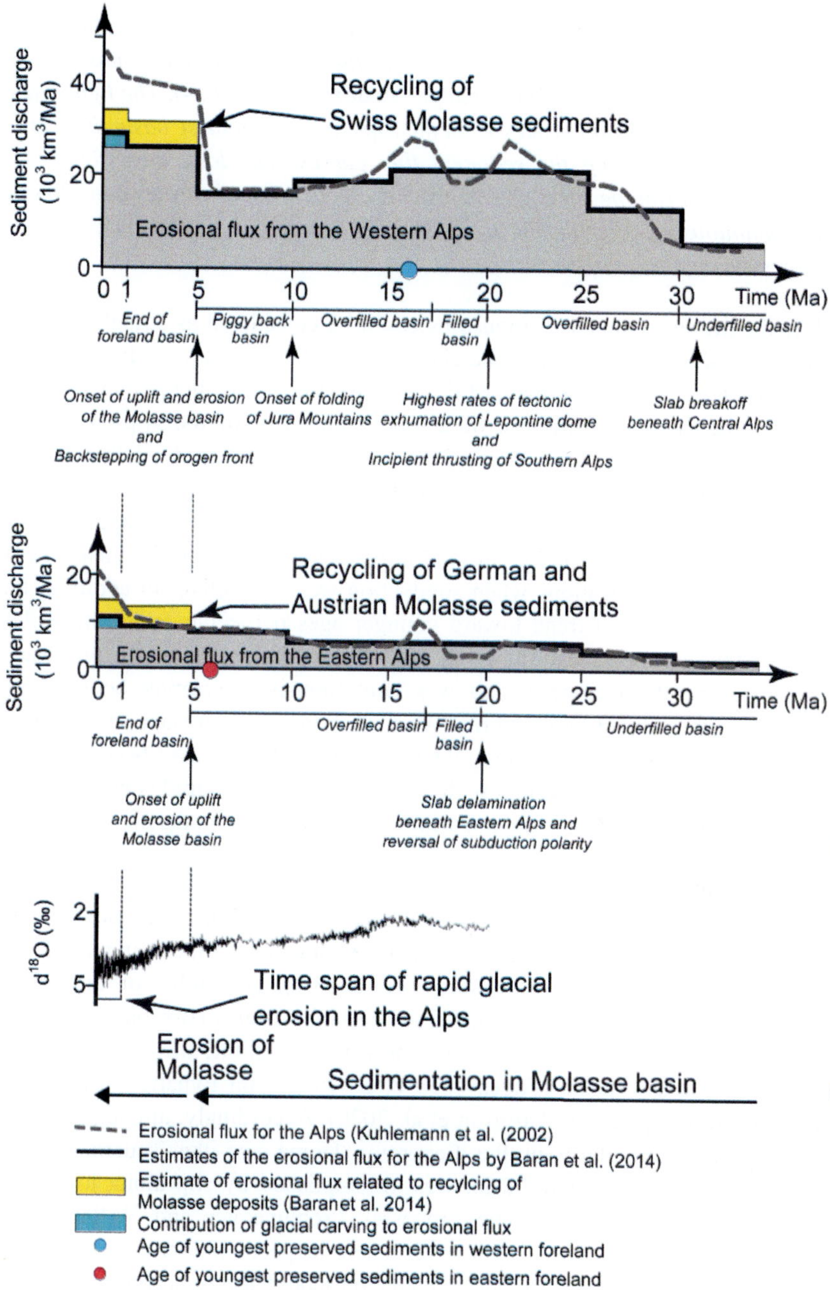

Figure 3.5. *Sediment flux to a) the Swiss Molasse basin and b) to the German and Austrian Molasse basin*

COMMENT ON FIGURE 3.5.– *The original data were collected by Kuhlemann (2000) and then reanalyzed by multiple authors. This figure illustrates the most recent modification of the Kuhlemann (2000) budget by Baran et al. (2014). The figure also illustrates the most important processes in the Alps and the Molasse basin that could be relevant for interpreting the sediment flux curves. The lowermost figure (c) displays the oxygen isotope curve of Zachos et al. (2001) that illustrates how climate became continuously cooler during the past ca. 5 Ma, which finally culminated in the ice ages during the quaternary.*

While the aforementioned relationships could be considered as straightforward, interpretations of the controls on these changes have been contested. In particular, whereas it has been accepted in the past decades that the 30 Ma increase in sediment flux reflects the erosional response of a topographic rise of the Alps after slab breakoff (Sinclair 1997; Baran et al. 2014; Schlunegger and Castelltort 2016) (section 3.7.1), the 30 Ma old slab breakoff hypothesis has just recently been contested by Ji et al. (2019). These authors used the zircon U-Pb ages and Hf isotopic compositions from the Periadriatic intrusives to document that magmatism started in the different locations when subduction was still active. In addition, the zircon U–Pb ages reveal a trend toward younger ages perpendicular to the strike-direction of the European slab. Ji et al. (2019) used these signals to propose a mechanism where the European slab was continuously steepening between the Eocene–Oligocene without breaking off. We note, however, that such a mechanism is not consistent with the stratigraphic development in the Molasse basin, because slab steepening would be associated with a period of ongoing rapid subsidence and continued Flysch sedimentation and not with a shift from basin underfill to overfill.

A larger controversary exists regarding the exhumation information from the Central Alps to explain the drop of sediment flux at 20 Ma. This is particularly the case because 20 Ma was also the time when exhumation in the Aar massif, the Lepontine area and the Tauern window (Figure 3.1), and uplift of the region surrounding the external massifs occurred at the highest rates (Boston et al. 2017; Herwegh et al. 2017, 2020). These regions, however, constitute the major sediment sources of the Molasse basin (Fügenschuh et al. 1997; von Eynatten et al. 1999; Kuhlemann and Kempf 2002; Anfinson et al. 2020). Accordingly, although most of this exhumation, particularly in the Lepontine Dome and the Tauern window (Figure 3.1; sections 3.2.1 and 3.2.2), was accomplished by slip along low angle normal faults and thus by tectonic processes (Schlunegger and Willett 1999; Kuhlemann et al. 2002), we would nevertheless expect an increase and not a decrease in sediment flux, since tectonic exhumation should also be accompanied by an equivalent increase in rock uplift to maintain an Alpine-type topography (Garefalakis and Schlunegger 2019).

The search for the controls on the increase in the sediment flux during the past 5 My has also been a topic of considerable debate in the literature. Some authors even noted that the interpretations of the underlying database, that is, the sediment volumes estimated by Kuhlemann (2000), may be biased by a Sadler (1999) effect (Willenbring and von Blanckenburg 2010). However, a recent analysis of the erosional flux that also considers the contribution of the post 5 Ma phase of Molasse erosion to the sediment budget has not disclosed a major bias in the way the post 5 Ma increase in sediment discharge has been determined (Baran et al. 2014). A further large divergence also exists in the interpretations of this flux increase, which includes the full range from an environmental to a tectonic driving mechanism (Kuhlemann et al. 2001, 2002; Cederbom et al. 2004, 2011; Molnar 2004; Willett et al. 2006; Genser et al. 2007; Willett, 2010; Schlunegger and Mosar 2011; Winterberg et al. 2019). In this context, Baran et al. (2014) proposed that possibly the largest portion of this increase could be explained by the recycling of previously deposited Molasse sediments (Figure 3.5).

In the same sense, although efforts have been undertaken to identify climate signals, at least in the patterns of surface erosion and exhumation rates (e.g. Cederbom et al. 2004), changes in sediment flux have been explained by a tectonic driving force, rather than by climate changes, mainly because sediment-flux driven progradations of alluvial megafans neither occured during periods of climate warming nor during phases of cooling (Schlunegger and Norton 2015; Schlunegger and Castelltort 2016). An exception, however, is the Pleistocene increase in the erosional mass flux that was related to global cooling (Baran et al. 2014) and glacial erosion (Herman et al. 2013), but this has again been contested recently (see discussion in Schildgen et al. (2018) and Willett et al. (2021)). Accordingly, the search for a climate control on surface erosion and sediment flux remains a yet not-fully resolved question, and it appears that we are still far from a consensus on how to interpret the inferred increase in sediment flux during the most recent history of the Alps/Molasse basin ensemble.

3.5.4. *Eustatic variations in sea level and stratigraphic response*

Accommodation space is not only formed or destroyed through changes in the deflection of the foreland plate and/or a modification in the supply rates of sediment, but also through variations in the eustatic sea level. Because the Molasse basin had a direct link to the global sea level either via the Paratethys in the northeast or the Tethys in the southwest (Allen et al. 1985; Kuhlemann and Kempf 2002), any changes in the elevation of the global sea level is expected to have had a direct and measurable response in the stratigraphic record of the Molasse basin, provided that other parameters such as subsidence rates and sediment supply remain constant (e.g. Romans et al. 2016; Castelltort et al. 2017). Therefore, it is not surprising that eustatic controls on the establishment of distinct depositional environments were also considered for the time when the Upper Marine Molasse (Keller 1989; Schaad et al. 1992; Jin et al. 1995; Sissingh 1998; Zweigel et al. 1998; Garefalakis and Schlunegger 2019) and the uppermost regressive part of the Lower Marine Molasse (Diem 1986) were deposited. In particular, and as outlined in section 3.5.2, changes in the eustatic sea level have been invoked to explain the first order stratigraphic changes in the German and also possibly in the Austrian Molasse basin (Jin et al. 1995; Zweigel et al. 1998). These include the transition from the Lower Marine to the Lower Freshwater Molasse, the Burdigalian transgression of the Upper Marine Molasse, and the change to the terrestrial conditions of the Upper Freshwater Molasse (Figure 3.2(b)). Related interpretations have been justified because the shifts from marine to terrestrial conditions and vice versa appear to coincide with high and/or low stands of the eustatic sea level (Haq et al. 1987).

For the Swiss Molasse basin, changes in the eustatic sea level have also been used to explain the transition from the Lower Marine Molasse to the Upper Freshwater Molasse (Diem 1986), and the Burdigalian transgression of the Upper Marine Molasse (Keller 1989). An eustatic control could also possibly be invoked to explain some of the unconformities identified within the Lower Freshwater Molasse deposits in the Swiss part of the basin when terrestrial conditions prevailed (e.g. Schlunegger et al. 1996; Kempf et al. 1999). This is the case because the downstream distance to the shoreline, which was situated near Munich at that time (Figure 3.3), was short enough (ca. 250–300 km from Zurich) so that any changes in the eustatic sea level are likely to have left a measurable signal in the stratigraphic record (Schlunegger and Kissling 2022). In particular, a drop of 20–40 m of the shoreline's elevation will not only initiate a phase of erosion of the former coast and a shift of the new coastal margin toward more distal sites, as documented by Jin et al. (1995) and Zweigel et al. (1998) based on onlap relationships in seismic data, but it will also result in the establishment of a new equilibrium longitudinal profile of the trunk stream and thus in an elevation adjustment of the topographic axis within the basin. The consequence is most likely a destruction of accommodation space or sediment bypass, even in a terrestrial environment. In the Swiss part of the Molasse basin, such an adjustment would explain the various hiati that have been identified not only in the USM (e.g. at 25 and 21 Ma; Schlunegger et al. 1997d; Kempf et al. 1999; Kälin and Kempf 2009), but also at the base of the OMM (Burdigalian unconformity) and in the OMM itself, mainly because they coincide with time periods of a low sea level (Zweigel et al. 1998; Figure 3.2(b)).

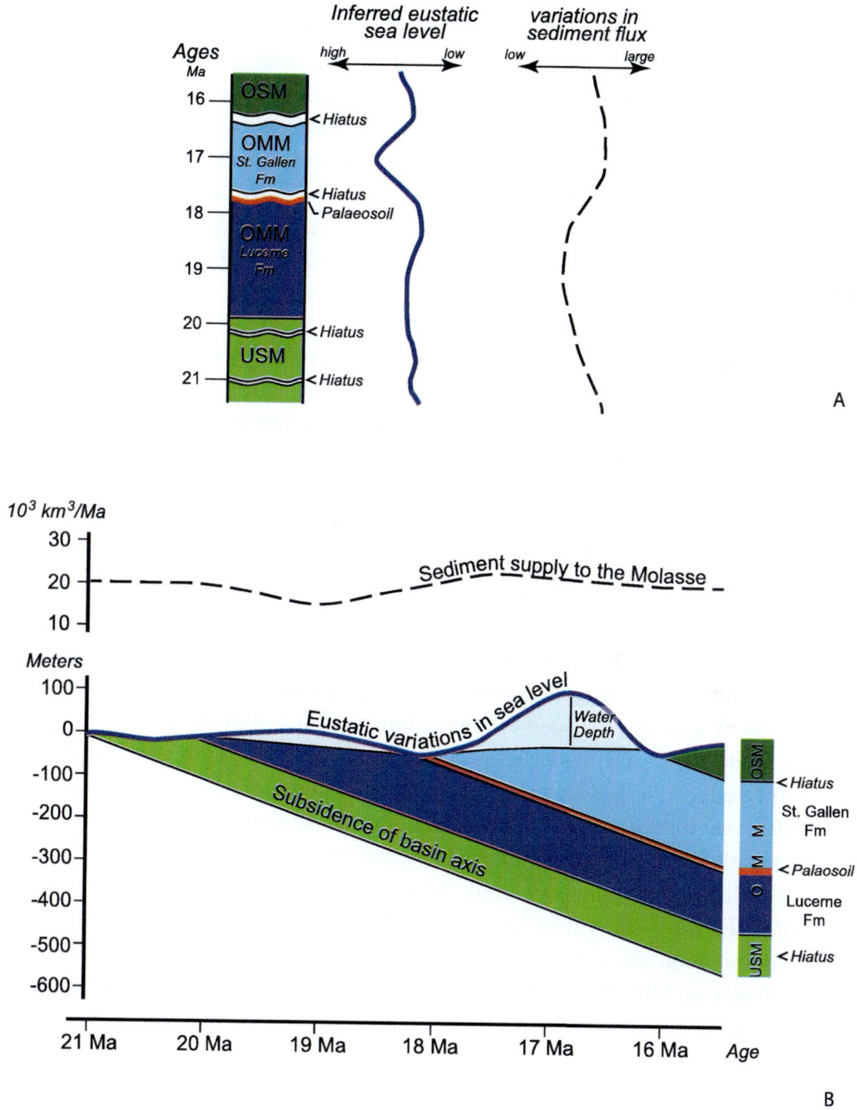

Figure 3.6. *Conceptual figures that illustrate how eustatic sea level variations could have affected the stratigraphic development of the Molasse basin*

COMMENT ON FIGURE 3.6.– *(a) Data from the Swiss Molasse basin, where multiple hiati (time spans of no sedimentation and/or erosion) were related to a drop in the eustatic sea level by Garefalakis and Schlunegger (2019). These authors used stable isotope data from the Atlantic Ocean by Miller et al. (1996) as a proxy for the eustatic sea level. The figure also shows the pattern of sediment supply rates to the Molasse basin that was taken from Kuhlemann (2000) (modified from Garefalakis and Schlunegger (2019)). (b) Sketch that illustrates, from a conceptual point of view, how variations in sediment supply rates and the eustatic sea level at a constant subsidence rate could have influenced the stratigraphic architecture of the Swiss Molasse basin. In case of a constant subsidence rate, a drop in sediment supply associated with an increase in the eustatic sea level is likely to explain the Burdigalian transgression, while a drop in the eustatic sea level paired with an increase in the sediment flux could explain the shift toward basin overfill during OSM times.*

If a modification in the eustatic sea level occurred at the same time when sediment supply to the basin also changed, then the resulting stratigraphic signals are likely to be non-unique, which complicates any interpretation. This is illustrated here for the case of the Burdigalian transgression at ca. 20 Ma in central Switzerland, leading to the deposition of the Upper Marine Molasse (Garefalakis and Schlunegger 2019) (Figure 3.6(a)). At that time, the transgression occurred contemporaneously with a decrease in sediment flux (Kuhlemann 2000), an increase in the eustatic sea level (although the magnitude is not known) and an inferred tectonically controlled widening of the basin, with the consequence that multiple controls (or a combination of them) can be invoked to explain this transgression (Garefalakis and Schlunegger 2019). The deposition of the OMM was also characterized by a short time span of non-sedimentation and the formation of a thick paleosoil, separating the underlying Lucerne Formation from the St. Gallen Formation (Keller 1989). The top of the OMM is additionally characterized by a short hiatus, as identified in magnetostratigraphic sections (Kählin and Kempf 2009). A simple model where subsidence is maintained at a constant rate, but where eustatic sea level variations occurred simultaneously with changes in sediment supply rates (Figure 3.6(b)) illustrates that such a complex stratigraphy could result from a combination of various non-related effects. In particular, Figure 3.6(b) illustrates that a ca. 30% decrease in sediment supply rates paired with an increase in the eustatic sea level could explain the Burdigalian transgression at ca. 20 Ma, while an increase in sediment supply at ca. 18 Ma paired with a drop in the eustatic sea level possibly explains the shift toward terrestrial conditions toward the end of the OMM. Also in this model, the formation of the paleosoil separating the Lucerne from the St. Gallen Formation could result from the combined effect of the 18 Ma old increase in sediment flux and a relatively low sea level. This model is by far simplistic, but it shows that interpretations of the controls on the stratigraphic records are definitely non-unique and complicated by changes in variables that are difficult to constrain.

We note that Sinclair et al. (1991) and Zweigel et al. (1998) propose a short phase of forebuldge uplift along the basin axis prior the Burdigalian transgression, which was considered to have been caused either by an episodic period of surface loading in the Alps (Sinclair et al. 1991), or by a visco-elastic relaxation after thrusting and loading (Zweigel et al. 1998). Accordingly, depending on whether an elastic or visco-elastic model is considered to explain the deflection of the Molasse foreland plate, the interpretation of a possible geodynamic scenario is different. We refrain from such an interpretation because the unconformities within the Lower Freshwater Molasse and the Upper Marine Molasse span nearly the same time span as the Burdigalian unconformity (Figure 3.2(a)). In addition, after each phases of erosion, sedimentation started at the proximal basin border and then progressed toward more distal sites, thus revealing a N-directed onlap on the erosional surface, as seen in seismic lines (Jin et al. 1995; Schlunegger et al. 1997d). Because these unconformities have the same characteristics both in the USM and the OMM, and could possibly be explained by changes in the eustatic sea level in combination with variations in sediment supply rates (Figure 3.6), we consider a surface (i.e. eustatic and sediment supply) rather than a tectonic control (see also above). As a result, we do not consider the occurrence of a forebulge along the basin axis at ca. 20 Ma as a plausible mechanism to explain the erosional unconformity between the Aquitanian and the Burdigalian Molasse sequences. However, an interpretation where a forelbulge uplift could have controlled the formation of the Northern Slope Unconformity (NSU, Figures 3.2(a) and 3.3(b)) on the northern margin of the Austrian Molasse basin is consistent with both subsidence data and the principles of an elastic response of a foreland plate to loading.

3.6. Theoretical concepts for exploring orogenesis and sedimentation

Because the construction of the Alps has mainly been controlled by the subduction of the European plate and not of the Adriatic plate (Holliger and Kissling 1992; Schmid et al. 1996; Rosenberg et al. 2015; Kissling and Schlunegger 2018), the analysis of the subsidence history of the Molasse basin has been the preferred focus of research in the past years, if the scope was to understand how the development of the Alps proceeded. In this context, interpretations of data on the basin's subsidence have commonly been guided by the results of numerical models that predict how foreland basins are expected to respond to orogenic processes. Our understanding of Alpine orogenesis and Molasse sedimentation thus hinges on the progress the modeling community has achieved. This section summarizes the most important results achieved by the modelers, and it explains how they have shaped and guided our view on the ensemble of the Alps and the Molasse basin.

3.6.1. *Elastic models explaining the subsidence history of a basin*

The work by Gunn (1943), Hetényi (1946) and Turcotte and Schubert (1982) offered the first quantitative explorations on how the formation of accommodation space in foreland basins can be understood as a function of crustal loading in the adjacent mountain belt. These authors particularly provided the theoretical framework to calculate the deflection of an elastic plate overlying a fluid substratum as a function of a distributed load. For foreland basins, first applications of these models were published by Jordan (1981) for an elastic plate, and Beaumont (1981) for a visco-elastic plate. These models are based on the generally accepted notion where a stack of thrust nappes represent the load on such a plate. A slab load component was not considered in these calculations (Sinclair et al. 1991; Sinclair 1996; Schlunegger et al. 1997c; Zweigel et al. 1998; Pfiffner et al. 2002). The flexural models were then coupled with a diffusion algorithm that simulated the re-distribution of surface loads from the mountain belt to the basin (Flemings and Jordan 1990; Jordan and Flemings 1991; Sinclair et al. 1991). These models became very powerful since they allowed us to explore a basin's response to thrusting and loading at the mountain front and to erosional redistribution of these loads from the orogen to the basin for the first time (Figure 3.7(a)). For a fully elastic plate, as has been inferred for most foreland plates, the theoretical investigations have shown that the wavelength of a flexural basin, and thus the width of a foreland basin, mainly depends on the flexural strength of the foreland plate, whereas the amplitude of the deflection and thus the depth of a basin is dependent on both the foreland plate's strength and the applied loads (Turcotte and Schubert 1982). These models also predict that any amplifications in the loads either on the surface through thrusting and loading, or at deeper crustal levels through the emplacement of thrust sheets, should have an immediate, measurable and predictable response in the foreland basin, expressed by an acceleration of sediment accumulation rates (Sinclair et al. 1991; Allen et al. 2013) (Figure 3.7(a)).

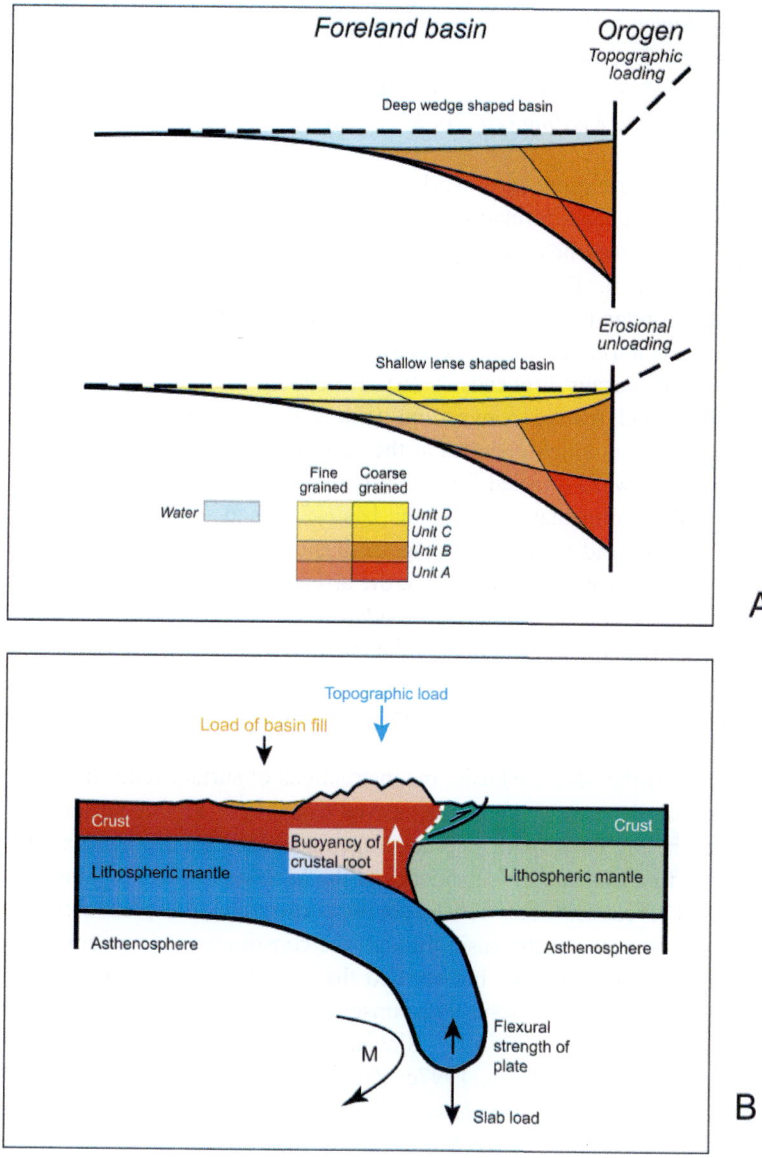

Figure 3.7. *Flexural models of foreland basin formation*

COMMENT ON FIGURE 3.7.– *(a) A period of topographic loading is expected to increase the deflection of the foreland plate and form a deep basin. The depocenters are situated close to the orogen front. A period of erosional unloading is considered to result in a lower surface load. As a consequence, the deflection of the basin is less, and the proximal depocenters migrate toward more distal sites (modified from Flemings and Jordan (1990)). (b) Conceptual model for a foreland basin (modified from Schlunegger and Kissling (2022)). M = Bending moment.*

The results of these pioneering works then converged in a conceptual model on loading and foreland basin formation that was developed by DeCelles and Gilles in 1996 and that has generally been accepted by the stratigraphic community until today (Figure 3.7(b)). Following these authors, the loads responsible for the formation of a foreland basin consist of four major components including (i) the topographic loads attributed to the pressure of upper crustal material that accreted and accumulated during orogenesis; (ii) a bending moment (M) due to the subducting slab and mantle convection; (iii) a vertical-directed shear force or concentrated line load which operates at the end of the plate, or somewhere within the plate, (iv) an upward-directed force that is related to the flexural strength of the plate, and (v) the sedimentary load of the basin-fill itself, which can result in the buckling of the foreland plate underlying the basin (DeCelles and Gilles 1996). In contrast to previous work (Figure 3.7(a)), the article by DeCelles and Gilles (1996) suggests that slab loads, or subduction loads, may also contribute to some of the deflection of a foreland plate and thus to the creation of a foreland trough. However, the buoyancy of the crustal root (Figure 3.7b) exerts an upward-directed force, which has generally not been considered in these models.

For the case of the Molasse basin, reconstructions of surface loads have been the primary focus until recently, if the scope was to explain the relationship between sedimentation in the basin with Alpine orogenic events. In this context, Pfiffner (1986) was the first to relate the subsidence of the Molasse basin to tectonic stacking at upper crustal levels. His study indeed set the cornerstone to explain the formation of accommodation space in the basin through the construction of topographic loads. Others used the same principle, but applied the theoretical models outlined above with an effort to identify the basin's response to episodic loading at or close to the thrust front (Sinclair et al. 1991; Sinclair and Allen 1992; Sinclair 1996; Schlunegger et al. 1997a, 1997b, 1997c; Zweigel et al. 1998; Kempf et al. 1999; Kempf and Matter 2002). While these relationships have been generally accepted, views where the subsidence of the Molasse basin was explained by the build up of topographic loads alone have been contested. In particular, Pfiffner et al. (2002) and Schlunegger and Kissling (2015) noted that changes in surface loads alone cannot be invoked to explain the continuous subsidence of the Molasse basin, mainly because of lack of evidence for a simultaneous and continuous growth of the Alpine topography (Kuhlemann 2007; Campani et al. 2012; Schlunegger and Kissling 2015) (Figure 3.4(c), section 3.5.2).

3.6.2. *Critical taper wedge models*

Critical taper wedge models, which are based on the principle that rocks can sustain a critical differential stress before failure (Figure 3.8(a)), have offered a further avenue for interpreting the relationships between Alpine orogenesis and the development of the Molasse basin (Mosar 1999; Willett et al. 2006; Schlunegger et al. 2007; Willett 2010; Willett and Schlunegger 2011). This was the case because these models allowed us to directly relate the erosional processes on the surface of a mountain belt with its structural development for the first time (Willett et al. 1993; Batt and Braun 1997). The fundamental research in this field was conducted by Davis et al. (1983) and Dahlen (1984) for 2D non-cohesive frictional Coulomb wedges, and by Koons (1994) for 3D wedges. Solutions for viscous and plastic wedges were developed by Platt (1986) and Chapple (1978), respectively. These models predict that mountain belts develop a taper toward the foreland (Figure 3.8(a)), thereby reaching a critical value of the sum $\alpha+\beta$ of their basal and upper slopes, respectively (e.g. Davis et al. 1983; DeCelles and Mitra 1995; Meigs and Burbank 1997). These models thus offer the possibility to explore the response of mountain belts to changes in surface erosion, as proposed, for example, for the Central Andes (Norton and Schlunegger 2011). For instance, enhanced denudation rates at constant rates of crustal thickening will lower the surface angle to subcritical. Critical taper wedge models then predict that the rear of the wedge will shorten and thicken in order to increase the surface slope to a critical value. Likewise, a decrease in erosional flux at constant convergence rates will cause an increase in the taper angle. According to critical taper wedge models, the mountain belt must then grow on the distal margin to re-establish a critically tapered geometry for the entire mountain belt. Because the erosional flux from the Alps to the Molasse basin has experienced large variations through time (Figure 3.5), research was conducted to explore whether such changes in surface mass flux had an influence on the development of the Alps. Using critical taper wedge models, for instance, Schlunegger and Willett (1999; Figure 3.8(b)) proposed that incipient thrusting in the Southern Alps at 20 Ma (Schönborn 1992) was stimulated by the contemporaneous decrease in the erosional mass flux (Figure 3.5(a)). In the same sense, Willett et al. (2006) considered that the Pliocene decrease in the cross-sectional width of the Alps, when the deformation front of the Alps stepped back from the vicinity of Milano to the core of the Alps between the Periadriatic lineament and the Basal Alpine thrust (Figure 3.1), occurred in response to the fast erosional unloading since 5 Ma (Figure 3.5(a)), driven by a lowered base level in a dry Mediterranean area. Finally, also based on the concepts offered by critical taper wedge models, Erdős et al. (2019) recently proposed that the 30 Ma increase in sediment flux (Figure 3.5), which occurred in response to the rise of the Alpine topography, could be considered as a possible reason why the propagation rate of the Alpine front slowed down (see also, Simpson (2006), for a similar explanation).

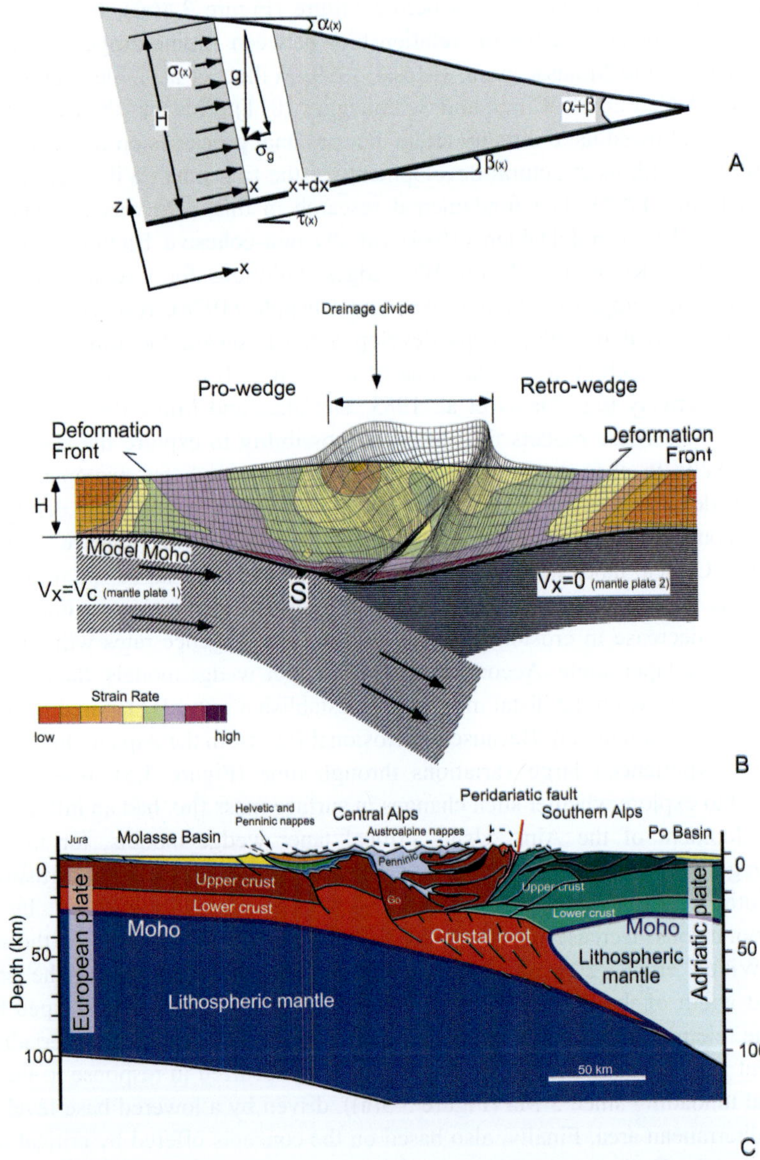

Figure 3.8. *Conceptual critical taper wedge model, coupled surface deformation and comparison of the model orogen and the Central Alps*

COMMENT ON FIGURE 3.8.– *(a) Conceptual framework that was used to develop the critical taper wedge models. The figure illustrates the body forces* σ_g*, the shear forces* τ*, and the compressive push* σ_x *that operate on a moving wedge. H is the thickness of the wedge at location x,* α *is the surface taper (or surface slope) and* β *is the basal slope (modified from Davis et al. (1983)). (b) Example of how a coupled surface deformation – surface erosion model was applied to the Alps. The white part of the mesh above the model orogen corresponds to the eroded material. This model is based on the concept of a critically tapered orogen. Subduction loads were not considered in these calculations. This model was employed to illustrate how an orogen widens and enters a constructive state when surface erosion rates decrease (modified from Schlunegger and Willett (1999)). (c) The comparison between the model orogen and the Central Alps illustrates that while the model is capable of predicting an orogen response to a reduction in surface mass flux, it fails to reproduce first-order features of the architecture of the Alps such as the crustal root (modified from Schlunegger and Kissling (2015)).*

Because these models predict that the erosional flux, and related to this the topographic development of a mountain belt, has a measurable impact on how an orogeny proceeds, stratigraphers have invested large efforts for reconstructing the dynamics of the Alpine topography based on stratigraphic information from the Molasse basin. This includes a re-analysis of provenance information and the collection of new compositional and isotope data from individual detrital grains (detrital zircon, garnet and mica) using novel techniques in geochemistry (von Eynatten et al. 1999, 2002; Spiegel et al. 2000, 2001, 2002, 2004; Malusà et al. 2016; Stutenbecker et al. 2019; Anfinson et al. 2020), a collection of new grain size data (Garefalakis and Schlunegger 2018), a reconsideration of sediment flux data (Willett 2010; Allen et al. 2013; Baran et al. 2014) and reconstructions of the Alpine paleo-topography using stratigraphic data and concepts from tectonic geomorphology (Kuhlemann 2007; Allen et al. 2013; Schlunegger and Norton 2013; Schlunegger and Kissling 2015, Winterberg et al. 2019; Figure 3.4(c)). However, as noted by Kissling and Schlunegger (2018), critical taper wedge models have failed to properly reproduce the structure and geometry of the Moho (e.g. Spada et al. 2013), which is situated at much deeper levels underneath the Alps than predicted by any of these models (Figure 3.8(c)). In the same sense, critical taper wedge models were not capable of reproducing the accretion of a crustal root (Kissling and Schlunegger 2018), which for the case of the Alps has been responsible for maintaining a constant Alpine elevation over millennia. Therefore, information on the deep crustal architecture offered by seismic tomography investigations (Lippitsch et al. 2003; Hetényi et al. 2018) have started to play an increasing role in exploring the causal links between sedimentation in the Molasse basin and the Alpine orogeny.

3.7. Slab load controls

Mechanical models to approximate elastic lithospheric bending due to loading resulted in the generally accepted notion that the width and depth of a foreland basin primarily depend on the interplay between the flexural rigidity of the foreland plate and the applied loads (Figure 3.7). These models also predict that any changes in these loads should have an immediate and predictable response in the foreland basin, expressed by a change in sediment accumulation rates. Although these concepts have largely been applied by the stratigraphic community (section 3.6.1), the exploration of the controls of slab loads on the deflection of the foreland plate and the formation of accommodation space in the basin has received less attention, despite early studies that highlighted their importance (Royden 1993; Carminati and Doglioni 2012). In the context of the Molasse basin, surface loads alone have not been able to explain why the foreland plate continued to subside despite a constant surface elevation (Figure 3.4(c)). As we will explain in detail below, this was also the main reason why Schlunegger and Kissling (2015, 2022) invoked the occurrence of slab loads as the most important driving force on the deflection of the European plate underlying the Molasse basin. The need for the occurrence of such a load was already proposed in the early 1990s, mainly because of the strong negative Bouguer anomaly in the region surrounding the Central Alps (Lyon-Caen and Molnar 1989; Kissling 1993, 2008), and this has subsequently been confirmed by the results of seismic tomography studies (Lippitsch et al. 2003, Figure 3.9(a)). The invoked slab loads add to the surface loads, and these have to be compensated by the buoyancy of a crustal root and an upward-directed elastic force (Figures 3.7(b) and 3.9(b)) in order to maintain the Alpine topography (Lin and Watts 2002; Fry et al. 2010; Schlunegger and Kissling 2015, 2022). If we accept these additional components as driving forces for the development of the Alps (Kissling and Schlunegger 2018; Zilio et al. 2020), then we are capable of addressing some first-order changes in the stratigraphic development of the Molasse basin, some of which, however, still await an explanation (section 3.7.3). These are outlined in the next section.

3.7.1. *Change from Flysch- to Molasse-type sedimentation*

Changes in the relative importance of slab loads versus surface loads were recently used to explain the shift from Flysch to Molasse sedimentation, which occurred in the Swiss Molasse basin ca. 32–30 Ma before present (Figure 3.2(a)). Prior to that time, referred to as the Flysch stage in the literature (Sinclair 1992; Sinclair and Allen 1992), the load of the European oceanic lithosphere, which subducted beneath the Adriatic continental plate, resulted in a slab pull, which flexed the European plate downward (Schlunegger and Kissling, 2022). At ca. 35–32 Ma, the buoyant continental lithosphere of the European plate entered the subduction channel (start of the continent–continent collision), and eventually thereafter the subducted oceanic lithosphere of the European plate broke off (Davies and von Blanckenburg 1995; Schmid et al. 1996). It was proposed that the consequence was a rebound of the European plate, a rise of the Alpine topography in the Central Alps and a larger sediment flux to the Swiss Molasse basin (Figure 3.5; Sinclair 1997; Garefalakis and Schlunegger 2018), which became overfilled between 32 and 30 Ma. East of Munich, however, debris flow deposits and proximal turbidites continued to accumulate within a ca. 1,000–1,500 m deep basin axis basin until approximately 20 Ma (Rögl et al. 1979; Fertig et al. 1991; Malzer et al. 1993; Hülscher et al. 2019), suggesting that the basin stayed in underfilled conditions until at least that time (Figures 3.2(a) and 3.3(a)). It is thus possible that slab-load forces were still downwarping the foreland plate beneath the Eastern Alps and creating a deep trough, whereas oceanic lithosphere slab break-off beneath the Central Alps caused a rebound of the foreland plate in Switzerland (Garefalakis and Schlunegger 2018). As a further implication, the stronger downward deflection of the European foreland plate beneath the Eastern Alps compared to the Central Alps could also explain the east-directed sediment transport prior to ca. 20 Ma (Figures 3.2(a) and 3.3(a) and (b)), but a confirmation of this hypothesis warrants further research on this topic. In the same sense, while these mechanisms are potentially capable of explaining the along-strike sedimentation pattern between the Swiss and the German Molasse basin, they cannot fully explain why the Austrian was not part of the Flysch basin prior to 30 Ma. We see the need for further research that particularly considers the transition between the German and the Austrian Moalsse basin. Such research also includes a focus on the basin situation at the thrust front, where sedimentation occurred in a wedge-top position, which in turn could have resulted in a partial disconnection between the Eastern Alps and the Molasse foredeep situated to the North of the wedge-top basin (e.g. Schlunegger and Kissling 2022).

3.7.2. *The 30 to 20 Ma old development of the Swiss Molasse basin*

We recently proposed a mechanism where slab loads played an increasingly important load on the subsidence history of the foreland plate underlying the Molasse basin (Schlunegger and Kissling 2015). The need for such a model was primarily given by first-order observations from the Swiss Molasse basin itself where (i) the spacing between the major alluvial fans decreased between ca. 30 and 20 Ma, and where (ii) the Molasse basin continued to subside (Figures 3.4(a) and (c)). We employed concepts derived from tectonic geomorphology (Allen et al. 2013) to invert the lateral distance between the depocenters into a paleo-elevation for the time the fans were formed (Figure 3.4(c)). We found that while the reconstructed paleo-elevation was growing between 30 and 20 Ma, and has remained stationary since then, the proximal basin border continued to subside (Figure 3.4(c)). Because the subsidence history of the basin was decoupled from the development of topographic loads, we concluded that surface loads alone are incapable of explaining the formation of accommodation space in the Molasse basin, and that subduction loads need to be invoked instead to explain the observed subsidence patterns.

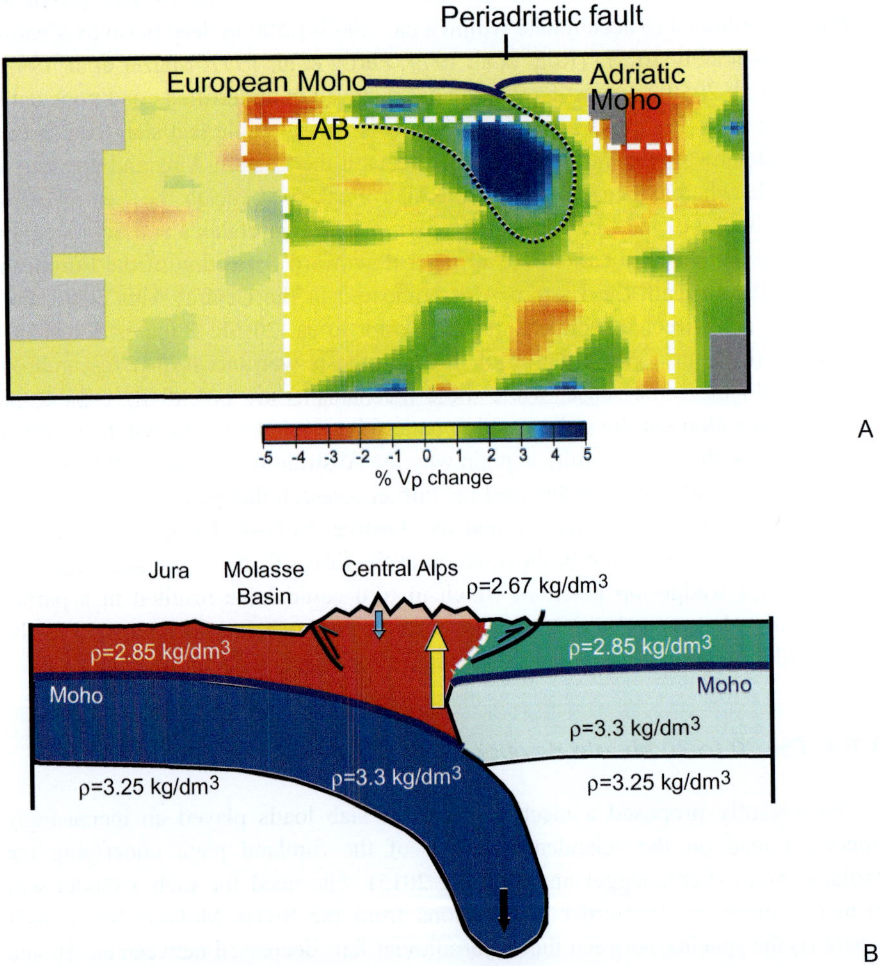

Figure 3.9. *a) Seismo-tomographic image of the Central Alps (modified from Lippitsch et al. (2003)) and b) simple geophysical model that considers the principle of isostacy (Kissling and Schlunegger 2018)*

We used these dataset to propose a mechanism, referred to as a rollback orogeny model (Kissling and Schlunegger 2018; Zilio et al. 2020), where a continuous rollback of the European plate, associated with delamination of crustal material, resulted in the ca. 160 km long subduction of the European mantle slab and in the corresponding shortening of the crustal material (Figure 3.9). This model also allows us to explain the stacking of a crustal root derived from delamination of lower crustal material from the European lithosphere (Figure 3.8(c)). The occurrence of a crustal root was imaged by seismic tomography (Fry et al. 2010) and is required to maintain the Alpine topography at the current elevation (Lyon-Caen and Molnar 1989; Holliger and Kissling 1992). We refer here to Kissling and Schlunegger (2018) and Zilio et al. (2020) for a detailed description of the related processes.

3.7.3. *The post 20 Ma development of the Molasse basin*

As with many foreland basins, the Molasse basin hosted an axial trunk stream collecting transverse drainage where streams with sources in the Alps supplied their material into the basin, thereby forming large radial fans (Figure 3.4(a)). Between 30 and 20 Ma, the axial material transport was to the east via braided to meandering streams. A coastline near Munich separated the terrestrial environment in the west from a deep marine trough farther east (Kuhlemann and Kempf 2002; section 3.4.2 and Figures 3.2 and 3.3) where a submarine basin-axial channel belt continuedly drained the marine Austrian Molasse basin toward the east (de Ruig and Hubbard 2006; Hubbard et al. 2009). After 20 Ma, however, heavy mineral assemblages together with orientations of cross beds in sandstones imply that material transport occurred to the west through the "Graupensandrinne" or "Glimmersandrinne" (Hofmann 1956; Allen et al. 1985; Kuhlemann and Kempf 2002; Berger et al. 2005). During the same time, the Austrian Molasse basin experienced a change from deep to shallow marine conditions (Kuhlemann and Kempf 2002). Garefalakis and Schlunegger (2019) related this drainage reversal between 20 and 18 Ma together with the shallowing-upward trend in the Austrian Molasse basin (Figures 3.2(a) and 3.3) to a change in slab loads underneath the Alps and to differences in subduction mechanisms between the Eastern and Central Alps. In particular, it is very likely that slab unloading beneath the Eastern Alps (Handy et al. 2015) (section 3.2.2) and ongoing subduction of the European lithospheric mantle beneath the Central Alps (section 3.2.1) could have caused a westward tilt of the foreland plate, which could explain this drainage reversal. The timing of this tilting can be further constrained by stratigraphic information from the Molasse basin. In particular, Strunck and Matter (2002) and Garefalakis and Schunegger (2019) showed that at ca. 20 Ma the transgression of the Lucerne Formation progressed from the East to the West, implying an east-directed tilt of the basin. In contrast, at 17.5 Ma, the transgression of the overlying St. Gallen Foramtion occurred in the opposite direction, that is, from the west to the east, which suggests a west-directed tilt of the basin. This indicates that the tilting occurred sometime between 20 and 17.5 Ma. Accordingly, this would be a further mechanism where changes in slab loads could be invoked to interpret a basin-ward shift in sedimentation patterns and material dispersion. This hypothesis (Schlunegger and Kissling 2022), however, is still at its infancy and warrants a more careful evaluation and further data.

Finally, the possible controls leading to the post 5 Ma and thus to the latest stage of uplift and erosional recycling of the Molasse basin are also a matter of current scientific debate (section 3.4.4). The proposed mechanisms include a driving force on the surface (Schlunegger and Mosar 2011; Winterberg and Willett, 2019), at upper crustal levels (uplift of the Jura; Kuhlemann and Kempf 2002), but also within mantle depths such as slab unloading beneath the Western Alps (Baran et al. 2014; Fox et al. 2015). These proposals for an explanation of the large-scale basin uplift and erosion are still speculative, and much work is needed to better understand the underlying mechanisms.

3.8. Conclusion

The Molasse basin situated on the northern side of the Alps has traditionally been the basin of choice if the scope was to improve our understanding of Alpine orogenic events based on sedimentary records of a basin. There is a geodynamic justification for this selection, because the Alpine orogeny was controlled by the subduction of the European plate beneath the Adriatic continental plate, and the Molasse basin sits on the European plate itself. Therefore, any changes in the subduction mechanisms not only resulted in a structural response in the Alps at crustal levels, but also in a modification of the deflection of the foreland plate and thus in a change of the subsidence history of the Molasse basin. The deflection and the subsidence history of the Po basin, situated on the southern side of the Alps, is mainly related to the rollback subduction of the Adriatic plate beneath the Apennines (Carminati and Doglioni 2012), but admittedly the Alps have been by far the most important sediment source. Accordingly, the Po basin mainly records changes in the provenance of the clastic material, and bears information on the erosional history and the drainage configuration within the Southern Alps. Yet the development of the Molasse basin bears a more complete picture on both the history of erosion on the surface of the Alps and that of the subduction processes at crustal levels as deep as the mantle.

Much research has been conducted in the Molasse basin, with a primary focus on improving the chronological framework of the deposits. While most of the sections and drillings have been successively dated with magnetopolarity-stratigraphy and biostratigraphy, reconstructions of the facies relationships still offer a major challenge because the depositional environments change over short distances, and so do assignments of names to the Molasse deposits. This has hampered a harmonization of the Molasse units across the basin for decades. Despite this hurdle, it is possible to identify a first-order sedimentation pattern across the entire Molasse basin, at least between Grenoble and Munich and partly as far east as Linz (Figure 3.1). In particular, the sedimentary architecture is quite homogeneous during deposition of the North Helvetic Flysch between ca. 35 and 32 Ma, and then again during deposition of the Upper Marine Molasse and Upper Freshwater Molasse after 20/18 Ma. The divergence in basin evolution mainly comprises the time interval between ca. 32 Ma and 20 Ma when the upper part of the Lower Marine Molasse and the entire Lower Freshwater Molasse were deposited. In particular, a terrestrial environment occupied the Swiss and German Molasse basin west of Munich during USM times, while Flysch-type deep marine sedimentation continued in the Bavarian and Austrian Molasse basin east of Munich. Also during USM times, the material transport in the basin axis mainly occurred toward the NE, while sediment was routed in the opposite direction after 18 Ma and during OSM times. The sediment dispersion then flipped again toward the NE at 10 Ma. From 5 Ma onwards, sediment dispersion occurred in nearly all directions (Mediterranean, North Sea, Black Sea/Paratethys).

The facies patterns in the Molasse basin mainly responded to changes in the relationship between accommodation space formation and sediment supply. Accommodation space is formed either through subsidence and/or an increase in the eustatic sea level. Shifts in sediment supply primarily occur in response to a change in the tectonic processes in the hinterland or to the exposure of lithologies with a different erodibility. A possible climate control on changes in sediment supply has been proposed by multiple authors and can therefore not be excluded at this stage. But a confirmation of related interpretations needs further in-situ climate proxy data (e.g. Berger 1992; Bruch 1998; Methner et al. 2020) and thus further research. The signals in the Molasse basin that are related to these mechanisms are often non-unique, possibly and mainly because sediment supply to the Molasse basin, and the long-term accommodation space formation in this basin are directly coupled through the same subduction processes. Eustatic variations in sea level exert an independent control on the formation of accommodation space. Although variations in the eustatic sea level are less important if the long-term development of the Molasse basin is considered, variations in eustacy are registered by signals such as hiati and unconformities that are also registered in the terrestrial sequences of the Lower Freshwater Molasse, and have thus confused stratigraphers for decades.

3.9. Acknowledgments

This chapter greatly benefitted from the constructive reviews by A. Berhnardt and A. Di Giulio. The support by C.L. Rosenberg during the review process is greatly acknowledged.

3.10. References

Affolter, T. and Gratier, J.P. (2004). Map view retrodeformation of an arcuate fold-and-thrust belt: The Jura case. *J. Geophys. Res Solid Earth,* 109, B03404.

Allen, P.A. (1984). Reconstruction of ancient sea conditions with an example from the Swiss Molasse. *Mar. Geol.*, 60, 455–473.

Allen, P.A. and Bass, J.P. (1993). Sedimentology of the upper marine molasse of the Rhône-Alp region. Eastern France: Implications for basin evolution. *Eclogae Geol. Helv.*, 86, 121–171.

Allen, P.A., Mange-Rajetky, A., Matter, A. (1985). Dynamic palaeogeography of the open Burdigalian seaway, Swiss Molasse basin. *Eclogae Geol. Helv.,* 78, 351–381.

Allen, P.A., Armitage, J.J., Carter, A., Duller, R., Michael, N.A., Sinclair, H.D., Whitchurch, A.L., Whittaker, A.C. (2013). The Qs problem: Sediment volumentric balance of proximal foreland basin systems. *Sedimentology,* 60, 102–103.

Amadori C., Garcia-Castellanos D., Toscani G., Sternai P., Fantoni R., Ghielmi M., Di Giulio A. (2018). Restored topography of the Po Plain Northern Adriatic region during the Messinian base-level drop – Implications for the physiography and compartmentalisation of the paleo-Mediterranean basin. *Basin Res.*, 30, 1247–1263.

Amadori C., Toscani G., Di Giulio A., Maesano F., D'Ambrogi C., Ghielmi M., Fantoni R. (2019). From cylindrical to non-cylindrical foreland basin: Pliocene-Pleistocene evolution of the Po Plain-Northern Adriatic Basin (Italy). *Basin Res.*, 31, 991–1015.

Amidon, W.H., Luna, L.V., Fisher, G.B., Burbank, D.W., Kylander-Clark, A.R.C., Alonso, R. (2017). Provenance and tectonic implications of Orán Group foreland basin sediments, Río Iruya canyon, NW Argentina (23° S). *Basin Res.,* 29, 96–112.

Anfinsion, O.A., Stockli, D.F., Miller, J.C., Möller, A., Schlunegger, F. (2020). Tectonic exhumation of the Central Alps recorded by detrital zircon in the Molasse basin, Switzerland. *Solid Earth,* 11, 2197–2220.

Argnani, A. (2012). Plate motion and the evolution of Alpine Corsica and Northern Apennines. *Tectonophysics*, 579, 207–219.

Argnani, A. and Ricci Lucchi, F. (2001). Tertiary silicoclasticturbidite systems of the northern Apennines, in anatomy of an orogen. In *The Apennines and Adjacent Mediterranean Basins*, Vai, G. and Martini, I.P.E. (eds). Springer, Dordrecht.

Bachmann, G.H. and Müller, M. (1992). Sedimentary and structural evolution of the German Molasse basin. *Eclogae Geol. Helv.,* 85, 519–530.

Baran, R., Friedrich, A., Schlunegger, F. (2014). The late Miocene to Holocene erosion pattern of the Alpine foreland reflects Eurasian slab unloading beneath the western Alps rather than global climate change. *Lithosphere,* 6, 124–131.

Barbieri, C., Bertotti, G., Di Giulio, A., Fantoni, R., Zoetermeijer, R. (2004). Flexural response of the Southalpine tectonics along the TRANSALP profile. *Terra Nova,* 16, 273–280.

Batt, G.E. and Braun, J. (1997). On the thermomechanical evolution of compressional orogens. *J. Geophys. Int.*, 128, 364–382.

Beaumont, C. (1981). Foreland basins. *Geophys. J. Int.*, 65, 291–329.

Berger, J.-P. (1992). Paléontologie de la Molasse de Suisse Occidentale, taxonomie, biostratigraphie, paléoecologie et paléoclimatologie. Habilitation Thesis, Université de Fribourg.

Berger, J.-P., Reichenbacher, B., Becker, D., Grimm., M., Grimm, K., Picot, L., Storni, A., Pirkenseer, C., Derer, C., Schäfer, A. (2005). Paleogeography of the Upper Rhine Graben (URG) and the Swiss Molasse basin (SMB) from Eocene to Pliocene. *Int. J. Earth Sci.,* 94, 697–710.

Bernhardt, A., Stright, L., Lowe, D.R. (2012). Channelized debris-flow deposits and their impact on turbidity currents: The Puchkirchen axial channel belt in the Austrian Molasse basin. *Sedimentology*, 59, 2042–2070.

Berra, F. and Carminati, E. (2009). Subsidence history from backstripping analysis of the Permo-Mesozoic succession of the Central Southern Alps (Northern Italy). *Basin Res.,* 22, 952–975.

Bertotti, G., Picotti, V., Cloething, S. (1998). Lithospheric weakening during "retroforeland" basin formation: Tectonic evolution of the central South Alpine foredeep. *Tectonics,* 17, 131–142.

Bigi, G., Castellarin, A., Coli, M., Dal Piaz, G.V., Sartori, M., Scandone, P., Vai, G.B. (1990). *Structural Model of Italy*. SELCA, Florence.

Boriani, A. and Sacchi, R. (1973). Geology of the junction between the Ivrea-Verbano and Strona-Ceneri Zones. *Mem. Inst. Geol. Min. Univ. Padova*, 28, 3–36.

Boston, K.R., Rubatto, J., Hermann, J., Engi, M., Amelin, Y. (2017). Geochronology of accessory allanite and monazite in the Barrovian metamorphic sequence of the Central Alps, Switzerland. *Lithos*, 286–287, 502–518.

Bridge, J.S. and Leeder, M.R. (1979). A simulation model of alluvial stratigraphy. *Sedimentology,* 26, 617–644.

Bruch, A.A. (1998). Palynologische Untersuchungen im Oligozän Sloweniens – Paläoumwelt und Paläoklime im Ostalpenraum. *Tüb. Mikropaleont. Mitt.,* 18, 193.

Brügel, A., Dunkl, I., Frisch, W., Kuhlemann, J., Balogh, K. (2003). Geochemistry and geochronology of gneiss pebbles from foreland Molasse conglomerates: Geodynamic and paleogeographic implications for the Oligo-Miocene evolution of the Eastern Alps. *J. Geology,* 111, 543–563.

Burbank, D.W., Engesser, B., Matter, A., Weidman, M. (1992). Magnetostratigraphic chronology, mammalian faunas and stratigraphic evolution of the Lower Freshwater Molasse, Haute Savoie, France. *Eclogae Geol. Helv*., 85, 399–431.

Burkhard, M. (1990). Aspects of the large-scale miocene deformation in the most external part of the Swiss Alps (Sub-Alpine Molasse to Jura Fold Belt). *Eclog. Geol. Helv*., 83, 559–583.

Burkhard, M. and Sommaruga, A. (1998). Evolution of the Western Swiss Molasse basin: Structural relations with the Alps and the Jura Belt. In *Cenozoic Foreland Basins of Western Europe*, Mascle, A., Puigdefabregas, C., Luterbacher, H.P., Fernandez, M. (eds). Geol. Soc. Spec. Publ. No. 134.

Buxtorf, A. (1916). Prognosen Und Befunde Beim Hauensteinbasis Und Grenchenbergtunnel Und Die Bedeutung Der Letzteren Für Die Geologie Der Jurabebirges. *Verh. Naturforsch. Ges. Basel*, 27, 185– 254.

Campani, M., Mulch, A., Kempf, O., Schlunegger, F., Mancktelow, N. (2012). Miocene paleotopography of the Central Alps. *Earth Planet. Sci. Lett.,* 337–338, 174–185.

Carminati, E. and Doglioni, C. (2012). Alps vs. Apennines: The paradigm of a tectonically asymmetric Earth. *Earth-Sci. Rev*., 112, 67–96.

Carminati, E., Lustrino, M., Doglioni, C. (2012). Geodynamic evolution of the central and western Mediterranean: Tectonics vs. igneous petrology constraints. *Tectonophysics*, 579, 173–192.

Castelltort, S., Honegger, L., Adatte, T., Clark, J.D., Puigdefàbregas, C., Spangenberg, J.E., Dykstra, M.L., Fildani, A. (2017). Detecting eustatic and tectonic signals with carbon istotopes in deep-marine strata, Eocene Ainsa Basin, Spanish Pyrenees. *Geology,* 45, 707–710.

Cederbom, C., Sinclair, H.D., Schlunegger, F., Rahn, M. (2004). Climate-induced rebound and exhumation of the European Alps. *Geology*, 32, 709–712.

Cederbom, C., van der Beek, P., Schlunegger, F., Sinclair, H., Oncken, O. (2011). Rapid extensive erosion of the North Alpine foreland basin at 5–4 Ma. *Basin Res.*, 23, 528–550.

Chapple, W.M. (1978). Mechanics of thin-skinned fold-and-thrust belts. *Geol. Soc. Am. Bull.*, 97, 1189–1198.

Dahlen, F.A. (1984). Noncohesive critical coulomb wedges: An exact solution. *J. Geophys. Res.*, 89, 10,125–10,133.

Dahlen, F.A. (1990). Critical taper model of fold-and-thrust belts and accretionary wedges. *Ann. Rev. Earth Planet. Sci.*, 18, 55–99.

Dando, B.D.E., Stuart, G.W., Houseman, G.A., Hegedüs, E., Brückl, E., Radovanović, S. (2011). Teleseismic tomography of the mantle in the Carpathian–Pannonian region of central Europe. *Geophys. J. Int.*, 186, 11–31.

Davis, J.H. and von Blanckenburg, F. (1994). Slab breakoff: A model of lithospheric detachment and its test in the magmatism and deformation of collisional orogens. *Earth Planet. Sci. Lett.*, 129, 85–102.

Davis, D.M., Suppe, J., Dahlen, F.A. (1983). Mechanics of fold-and-thrust belts and accretionary wedges. *J. Geophys. Res.*, 88, 1153–1172.

Daxner-Höck, G. and Höck, E. (2009). New data on Eomyidae and Gliridae (Rodentia, Mammalia) from the Late Miocene of Austria. *Ann. Naturhist. Mus. Wien,* 111A, 375–444.

De Ruig, M.J. and Hubbard, S.M. (2006). Seismic facies and reservoir characteristics of a deep-marine channel belt in the Molasse foreland basin. Puchkirchen Formation, Austria. *AAPG Bull.*, 90, 735–752.

DeCelles, P.G. and Giles, K.A. (1996). Foreland basin systems. *Basin Res.,* 8, 105–123.

DeCelles, P.G. and Mitra, G. (1995). History of the Sevier orogenic wedge in terms of critical taper models, northeast Utah and southwest Wyoming. *Geol. Soc. Am. Bull.*, 107, 454–462.

DeCelles, P.G., Gehrels, G.E., Quade, J., Ojha, T.P. (1998). Eocene-early Miocene foreland basin development and the history of Himalayan thrustiing, western and central Nepal. *Tectonics*, 5, 741–765.

Dèzes, P., Schmid, S.M., Ziegler, P.A. (2004). Evolution of the European Cenozoic Rift system: Interaction of the Alpine and Pyrenean orogens with their foreland lithosphere. *Tectonophysics*, 389, 1–33.

Di Giulio, A., Carrapa, B., Fantoni, R., Gorla, L., Valdisturlo, A. (2001). Middle Eocene-Early Miocene sedimentary evolution of the Western Lombardy South Alpine foredeep (Italy). *Int. J. Earth Sci.,* 90, 534–548.

Diem, B. (1986). Die Untere Meeresmolasse zwischen der Saane (Westschweiz) und der Ammer (Oberbayern). *Eclogae Geol. Helv.,* 79, 493–559.

Elliott, T., Apps, G., Davies, H., Evans, M., Ghibaudo, G., Graham, R.H. (1985). A structural and sedimentological traverse through the Tertiary foreland basin of the external Alps of South-East France. In *International Symposium on Foreland Basins*, Allen, P., Homewood, P., Williams, G. (eds). Int. Assoc.of Sedimentol., Excursion guide book, Fribourg.

Engesser, B. (1987). New Eomyidae, Dipodidae, and Cricetidae (Rodentia, Mammalia) of the lower freshwater molasse of Switzeraland and Savoy. *Eclogae Geol. Helv.,* 80, 943–994.

Engesser, B. and Kälin, D. (2017). Eomys helveticus n.sp. and Eomys schlunegger in.sp., two new small eomyids of the Chattian (MP25/MP26) subalpine lower freshwater molasses of Switzerland. *Fossil Imprint*, 73, 213–224.

Erdös, Z., Huismans, R.S., van der Beek, P. (2019). Control of increased sedimentation on orogenic fold-and-thrust belt structure – Insights into the evolution of the Western Alps. *Solid Earth,* 10, 391–404.

Fertig, J., Graf, R., Lohr, H., Mau, J., Müller, M. (1991). Seismic sequences and facies analysis of the Puchkirchen Formation, Molasse basin, south-east Bavaria, Germany. *Gener. Accumulation, Prod. Eur. Hydrocarb* : *Eur. Assoc. Pet. Geosci. Spec. Publ.*, 1, 277–287.

Flemings, P.B. and Jordan, T.E. (1990). Stratigraphic modeling of foreland basins: Interpreting thrust deformation and lithosphere rheology. *Geology*, 18, 430–434.

Ford, M. and Lickorish, H. (2004). Forleand basin evolution around the western Alpine arc. In *Deep-Water Sedimentation in the Alpine Basin of SE France. New Perspectives on the Grbs d'Annot and Related Systems,* Joseph, P. and Lomas, S.A. (eds). Geol. Soc. London Spec. Publ. No. 221.

Ford, M., Duchêne, S., Gasquet, D., Vanderhaeghe, O. (2006). Two-phase orogenic convergence in the external and internal SW Alps. *J. Geol. Soc. London,* 163, 815–826.

Fountain, D.M. (1976). The Ivrea-Verbano and Strona-Ceneri zones, Northern Italy: A cross-section of the continental crust: New evidence from seismic velocities of rock samples. *Tectonophysics*, 33, 145–165.

Fox, M., Herman, F., Kissling, E., Willett, S.D. (2014). Rapid exhumation in the Western Alps driven by slab detachment and glacial erosion. *Geology,* 43, 379–382.

Frisch, W., Kuhlemann, J., Dunkl, I., Székely, B. (2001). The Dachstein paleosurface and the Augenstein Formation in the Northern Calcareus Alps – A mosaic stone in the geomorphological evolution of the Eastern Alps. *Int. J. Earth Sci.,* 90, 500–518.

Fry, B., Deschamps, F., Kissling, E., Stehly, L., Giardini, D. (2010). Layered azimuthal anisotropy of Rayleigh wave phase velocities in the European Alpine lithosphere inferred from ambient noise. *Earth Planet. Sci. Lett.*, 297, 95–102.

Füchtbauer, H. (1964). Sedimentpetrographische Untersuchungen in der älteren Molasse nördlich der Alpen. *Eclogae Geol. Helv.*, 61, 157–298.

Fügenschuh, B., Seward, D., Mancktelow, N. (1997) Exhumation in a convergent orogen: The western Tauern window. *Terra Nova*, 9, 213–217.

Fügenschuh, B., Seward, D., Mancktelow, N. (2006). Exhumation in a convergent orogen: The western Tauern window. *Terra Nova,* 9, 213–217.

Garcia-Castellanos, D., Fernàndez, M., Torne, M. (2002). Modeling the evolution of the Guadalquivir foreland basin (southern Spain). *Tectonics*, 21, 1018.

Garefalakis, P. and Schlunegger, F. (2018). Link between concentrations of sediment flux and deep crustal processes beneath the European Alps. *Sci. Rep.,* 8, 183.

Garefalakis, P. and Schlunegger, F. (2019). Tectonic processes, variations in sediment flux, and eustatic sea level recorded by the 20 Myr old Burdigalian transgression in the Swiss Molasse basin. *Solid Earth,* 10, 2045–2072.

Genser, J., Cloething, S.A.P.L., Neubauer, F. (2007). Late orogenic rebound and oblique Alpine convergence: New constraints from subsidence analysis of the Austrian Molasse basin. *Global Planet. Change*, 58, 214–223.

Grool, A.R., Ford, M., Vergés, J., Huismans, R.S., Christophoul, F., Dielforder, A. (2018). Insights into the crustal-scale dynamics of a doubly vergent orogen from a quantitative analysis of its forelands: A case study of the eastern Pyrenees. *Tectonics*, 37, 450–476.

Grunert, P., Hinsch, R., Sachsenhofer, R.F., Bechtel, A., Ćorić, S., Harzhauser, M., Piller, W.E., Sperl, H. (2013). Early Burdigalian infill of the Puchkirchen Trough (North Alpine Foreland Basin, Central Paratethys): Facies development and sequence stratigraphy. *Marine Petrol. Geol.,* 39, 164–186.

Grunert, P., Auer, G., Harzhauser, M., Piller, W.E. (2015). Stratigraphic constraints for the upper Oligocene to lower Miocene Puchkirchen Group (North Alpine Foreland Basin. Central Paratethys). *Newslett. Stratigr.*, 48, 111–133.

Gunn, R. (1943). A quantitatie evaluation of the influence of the lithosphere on the anomalies of gravity. *J. Franklin. Inst.,* 236, 373–377.

Gusterhuber, J., Dunkl, I., Hinsch, R., Linzer, H.-G., Sachsenhofer, R.F. (2012). Neogene uplift and erosion in the Alpine Foreland Basin (Upper Austria and Salzburg). *Geol. Carp.,* 63, 295–305.

Handy, M., Schmid, S., Bousquet, R., Kissling, E., Bernoulli, D. (2010). Reconciling plate-tectonic reconstructions of Alpine Tethys with the geological-geophysical record of spreading and subduction in the Alps. *Earth-Sci. Rev.*, 102, 121–158.

Handy, M.R., Ustaszewski, K., Kissling, E. (2015). Reconstructing the Alps-Carpathians-Dinarides as key to understanding switches in subduction polarity, slab gaps and surface motion. *Int. J. Earth Sci.*, 104, 1–26.

Handy, M.R., Schmid, S.M., Paffrath, M., Friederich, W., AlpArray Working Group (2021). Orogenic lithosphere and slabs in the greater Alpine area–interpretations based on teleseismic P-wave tomography. *Solid Earth*, 12, 2633–2669.

Hänni, R. and Pfiffner, O.A. (2001). Evolution and internal structure of the Helvetic nappes in the Bernese Oberland. *Eclogae Geol. Helv.*, 94, 161–171.

Haq, B., Hardenbol, J., Vail, P. (1987). Chronology of fluctuating sea levels since the Triassico. *Science,* 235, 1156–1167.

Hay, W.W., Wold, C.N., Herzog, J.M. (1992). Preliminary mass-balanced 3D reconstructions of the Alps and surrounding areas during the Miocene. In *Computer Graphics in Geology, Three-Dimensional Computer Graphics in Modeling Geologic Structures and Simulating Geologic Processes,* Pflug, R. and Harbaugh, J.W. (eds). Lecture Notes Earth Science, 41, 99–100.

Herman, F., Seward, D., Valla, P.G., Carter, A., Kohn, B., Willett, S.D., Ehlers, T.A. (2013). Worldwide acceleration of mountain erosion under a cooling climate. *Nature,* 504, 423–426.

Herwegh, M., Berger, A., Baumberger, R., Wehrens, P., Kissling, E. (2017). Large-scale crustal-block-extrusion during Late Alpine collision. *Sci. Rep.*, 7, 413.

Herwegh, M., Berger, A., Glotzbach, C., Wangenheim, C., Mock, S., Wehrens, P., Baumberger, R., Egli, D., Kissling, E. (2020). Late stages of continent-continent collision: Timing, kinematic evolution, and exhumation of the Northern rim (Aar Massif) of the Alps. *Earth-Sci. Rev.,* 200, 102959.

Hetényi, M. (1946). *Beams on Elastic Foundation.* University of Michigan Press, Ann Arbor.

Hetényi, G., Plomerová, J., Bianchi, I., Kampfová Exnerová, H., Bokelmann, G., Handy, M.R., Babuška, V. (2018). From mountain summits to roots: Crustal structure of the Eastenr Alps and Bohemian Massif along longitude 13.3°E. *Tectonophysics,* 744, 239–255.

Hinderer, M., Kastowski, M., Kamelger, A., Bartolini, C., Schlunegger, F. (2013). River loads and modern denudation of the Alps – A review. *Earth-Sci. Rev.*, 118, 11–44.

Hinsch, R. (2013). Laterally varying structure and kinematics of the molasse fold and thrust belt of the central eastern alps: Implications for exploration. *AAPG Bull.,* 97, 1805–1831.

Hofmann, F. (1956). Die Obere Süsswassermolasse in der Ostschweiz und im Hegau. *Bull. Ver. Schweiz. Pet-Geol. Ing.*, 23, 23–34.

Hofmayer, F., Kirscher, U., Sant, K., Krijgsman, W., Fritzer, T., Jung, D., Weissbrodt, V., Reichenbacher, B. (2019). Three-dimensional geological modeling supports a revised Burdigalian chronostratigraphy in the North Alpine Foreland Basin. *Int. J. Earth Sci.*, 108, 2627–2651.

Holliger, K. and Kissling, E. (1992). Gravity interpretation of a unified 2-D acoustic image of the central Alpine collision zone. *Geophys. J. Int.*, 111, 213–225.

Homewood, P. and Allen, P.A. (1981). Wave-, tide-, and current-controlled sandbodies in miocene molasse, western Switzerland. *Am. Assoc. Petr. Geol. B.*, 65, 2534–2545.

Homewood, P.W. and Carbon, C. (1982). Flysch of the western Alps. In *Mountain Building Processes,* Hsü, K.J. (ed.). Academic Press, London.

Homewood, P., Allen, P.A., Williams, G.D. (1986). Dynamics of the Molasse basin of western Switzerland. *Spec. Publs. Int. Ass. Sediment.*, 8, 199–217.

Horton, B.K. (2018). Sedimentary record of Andean mountain building. *Earth Sci. Rev.,* 178, 279–309.

Hubbard, S.M., de Ruig, M.J., Graham, S.A. (2009). Confined channel-levee complex development in an elongate depo-center: Deep-water tertiary strata of the Austrian Molasse basin. *Mar. Pet. Geol.*, 26, 85–112.

Hülscher, J., Fischer, G., Grunert, P., Auer, G., Bernhardt, A. (2019). Selective recording of the tectonic forcings in an Oligocene/Miocene submarine channel system: Insights from the wage constraints and sediment volumes from the Austrian northern Alpine forleand basin. *Front. Earth Sci.,* 7, 302.

Hurford, A.J. (1986). Cooling and uplift patterns in the Lepontine Alps, south central Switzerland and an age of vertical movement on the Insubric fault line. *Contrib. Mineral. Petr.*, 93, 413–427.

Ji, W.-Q., Malusà, M.G., Tiepolo, M., Langone, A., Zhao, L., Wu, F.-Y. (2019). Synchronous Periadriatic magmatism in the Western and Central Alps in the absence of slab breakoff. *Terra Nova,* 31, 120–128.

Jin, J., Aigner, T., Luterbacher, H.P., Bachmann, G.H., Müller, M. (1995). Sequence stratigraphy and depositional history in the south-eastern German Molasse basin. *Marine Petrol. Geol.,* 12, 929–936.

Jordan, T.E. (1981). Thrust loads and foreland basin evolution, Cretaceous, western United States. *AAPG Bull.*, 65, 2506–2520.

Jordan, T.E. and Flemings, P.B. (1991). Large-scale stratigraphic architecture, eustatic variation, and unsteady tectonism: A theoretical evaluation. *J. Geophys. Res.-Sol. Earth*, 96, 6681–6699.

Jost, J., Kempf, O., Kälin, D. (2016). Stratigraphy and palaeoecology of the Upper Marine Molasse (OMM) of the central Swiss Plateau. *Swiss J. Geosci.*, 109, 149–169.

Kälin, D. (1997). Litho und Biostratigraphie der mittel bis obermiozänen Bois De Raube Formation (Nordwestschweiz). *Eclog. Geol. Helv.*, 90, 97–114.

Kälin, D. and Kempf, O. (2009). High-resolution stratigraphy from the continental record of the Middle Miocene Northern Alpine Foreland Basin of Switzerland. *N. Jb. Geol. Paläont. Abh.*, 254, 177–235.

Karousova, H., Plomerova, J., Babuska, V. (2013). Upper-mantle structure beneath the southern Bohemian Massif and its surroundings imaged by high-resolution tomography. *Geophys. J. Int.*, 194, 1203–1215.

Keller, B. (1989). Fazies und Stratigraphie der Oberen Meeresmolasse (Unteres Miozän) zwischen Napf und Bodensee. PhD Thesis, University of Bern.

Kempf, O. and Matter, A. (1999). Magnetostratigraphy and depositional history of the Upper Freshwater Molasse (OSM) of eastern Switzerland. *Eclogae Geol. Helv.*, 92, 97–103.

Kempf, O. and Pfiffner, O.A. (2004). Early tertiary evolution of the North Alpine Foreland Basin of the Swiss Alps and adjoining areas. *Basin Res.,* 16, 549–567.

Kempf, O., Matter, A., Burbank, D.W., Mange, M. (1999). Depositional and structural evolution of a foreland basin margin in a magnetostratigraphic framework: The eastern Swiss Molasse basin. *Int. J. Earth Sci.*, 88, 253–275.

Kirscher, U., Prieto, J., Bachtadse, V., Abdul Aziz, H., Doppler, G., Hagmaier, M., Böhme, M. (2016). A biochronologic tie-point for the base of the Tortonian stage in European terrestrial settings: Magnetostratigraphy of the topmost Upper Freshwater Molasse sediments of the North Alpine Foreland Basin in Bavaria (Germany). *Newslett. Stratigraphy,* 49, 445–467.

Kissling, E. (1993). Deep structure of the Alps – What do we really know? *Phys. Earth Planet. Int.*, 79, 87–112.

Kissling, E. (2008). Deep structure and tectonics of the Valais and the rest of the Alps. *Bull. Angew. Geol.*, 13, 3–10.

Kissling, E. and Schlunegger, F. (2018). Rollback orogeny model for the evolution of the Swiss Alps. *Tectonics*, 37, 1097–1115.

Kissling, E., Schmid, S., Lippitsch, R., Ansorge, J., Fuegenschuh, B. (2006). Lithosphere structure and tectonic evolution of the Alpine arc: New evidence from high-resolution teleseismic tomography. *Geol. Soc. London Mem.,* 32, 129–145.

Koons, P.O. (1994). Three-dimensional critical wedges: Tectonics and topography in oblique collisional orogens. *J. Geophys. Res.*, 99, 12301–12315.

Kuhlemann, J. (2000). Postcollisional sediment budget of circum-Alpine basins (central Europe). *Mem. Sci. Geol. Univ. Padova*, 52, 1–91.

Kuhlemann, J. (2007). Paleogeographic and paleotopographic evolution of the Swiss and Eastern Alps since the Oligocene. *Global Planet. Change,* 58, 224–236.

Kuhlemann, J. and Kempf, O. (2002) Post-Eocene evolution of the North Alpine Foreland Basin and its response to Alpine tectonics. *Sediment. Geol.*, 152, 45–78.

Kuhlemann, J., Frisch, W., Dunkl, I., Skékely, B. (2001). Quantifying tectonic versus erosive denudation by the sediment budget: The Miocene core complexes of the Alps. *Tectonophysics*, 330, 1–23.

Kuhlemann, J., Frisch, W., Székely, B., Kunkl, I., Kázmér, M. (2002). Post-collisional sediment budget history of the Alps: Tectonic versus climatic control. *Int. J. Earth Sci.,* 91, 818–837.

Kühni, A. and Pfiffner, O.A. (2001a). The relief of the Swiss Alps and adjacent areas and its relation to lithology and structure: Topographic analysis from a 250-m DEM. *Geomorphology*, 41, 285–307.

Kühni, A. and Pfiffner, O.A. (2001b). Drainage patterns and tectonic forcing: A model study for the Swiss Alps. *Basin Res.*, 13, 169–197.

Lammerer, H., Gebrande, E., Lüschen, E., Veselá, P. (2008). A crustal-scale cross-section through the Tauern Window (eastern Alps) from geophysical and geological data. *Geol. Soc. London Spec. Publ.,* 298, 219–229.

Lateltin, O. (1988). Les dépôts turbiditiques Oligocènes d'avant-pays entre Annecy (Haute Savoie) et la Sanetsch (Suisse). Unpublished PhD. Thesis, Université de Fribourg.

Laubscher, H.P. (1988) The arcs of the Western Alps and the Northern Apennines: An updated view. *Tectonophysics*, 146, 67–78.

Laubscher, H. (1992). Jura Kinematics and the Molasse basin. *Eclog. Geol. Helv.*, 85, 653– 675.

Lemcke, K. (1974). Vertikalbewegungen des vormesozoischen Sockels im nördlichen Alpen-vorland Perm bis zur Gegenwart? *Eclogae Geol. Helv.,* 67, 121–133.

Lihou, J.C. and Allen, P.A. (1996). Importance of inherited rift margin structures in the early North Alpine Foreland Basin, Switzerland. *Basin Res.*, 8, 425–442.

Lin, A.T. and Watts, A.B. (2002). Origin of the West Taiwan basin by orogenic loading and flexure of a rifted continental margin. *J. Geophys. Res. – Solid Earth*, 107, ETG 2-1 – ETG 2-19.

Lippitsch, R., Kissling, E., Ansorge, J. (2003). Upper mantle structure beneath the Alpine orogen from high-resolution teleseismic tomography. *J. Geophys. Res.*, 108, 2376.

Loget, N., Van Den Driessche, J., Davy, P. (2005). How did the Messinian salinity crisis end? *Terra Nova,* 17, 414–419.

Lu, G., Winkler, W., Rahn, M., von Quadt, A., Willett, S.D. (2018). Evaluating igneous sources of the Taveyannaz formation in the Central Alps by detrital zircon U–Pb age dating and geochemistry. *Swiss J. Geosci.*, 111, 399–416.

Lyon-Caen, H. and Molnar, P. (1989). Constraints on the deep structure and dynamic processes beneath the Alps and adjacent regions from an analysis of gravity anomalies. *Geophys. J. Int.*, 99, 19–32.

Madritsch, H., Preusser, F., Fabbri, O., Bichet, V., Schlunegger, F., Schmid, S. (2010). Late Quaternary folding in the Jura Mountains: Evidence from syn-erosional deformation of fluvial meanders. *Terra Nova,* 22, 147-154.

Malusà, M.G., Anfinson, O.A., Dafov, L.N., Stöckli, D.F. (2016). Tracking Adria indentation beneath the Alps by detrital zircon U-Pb geochronology: Implications for the Oligocene-Miocene dynamics of the Adriatic microplate. *Geology*, 44, 155–158.

Malz, A., Madritsch, H., Jordan, P., Meier, B., Kley, J. (2019). Along-strike variation of thin-skinned thrusting style controlled by pre-existing basement structure in the easternmost Jura Mountains (Northern Switzerland). In *Fold and Thrust Belts: Structural Style, Evolution and Exploration*, Hammerstein, J.A., di Cuia, R., Cottam, M.A., Zamora, G., Butler, R.W.H. (eds). Geol. Soc. London Spec. Publ. No. 490.

Malzer, O., Rögl, F., Seifert, P., Wagner, L., Wessely, G., Brix, F. (1993). Die Molassezone und deren Untergrund. *Erdöl Erdgas Öster.*, 2, 281–358.

Mancktelow, N.S. (1985). The Simplon Line: A major displacement zone in the western Lepontine Alps. *Eclogae Geol. Helv.,* 78, 73–96.

Mancktelow, N.S. and Grasemann, B. (1997). Time-dependent effects of heat advection and topography on cooling histories during erosion. *Tectonophysics*, 270, 167–195.

Martinius, A.W., Elfenbein, C., Keogh, K.J. (2014). Applying accommodation versus sediment supply ratio concepts to stratigraphic analysis and zonation of a fluvial reservoir. *Int. Assoc. Sedimentol. Spec. Publ.,* 46, 101–126.

Matter, A., Homewood, P., Caron, C., Rigassi, D., van Stuijvenberg, J., Weidmann, M., Winkler, W. (1980). *Flysch and Molasse of Western and Central Switzerland.* Excursion Guidebook No. 126A of the 26th International Geological Congress. Schweizerische Geologische Kommission, Wepf & Co. Publishers, Basel.

Mazurek, M., Hurford, A., Leu, W. (2006). Unravelling the multi-stage burial history of the Swiss Molasse basin: Integration of apatite fission track, vitrinite reflectance and biomarker isomerization analysis. *Basin Res.,* 18, 27–50.

Meigs, A.J. and Burbank, D.W. (1997). Growth of the South Pyrenean orogenic wedge. *Tectonics,* 16, 239–258.

Methner, K., Campani, M., Fiebig, J., Löffler, N., Kempf, O., Mulch, A. (2020). Middle Miocene long-term continental temperature change in and out of pace with marine climate records. *Sci. Rep.,* 10, 7989.

Mey, J., Scherler, D., Wickert, A.D., Egholm, D.L., Tesauro, M., Schildgen, T.F., Strecker, M. (2016). Glacial isostatic uplift of the European Alps. *Nat. Comm.,* 7, 13382.

Miller, K.G. and Mountain, G.S. (1996). The leg 150 shipboard party, and members of the New Jersey Coastal Plain Drilling project: Drilling and dating New Jersey Oligocene–Miocene sequences: Ice volume, global sea level, and Exxon records. *Science*, 271, 92–94.

Mitterbauer, U., Behm, M., Brückl, E., Lippitsch, R., Guterch, A., Keller, G.R., Koslovskaya, E., Rumpfhuber, E.M., Sumanovac, F. (2011). Shape and origin of the East-Alp slab constrained by the ALPASS teleseismic model. *Tectonophysics*, 510, 195–206.

Mock, S. and Herwegh, S., (2017). Tectonics of the central Swiss Molasse basin: Post-Miocene transition to incipient thick-skinned tectonics? *Tectonics,* 36, 1699–1723.

Mojon, P.-O., De Kaenel, E., Kälin, D., Becker, D., Pirkenseer, C.M., Rauber, G., Ramseyer, K., Hostettler, B., Weidmann, M. (2018). New data on the biostratigraphy (charophytes, nannofossils, mammals) and lithostratigraphy of the Late Eocene to Early Late Miocene deposits in the Swiss Molasse Basin and Jura Mountains. *Swiss J. Paleontol.,* 137, 1–48.

Molnar, P. (2004). Late Cenozoic increase in accumulation rates of terrestrial sediment: How might climate change have affected erosion rates? *Ann. Rev. Earth Planet. Sci.,* 32, 67–89.

Mosar, J. (1999). Present-day and future tectonic underplating in the western Swiss Alps: Reconciliation of basement/wrench-faulting and decollement folding of the Jura and Molasse basin in the Alpine foreland. *Earth Planet Sci. Lett.,* 173, 143–155.

Naylor, M. and Sinclair, H.D. (2008). Pro-vs. retro-foreland basins. *Basin Res.,* 20, 285–303.

Norton, K.P. and Schlunegger, F. (2011). Migrating deformation in the Central Andes from enhanced orographic rainfall. *Nat. Comm., 2,* 548.

Ortner, H. and Stingl, V. (2001). Facies and basin development of the Oligocene in the Lower Inn Valley, Tyrol/Bavaria. In *Paleogene of the Eastern Alps,* Piller, W.E. and Rasser, M.W. (eds). *Erdwiss. Komm. Österr. Akad. Wiss.,* 14, 153–198.

Ortner, H., Aicholzer, S., Zerlauth, M., Pilser, R., Fügenschuh, B. (2015). Geometry, amount, and sequence of thrusting in the Subalpine Molasse of western Austria and southern Germany, European Alps. *Tectonics*, 34, 1–30.

Pfiffner, O.A. (1986). Evolution of the north Alpine foreland basin in the Central Alps. In *Foreland Basins,* Allen, P.A. and Homewood, P. (eds). Spec. Publ. Int. As. Sed. No. 8.

Pfiffner, O.A., Erard, P.-F., Stäuble, M. (1997). Two cross sections through the Swiss Molasse Basin (lines E4-E6, W1, W7-W10). In *Deep Structure of the Swiss Alps: Results of NRP 20,* Pfiffner, O.A., Lehner, P., Heitzmann, P., Müller, S. (eds). Birkhäuser, Basel.

Pfiffner, O.A., Schlunegger, F., Buiter, S.J.H. (2002). The Swiss Alps and their peripheral foreland basin: Stratigraphic response to deep crustal processes. *Tectonics*, 6, 1054.

Philippe, Y., Colletta, B., Deville, E., Mascle, A. (1996). The Jura fold–and–thrust belt: A kinematic model based on map–balancing. In *Peri–Tethys Memoir 2: Structure and Prospects of Alpine Basins and Forelands*, Ziegler P.A. and Horvath, F. (eds). Mem. Mus. Natnl. Hist. Nat., 170, 235– 261.

Pieri, M. and Groppi, G. (1981). Subsurface geological structures of the po plain. *Prog. Fin. Geodin.*, 414, 1–11.

Piromallo, C. and Morelli, A. (2003). P-wave tomography of the mantle under the Alpine–Mediterranean area. *J Geophys Res*, 108, 2065.

Platt, J.P. (1986). Dynamics of orogenic wedges and the uplift of high-pressure metamorphic rocks. *Geol. Soc. Am. Bull.*, 97, 1037–1053.

Platt, N.H. and Keller, B. (1992). Distal alluvial deposits in a foreland basin setting – The Lower Freshwater Miocene (Lower Miocene), Switzerland: Sedimentology, architecture and palaeosols. *Sedimentology*, 39, 545–565.

Reichenbacher, B., Krijgsman, W., Lataster, Y., Pipperr, M., Van Baak, C.G., Chang, L., Kälin, D., Jost, J., Doppler, G., Jung, D. et al. (2013). A new magnetostratigraphic framework for the Lower Miocene (Burdigalian/Ottnangian, Karpatian) in the North Alpine Foreland Basin. *Swiss J. Geosci.*, 106, 309–334.

Rögl, F., Hochuli, P., Müller, C. (1979). Oligocene-early miocene stratigraphic correlations in the molasse basin of Austria. *Ann. Geol. Pay. Helleniques. Tome Hors Series*, 30, 1045–1050.

Romans, B.W., Castelltort, S., Cavault, J.A., Fildani, A., Walsh, J.P. (2016). Environmental signal propagation in sedimentary systems across timescales. *Earth-Sci. Rev.,* 153, 7–29.

Rosenberg, C.L., Berger, A., Bellahsen, N., Bousquet, R. (2015). Relating orogen width to shortening, erosion, and exhumation during Alpine collision. *Tectonics*, 34, 1306–1328.

Rosenberg, C.L., Schneider, S., Scharf, A., Bertrand, A., Hammerschmidt, K., Rabaute, A., Brun, J.-P. (2018). Relating collisional kinemtatics to exhumation processes in the Eastern Alps. *Earth-Sci. Rev.*, 176, 311–344.

Royden, L.H. (1993). The tectonic expression slab pull at continental convergent boundaries. *Tectonics*, 12, 303–325.

Sachsenhofer, R.F., Leitner, B., Linzer, H.G., Bechtel, A., Ćorić, S., Gratzer, R., Reischenbacher, D., Soliman, A. (2010). Deposition, erosion and hydrocarbon source potential of the Oligocene Eggerding Formation (Molasse Basin, Austria). *Aust. J. Earth Sci.,* 103, 76–99.

Sadler, P.M. (1999). The influence of hiatuses on sediment accumulation rates. *GeoRes. Forum,* 5, 15–40.

Sant, K.V., Palcu, D., Mandic, O., Krijgsman, W. (2017). Changing seas in the Early–Middle Miocene of Central Europe: A Mediterranean approach to Paratethyan stratigraphy. *Terra Nova*, 29, 273–281.

Schaad, W., Keller, B., Matter, A. (1992). Die Obere Meeresmolasse (OMM) am Pfänder: Beispiel eines Gilbert-Deltakomplexes. *Eclogae Geol. Helv*., 85, 145–168.

Schegg, R. and Leu, W. (1998). Analysis of erosion ebents and paleogeothermal gradients in the North Alpine Foreland Basin of Switzerland. In *Basin Modelling: Practice and Progress*, Düppenbecker, D.J. and Iliffe, J.E. (eds). Geol. Soc. London Spec. Publ. No. 141.

Schegg, R., Leu, W., Conford, C., Allen, P. (1997). New coalification profiles in the molasse basin of western Switzerland: Implifications for the thermal and geodynamic evolution of the Alpine foreland. *Eclogae Geol. Helv.,* 90, 79–96.

Schildgen, T.F., van der Beek, P.A., Sinclair, H.D., Thiede, R.C. (2018). Spatial correlation bias in late-Cenzoic erosion histories derived from thermochronology. *Nature,* 559, 89–93.

Schlunegger, F. (1999). Controls of surface erosion on the evolution of the Alps: Constraints from the stratigraphies of the adjacent foreland basins. *Int. J. Earth Sci.,* 88, 285–304.

Schlunegger, F. and Castelltort, S. (2016). Immediate and delayed signal of slab breakoff in Oligo/Miocene Molasse deposits from the European Alps. *Sci. Rep*., 6, 31010.

Schlunegger, F. and Kissling, E. (2015). Slab rollback orogeny in the Alps and evolution of the Swiss Molasse basin. *Nat. Commun*., 6, 8605.

Schlunegger, F. and Kissling, E. (2022). Slab load controls beneath the Alps on the source-to-sink sedimentary parthways in the Molasse basin. *Geosciences*, 12, 226.

Schlunegger, F. and Mosar, J. (2011). The last erosional stage of the Molasse basin and the Alps. *Int. J. Earth Sci. (Geol. Rundsch.),* 100, 1147–1162.

Schlunegger, F. and Norton, K.P. (2013). Headward retreat of streams in the Late Oligocene to Early Miocene Swiss Alps. *Sedimentology,* 60, 85–101.

Schlunegger, F. and Norton, K.P. (2015). Climate versus tectonics: The competing roles of Late Oligocene warming and Alpine orogensis in constructing alluvial megafan sequences in the North Alpine foreland basin. *Basin Res.,* 27, 230–245.

Schlunegger, F. and Willett, S. (1999). Spatial and temporal variations in exhumation of the central Swiss Alps and implications for exhumation mechanisms. *Geol. Soc. Lond. Spec. Publ.*, 154, 157–179.

Schlunegger, F., Matter, A., Mange, M. (1993). Alluvial fan sedimentation and structure of the southern Molasse basin margin, Lake Thun area, Switzerland. *Eclogae Geol. Helv.*, 86, 717–750.

Schlunegger, F., Burbank, D.W., Matter, A., Engesser, B., Mödden, C. (1996). Magnetostratigraphic calibration of the Oligocene to Middle Miocene (30–15 Ma) mammal biozones and depositional sequences of the Swiss Molasse Basin. *Eclogae Geol. Helv.*, 89, 753–788.

Schlunegger, F., Leu, W., Matter, A. (1997a). Sedimentary sequences, seismic facies, subsidence analysis, and evolution of the Burdigalian upper marine Molasse group, Central Switzerland. *AAPG Bull.*, 81, 1185–1207.

Schlunegger, F., Jordan, T.E., Klaper, E.M. (1997b). Controls of erosional denudation on foreland basin evolution: The Oligocene central Swiss Molasse Basin as an example. *Tectonics*, 16, 823– 840.

Schlunegger, F., Jordan, T.E., Klaper, E.M. (1997c). Controls of erosional denudation in the orogeny on foreland basin evolution: The Oligocene central Swiss Molasse Basin as an example. *Tectonics,* 16, 823–840.

Schlunegger, F., Matter, A., Burbank, D.W., Leu, W., Mange, M., Màtyàs, J. (1997d). Sedimentary sequences, seismofacies and evolution of depositional systems of the Oligo/Miocene Lower Freshwater Molasse Group, Switzerland. *Basin Res.,* 9, 1–26.

Schlunegger, F., Melzer, J., Tucker, G.E. (2001). Climate, exposed source-rock lithologies, crustal uplift and surface erosion: A theoretical analysis calibrated with data from the Alps/North Alpine Foreland Basin system. *Int. J. Earth Sci.,* 90, 484–499.

Schlunegger, F., Rieke-Zapp, D., Ramseyer, K. (2007). Possible environmental effects on the evolution of the Alps-Molasse Basin system. *Swiss J. Geosci.,* 100, 383–405.

Schmid, S.M., Pfiffner, O.A., Froitzheim, N., Schönborn, G., Kissling, E. (1996). Geophysical-geological transect and tectonic evolution of the Swiss-Italian Alps. *Tectonics*, 15, 1036–1064.

Schmid, S.M., Fügenschuh, B., Kissling, E., Schuster, R. (2004). Tectonic map and overall architecture of the Alpine orogen. *Eclogae Geol. Helv.*, 97, 93–117.

Schmid, S.M., Kissling, E., Diehl, T., van Hinsbergen, D.J.J., Molli, G. (2017). Ivrea mantle wedge, arc of the Western Alps, and kinematic evolution of the Alps-Apennines orogenic system. *Swiss J. Geosci.*, 2017, 581–612.

Schönborn, G. (1992). Alpine tectonics and kinematic models of the Central Southern Alps. *Mem. Soc. Geol.,* 44, 229–393.

Scrocca, D., Doglioni, C., Innocenti, F. (2003). Constraints for an interpretation of the Italian geodynamics: A review. *Mem. Descr. Carta Geol. d'It.,* LXII, 15–46.

Simpson, G. (2006). How and to what extent does the emergence of orogens above sea level influence their tectonic development? *Terra Nova,* 18, 447–451.

Sinclair, H.D. (1992). Turbidite sedimentation during Alpine thrusting: The Taveyannaz sandstones of eastern Switzerland. *Sedimentology,* 39, 837–856.

Sinclair, H.D. (1996). Plan-view curvature of foreland basins and its implications for the palaeostrength of the lithosphere underlying the western Alps. *Basin Res.*, 8, 173–182.

Sinclair, H.D. (1997). Flysch to molasse transition in peripheral foreland basins: The role of the passive margin versus slab breakoff. *Geology*, 25, 1123–1126.

Sinclair, H.D. and Allen, P.A. (1992). Vertical versus horizontal motions in the Alpine orogenic wedge: Stratigraphic response in the foreland basin. *Basin Res.*, 4, 215–232.

Sinclair, H.D. and Naylor, M. (2012). Foreland basin subsidence driven by topographic growth versus plate subduction. *Geol. Soc. Amer. Bull.,* 124, 368–379.

Sinclair, H.D., Coakley, B.J., Allen, P.A., Watts, A.B. (1991). Simulation of foreland basin stratigraphy using a diffusion model of mountain belt uplift and erosion: An example from the Central Alps, Switzerland. *Tectonics*, 10, 599–620.

Sissingh, W. (1997). Tectonostratigraphy of the North Alpine Foreland Basin: Correlation of tertiary depositional cycles and orogenic phases. *Tectonophysics*, 282, 223–256.

Sissingh, W. (1998). Comparative tertiary stratigraphy of the Rhine Graben, Bresse Graben and Molasse basin: Correlation of Alpine foreland events. *Tectonophysics*, 300, 249–284.

Sommaruga, A. (1999). Decollement tectonics in the Jura foreland fold-and-thrust belt. *Mar. Petrol. Geol.*, 16, 111–134.

Sommaruga, A., Eichenbeger, U., Marillier, F. (2012). *Seismic Atlas of the Swiss Molasse Basin.* Swiss Geophysical Commission [Online]. Available at: https://www.geologieportal.ch/en/themes/fundamentals-of-geology/geophysics/seismic-atlas.html [Accessed 23 May 2022].

Spada, M., Bianchi, I., Kissling, E., Agostinetti, N.P., Wiemer, S. (2013). Combining controlled-source seismology and receiver function information to derive 3-D Moho topography for Italy. *Geophys. J. Int.*, 194, 1050–1068.

Speranza, F., Villa, I.M., Sagnotti, L., Florindo, F., Cosentino, D., Cipollari, P. (2002). Age of the Corsica-Sardinia rotation and Liguro-Provençal Basin spreading: New paleomagnetic and Ar/Ar evidence. *Tectonophysics,* 347, 231–251.

Spicher, A. (1980). *Tectonic Map of Switzerland.* Schweiz. Geol. Komm., Basel.

Spiegel, C., Kuhlemann, J., Dunkl, I., Frisch, W., von Eynatten, H., Balogh, K. (2000). The erosion history of the Central Alps: Evidence from zircon fission track data of the foreland basin sediments. *Terra Nova*, 12, 163–170.

Spiegel, C., Kuhlemann, J., Dunkl, I., Frisch, W. (2001). Paleogeography and catchment evolution in a mobile orogenic belt: The Central Alps in Oligo–Miocene times. *Tectonophysics*, 341, 33–47.

Spiegel, C., Siebel, W., Frisch, W., Berner, Z. (2002). Nd and Sr isotopic ratios and trace element geochemistry of epidote from the Swiss Molasse Basin as provenance indicators: Implications for the reconstruction of the exhumation history of the Central Alps. *Chem. Geol.*, 189, 231–250.

Spiegel, C., Siebel, W., Kuhlemann, J., Frisch, W. (2004). Toward a comprehensive provenance analysis: A multi-method approach and its implications for the evolution of the Central Alps. *Geol. Soc. Am. Spec. Paper*, 378, 37–50.

Stampfli, G.M. and Borel, G.D. (2002). A plate tectonic model for the Paleozoic and Mesozoic constrained by dynamic plate boundaries and restored synthetic oceanic isochrons. *Earth Planet. Sci. Lett.*, 196, 17–33.

Strunck, P. and Matter, A. (2002). Depositional evolution of the western Swiss Molasse. *Eclogae Geol. Helv*., 95, 197–222.

Stutenbecker, L., Tollan, P., Madella, A., Lanari, P. (2019). Miocene basement exhumation in the Central Alps recorded by detrital garnet geochemistry in foreland basin deposits. *Solid Earth,* 5, 1581–1595.

Tesauro, M., Kaban, M.K., Cloetingh, S.A.P.L. (2009). A new thermal and rheological model of the European lithosphere. *Tectonophysics*, 476, 478–495.

Tessier, B. and Gigot, P. (1989). A vertical record of different tidal cyclicities: An example from the Miocene Marine Molasse of Digne (Haute Provenace, France). *Sedimentology,* 36, 767–776.

Toscani, G., Bonini, L., Ahmad, M.I., Di Bucci, D., Di Giulio, A., Seno, S., Galuppo, C. (2014). Opposite verging chains sharing the same foreland: Kinematics and interactions through analogue models (Central Po Plain, Italy). *Tectonophysics,* 633, 268–282.

Turcotte, D.L. and Schubert, G. (1982). *Geodynamics: Application of Continuum Physics to Geological Problems*. Wiley, New York.

Ustaszewski, K., Schmid, S.M, Fügenschuh, B., Tischler, M., Kissling, E., Spakman, W. (2008). A map-view restoration of the Alpine-Carpathian-Dinaridic system for the Early Miocene. *Swiss J. Geosci.*, 101, 273–294.

Van der Boon, A., Beniest, A., Ciurej, A., Gaździcka, E., Grothe, A., Sachsenhofer, R.F., Langereis, C.G., Krijgsman, W. (2018). The Eocene-Oligocene transition in the North Alpine Foreland Basin and subsequent closure of a Paratethys gateway. *Global Planet. Change,* 162, 101–119.

Von Eynatten, H. and Wijbrans, J.R. (2003). Precise tracing of exhumation and provenance using $^{40}Ar/^{39}Ar$ geochronology of detrital white mica: The example of the Central Alps. In *Tracing Tectonic Deformation Using the Sedimentary Record,* McCann, T. and Saintot, A. (eds). Geol. Soc. London Spec. Publ. No. 208.

Von Eynatten, H., Schlunegger, F., Gaupp, R., Wijbrans, J.R. (1999). Exhumation of the Central Alps: Evidence from $^{40}Ar/^{39}Ar$ laserprobe dating of detrital white micas from the Swiss Molasse Basin. *Terra Nova,* 11, 284–289.

Waibel, A.F. (1990). Sedimentology, petrographic variability, and very-low-grade metamorphism of the Champsaur sandstone (Paleogene, Hautes- Alpes, France). Evolution of volcaniclastic foreland turbidites in the external Western Alps. PhD Thesis, University of Geneva.

Waschbusch, P.J. and Royden, L.H. (1992). Spatial and temporal evolution of foredeep basins: Lateral strength variations and inelastic yielding in continental lithosphere. *Basin Res.*, 4, 179–196.

Willenbring, J.K. and von Blanckenburg, F. (2010). Long-term stability of global erosion rates and weathering during late-Cenozoic cooling. *Nature,* 465, 211–214.

Willett, S.D. (2010). Late Neogene erosion of the Alps: A climate driver? *Annu. Rev. Earth Planet. Sci.*, 38, 411–437.

Willett, S.D. and Schlunegger, F. (2011). The last phase of deposition in the Swiss Molasse Basin: From foredeep to negative-alpha basin. *Basin Res.,* 22, 623–639.

Willett, S., Beaumont, C., Fullsack, P. (1993). Mechanical model for the tectonics of doubly vergent compressional orogens. *Geology*, 21, 371–374.

Willett, S.D., Schlunegger, F., Picotti, V. (2006). Messinian climate change and erosional destruction of the central European Alps. *Geology,* 34, 613–616.

Winterberg, S. and Willett, S.D. (2019). Greater Alpine river network evolution, interpretations based on novel drainage analysis. *Swiss J. Geosci.,* 112, 3–22.

Wolff, R., Hetzel, R., Dunkl, I., Anczkiewicz, A.A., Pomella, H. (2020). Fast cooling of normal-fault footwalls: Rapid fault slip or thermal relaxation? *Geology,* 4, 333–337.

Wolff, R., Hetzel, R., Dunkl, I., Anczkiewicz, A.A. (2021). New constraints on the exhumation history of the western Tauern Window (European Alps) from thermochronology, thermokinematic modeling, and topographic analysis. *Int. J. Earth Sci.,* 110, 2955–2977.

Zachos, J., Pagani, M., Sloan, L., Thomas, E., Billups, K. (2001). Trends, rhythms, and aberrations in global climate 65 Ma to present. *Science*, 292, 686–693.

Ziegler, P.A. and Fraefel, M. (2009) . Response of drainage systems to Neogene evolution of the Jura fold-thrust belt and Upper Rhein Graben. *Swiss J. Geosci.*, 102, 57–75.

Zilio, L.D., Kissling, E., Gerya, T., van Dinther, Y. (2020). Slab rollback orogeny model: A test of concept. *Geophys. Res. Lett.,* 18, e2020GL089917.

Zingg, A., Handy, M.R., Hunziker, J.C., Schmid, S.M. (1990). Tectonometamorphic history of the Ivrea Zone and its relationship to the evolution of the Southern Alps. *Tectonophysics*, 182, 169–192.

Zweigel, J., Aigner, T., Luterbacher, H. (1998). Eustatic versus tectonic controls on Alpine foreland basin fill: Sequence stratigraphy and subsidence analysis in the SE German Molasse. In *Cenzoic Foreland Basins of Western Europe,* Mascle, A., Puigdefabregas, C., Luterbacher, H.P., Fernandez, M. (eds). Geol. Soc. London Spec. Publ. No. 134.

4

The Detrital Thermochronological Record of Exhumation of the European Alps since the Early Oligocene

Matthias BERNET[1], Peter VAN DER BEEK[1,2], Pierre G. VALLA[1,3] and Christoph GLOTZBACH[4]

[1]CNRS, ISTerre, Université Grenoble Alpes, France

[2]Institute of Geosciences, University of Potsdam, Germany

[3]Institute of Geological Sciences & Oeschger Center for Climate Research, University of Bern, Switzerland

[4]Department of Geosciences, University of Tübingen, Germany

Low-temperature thermochronology of detrital minerals is a widely used tool for tracing the exhumation of crystalline rocks in convergent mountain belts. Detrital minerals are by definition the result of bedrock erosion and sediment transport and they are preserved in river sediments as well as modern and ancient clastic deposits in sedimentary basins adjacent to the evolving mountain belt. Detrital minerals obtained from sediments and sedimentary rocks provide a record of present-day and past exhumation rates on million-year time scales. In this respect, the European Alps, the Western and Central Alps in particular, are one of the most studied mountain belts in the world. Data are available from different thermochronological dating techniques on detrital minerals such as white mica $^{40}Ar/^{39}Ar$, as well as zircon

Geodynamics of the Alps 3,
coordinated by Claudio L. ROSENBERG and Nicolas BELLAHSEN.

and apatite fission-track. Here we review the currently available detrital thermochronological record of exhumation for the European Alps. Unfortunately, the same type and amount of data are not available for all Alpine regions, but nonetheless, they allow tracing and discussing temporal and spatial patterns of exhumation across the Alps. The higher-temperature white mica $^{40}Ar/^{39}Ar$ ages, if considered as cooling ages, depict phases of very rapid exhumation, on the order of 3 ± 1 km/Myr, in the Western Alps during the Early Oligocene, and somewhat slower exhumation of ~1 km/Myr in the Central Alps during the Late Oligocene–Early Miocene. Detrital zircon fission-track data indicate a relatively steady exhumation signal since the Late Oligocene for the Central Alps, whereas detrital apatite fission-track data emphasize regional differences in exhumation rates since at least Miocene times. The recently suggested regional increases in exhumation rates over the past 2 million years, as inferred from bedrock thermochronology data in the Western Alps, are not supported by the detrital record using low-temperature thermochronology.

4.1. Introduction

The European Alps are a convergent mountain belt that can be divided geographically into four main parts, the Western, Central, Southern and Eastern Alps (Figure 4.1; Pfiffner 2016). As outlined in the other chapters of this book, the evolution of the Alpine orogen was a long-term process, which started with the opening of the Alpine Tethys ocean basin during the Late Triassic to Early Jurassic and continued for millions of years after the Late Eocene to Oligocene continent–continent collision between the European margin and the Adriatic plate, the propagation of nappes, and the outward growth of the mountain belt until today. This evolution has been punctuated by tectonic events such as postulated breakoff of the European slab beneath the Central/Southern Alps and limited magmatism between about 42 and 30 Ma (e.g. Adamello, Bergell and Biella massifs, in Figure 4.1; von Blanckenburg and Davies 1995), or the reorientation of the convergence direction in the Western Alps during the Oligocene, resulting in the arc shape of the Western Alps (Figure 4.1; Dumont et al. 2012). Once the Alpine orogenic system had entered its main collisional phase at around 35 Ma, mountainous topography started to develop rapidly, driven by crustal thickening (e.g. Fauquette et al. 2015; Pfiffner 2016). As soon as topography grew above sea level, erosion set in, causing the removal of overburden and the exhumation of deep-seated metamorphic rocks towards the surface (Ring et al. 1999).

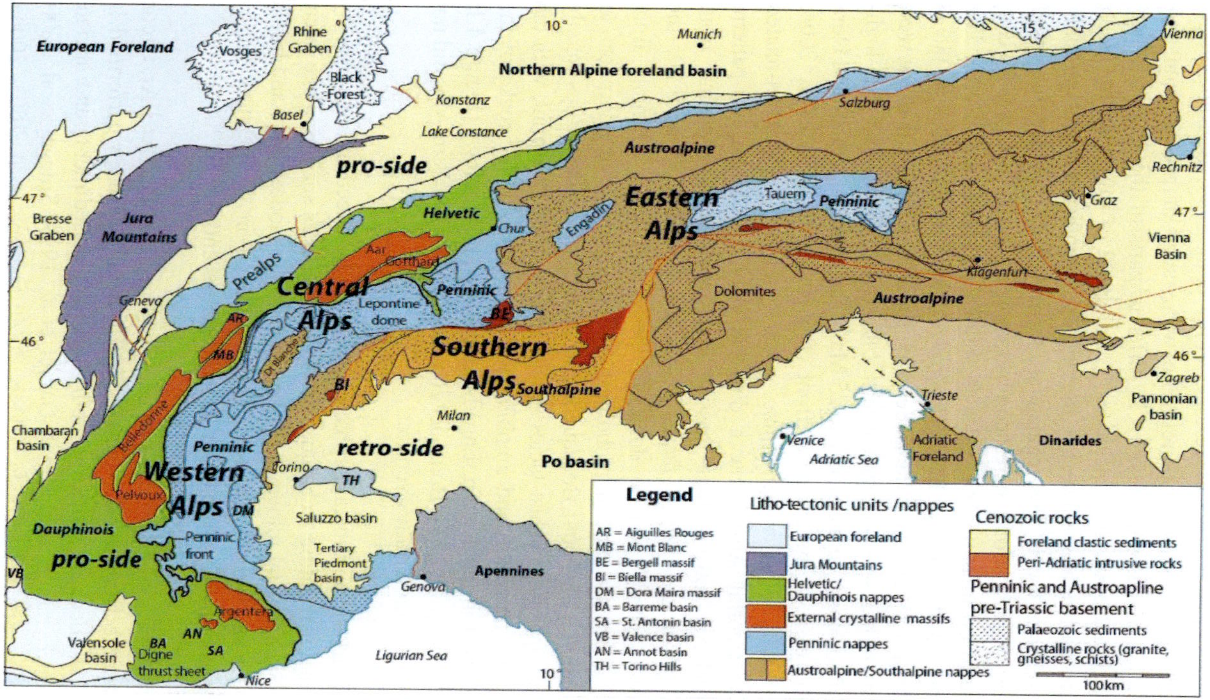

Figure 4.1. *Tectonic map of the Alps, modified from Pfiffner (2016). Shown are the main litho-tectonic units and nappes of the Western, Central and Eastern Alps. In the Western Alps, the Penninic nappes once covered the external crystalline massifs, and in the Central and internal Western Alps the Penninic nappes were once covered by the Austroalpine units, which are well preserved in the Eastern Alps*

As different parts of the Alps experienced different exhumation histories and styles, a variety of sedimentary cover units, plutonic rocks and metamorphic rocks are exposed throughout the Alps at the surface today (Frey et al. 1999; Bousquet et al. 2012). The exposed metamorphic rocks reflect a wide range of pre-collision and collisional metamorphic grades and histories in different parts of the mountain belt (Frey et al. 1999; Schmid et al. 2004; Chapter 7, Volume 2 of this book by Jacob et al.). Here we are particularly interested in the Alpine collisional metamorphic overprint, which is relevant for the low-temperature thermochronology data discussed below. In the Western Alps, a strong increase in metamorphic grade from west to east can be observed based on metamorphic index minerals (Bousquet et al. 2012). The Alpine metamorphic grade ranges from sub-greenschist and greenschist facies metamorphism at around 25–20 Ma in the Belledonne and Pelvoux massifs to the west of the Penninic front, to high-pressure greenschist, blueschist, eclogite and ultra-high-pressure facies metamorphism between 42 and 25 Ma of the Penninic nappes and Dora Maira massif to the east of the Penninic front (Figure 4.1; Schmid et al. 2004; Bousquet et al. 2012). This metamorphic pattern reflects subduction-related metamorphism. The Central Alps show a different Alpine metamorphic pattern. Whereas the external Aar and Gotthard massifs to the north of the Penninic front reached lower to upper greenschist facies conditions between 32 and 20, the metamorphic grade of the Lepontine dome to the south of the Penninic front experienced mainly Barrovian-type amphibolite to granulite facies metamorphism at around 35 Ma (Schmid et al. 2004; Bousquet et al. 2012). The Eastern Alps mainly show a very-low to upper greenschist facies Alpine metamorphic overprint, with the greenschist-amphibolite transition and amphibolite facies metamorphic rocks of the Penninic nappes exposed in the Tauern, Engadin and Rechnitz Windows (Figure 4.1; Frey et al. 1999; Schmid et al. 2004; Bousquet et al. 2012). The Southern Alps did not experience any Alpine metamorphic overprint.

Exhumation of plutonic and metamorphic bedrock can be traced through time by using thermochronological dating techniques on detrital minerals such as white mica, zircon and apatite. Such detrital minerals can be found in the modern river and beach sediments or ancient sandstones that have been deposited in sedimentary basins adjacent to the mountain belt (e.g. Garver et al. 1999; Bernet and Garver 2005).

Numerous detrital thermochronological studies with white mica $^{40}Ar/^{39}Ar$, zircon fission-track and apatite fission-track and (U-Th)/He data have been performed in the Alps over the past decades, targeting both modern river sediments and ancient sandstones (e.g. von Eynatten et al. 1999; Garver et al. 1999; Spiegel et al. 2000, 2001, 2007; Bernet et al. 2001, 2004a, 2004b, 2009; Carrapa et al. 2003, 2004a, 2004b, 2016; Kuhlemann et al. 2006; Morag et al. 2008; Malusà et al. 2009, 2016, 2017; Glotzbach et al. 2011, 2017; Bernet 2013; Jourdan et al. 2013, 2018;

Reiter et al. 2013; Mark et al. 2016). In some Alpine research projects, detrital thermochronology was combined with sediment-petrographic, heavy-mineral and geochemical analyses in order to trace both the exhumation of crustal rocks, their geological/geographical provenance (Malusà et al. 2009, 2016, 2017; Schwartz et al. 2012; Jourdan et al. 2012, 2013, 2018), and/or the evolution of the foreland basins (e.g. Schegg and Leu 1998; Cederbom et al. 2004, 2011; Mazurek et al. 2006). Recent developments in LA-ICP-MS analysis provide the possibility of double- and triple-dating of single apatite and zircon grains by combining fission-track, (U-Th)/He and U-Pb dating, which has greatly improved the opportunity for obtaining more detailed insights into the evolution of mountain belts such as the Alps (e.g. Jourdan et al. 2013; Mark et al. 2016) or the Andes (Carrapa et al. 2009).

In this chapter, we outline some basic principles of detrital thermochronology and focus on reviewing the detrital thermochronology data currently available for the Alps. We discuss the detrital record of Alpine exhumation since the Early Oligocene, the time when the first sediments derived from the emerging doubly-vergent Alpine orogenic wedge were deposited in the evolving foreland basins on both sides of the mountain belt (e.g. Sinclair 1997a, 1997b; Ford et al. 1999; Kuhlemann et al. 2006).

4.2. Principles of detrital thermochronology

The term "detrital thermochronology" generally refers to the analysis of detrital apatite and zircon grains with the fission-track and (U-Th)/He methods and of detrital white mica with $^{40}Ar/^{39}Ar$ dating, although higher-temperature methods such as U-Pb dating of detrital apatite and rutile are becoming more common (e.g. Carrapa et al. 2009; Mark et al. 2016; Govin et al. 2020). Apatite (U-Th)/He studies are mainly aimed at studying basin evolution and they are not further considered here, and we will mainly focus on fission-track dating, as detrital apatite and zircon fission-track data present the largest detrital thermochronological datasets available for the Alps today. The goal of thermochronological analyses of modern river sediment or ancient sandstone deposits is to determine (A) the long-term cooling history of bedrock exposed in the mountain belt, which can be translated into the timing and rates of exhumation allowing for a number of assumptions, (B) the determination of sediment provenance within the orogenic drainage system, with implications for the paleogeographic evolution of the area, and (C) the thermal evolution of the sedimentary basins adjacent to the mountain belt, as these basins may have been partially inverted when incorporated into the outward-propagating orogenic wedge through thrust faulting. Figure 4.2(A) exemplarily for all other thermochronological systems schematically summarizes how detrital fission-track thermochronology is used for estimating rates of exhumation with the lag-time concept. Lag time is defined as the difference between the apparent mineral cooling age and the time of sediment deposition (e.g. Garver et al. 1999). Following England and Molnar (1990), exhumation is here defined as the movement of deep-seated crustal rocks toward the topographic surface. Rock uplift is the upward vertical movement of a rock particle with respect to a reference point, such as the geoid or sea level. Surface uplift in turn is the increase in (average) surface elevation with respect to a fixed reference. Surface uplift is equal to rock uplift minus exhumation (England and Molnar 1990). In general, the exhumation of upper crustal rocks in convergent mountain belts is driven by erosion and normal faulting (Ring et al. 1999). Rocks cool during exhumation, and this is why thermochronological dating techniques provide quantitative information on the rates and timing of exhumation. The time for sediment transport from mountainous massifs into main rivers or peripheral basins, such as the European Alps, is generally considered negligible when dealing with low-temperature thermochronometers, as the transport time is commonly shorter than the error range of the apparent mineral cooling ages linked to the dating system-specific closure temperature and cooling rate (Figure 4.2(B); Bernet et al. 2004a, 2004b).

When analyzing detrital samples, it is recommended to analyze about 100 individual grains per sample (e.g. Vermeesch 2004), in order to capture the full grain-age spectrum of the bedrock exposed within the drainage area. Unfortunately, this is not always possible when working with detrital apatite and zircon because of potentially limited availability of sample material, the limited yield of these minerals, and/or complications such as fractures, strong uranium zonation, inclusions, accumulated radiation damage, etc., that render individual grains which are not datable. Malusà et al. (2013, 2016) and Malusà and Fitzgerald (2020) have described in detail the potential bias that may be introduced in detrital geo- and thermochronology studies because of lithological bias (i.e. mineral fertility) in the sediment source areas. Even more important than the fertility alone is the mineral yield, which is a function of source-rock mineral fertility, surface-area distribution of the source rock and the respective erosion rates of these lithologies, as outlined in Bernet et al. (2004a, 2004b). Another challenge is the potential detection bias, particularly in detrital zircon fission-track studies, as fission tracks cannot be etched equally in all zircon grains and the risk exists that certain source areas may be missed or over-represented. Particularly zircons with very old apparent cooling ages (>100 Ma), high amounts of accumulated radiation damage, and very high track densities can be difficult or impossible to analyze (Bernet and Garver 2005). In contrast, apatite and zircon with very young apparent cooling ages (<1 Ma for apatite and 4 Ma for zircon) and low uranium concentrations may be challenging to analyze because of a lack of fission tracks (e.g. Huyghe et al. 2020). Nonetheless, in a mountain belt such as the Alps, characterized by moderate exhumation rates and by detrital apatite and zircon derived from sedimentary, meta-sedimentary, or igneous rocks that all contain apatite and zircon with a wide range of uranium concentrations, such detection problems are negligible. Using a multi-mount etching technique for zircon fission-track dating, the detrital grain-age distributions faithfully represent the range of bedrock cooling ages in the drainage areas (Garver et al. 1999; Bernet et al. 2004a, 2004b; Malusà et al. 2013). As etching procedures are standardized for apatite fission-track dating (Donelick et al. 2005), a single grain mount is sufficient in detrital apatite fission-track studies to reflect the bedrock age distribution (e.g. Bermúdez et al. 2013).

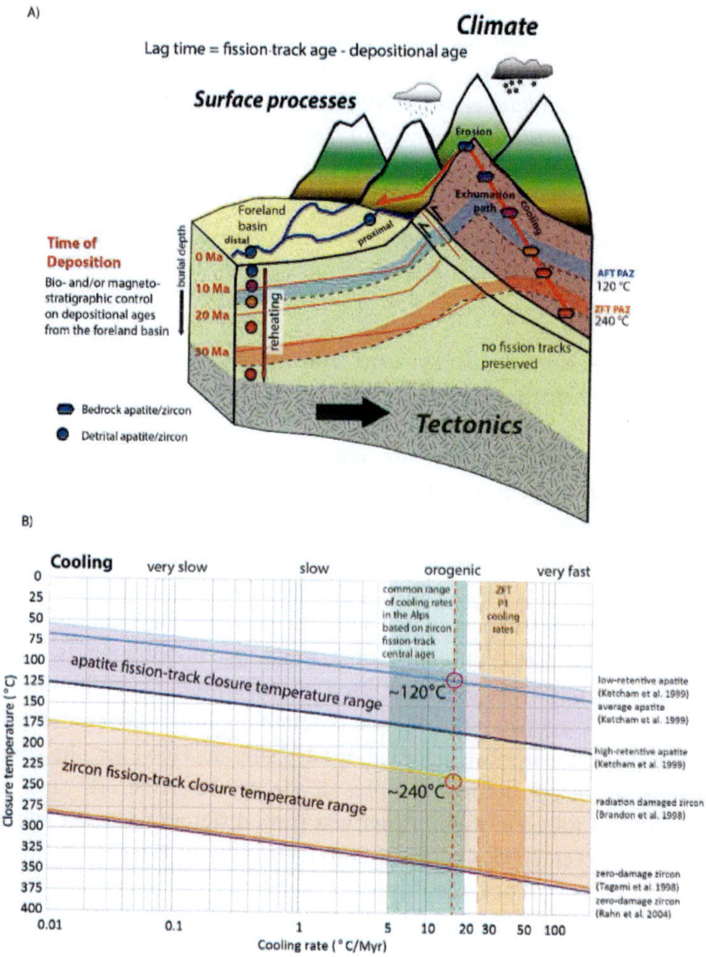

Figure 4.2. *Schematic diagram of detrital thermochronology and the lag-time concept, and graph of apatite and zircon fission-track closure temperature ranges*

COMMENTS ON FIGURE 4.2.– *(A) Schematic diagram of detrital thermochronology and the lag-time concept. Lag time is defined as the difference between the thermochronological age and the time of deposition. The colors of the bedrock (diamonds) and detrital (circles) grains symbolize the cooling during exhumation or reheating during burial. Before apatites and zircons with orogenic cooling ages are exposed at the surface, the rocks of the cover units containing non- and or partially annealed apatite or zircon need to be removed by erosion and/or normal faulting. (B) Apatite and zircon fission-track closure temperature ranges in relation to cooling rate, chemical composition (apatite) and radiation damage (zircon). The closure temperature can be applied in the case of monotonic cooling since peak temperatures are higher than the closure temperature. The closure temperatures given here are based on annealing models published by Brandon et al. (1998), Tagami et al. (1998), Ketcham et al. (1999), and Rahn et al. (2004). The red dashed line indicates an orogenic cooling rate of 15°C/Myr, common for the European Alps, resulting in an apatite fission-track closure temperature of ~120°C for apatite of average chemical composition (Brandon et al. 1998), and a zircon fission-track closure temperature of ~240°C for natural radiation–damaged zircons (Hurford 1986; Brandon et al. 1998; Bernet 2009).*

4.2.1. *Single-age distributions and mean age estimates of detrital thermochronology data*

Detrital thermochronology data are commonly presented in radial plots for fission-track data and frequency plots (either Kernel density estimate (KDE) plots or probability density (PD) plots) for both $^{40}Ar/^{39}Ar$ and fission-track data (Figure 4.3; e.g. Vermeesch 2019). In bedrock studies, the fission-track age is generally reported as either a pooled age or a central age (e.g. Galbraith and Laslett 1993; Galbraith 2005; Vermeesch 2019). The pooled age basically pools all track counts into a single age estimate, whereas the central age of a grain-age distribution is an iterative estimate of the mean age, for over-dispersed age distributions (Galbraith 2005). If the single-grain age dispersion of a sample is relatively low (<20%), the central-age and the pooled-age estimates are identical or very similar. However, the single-grain age dispersion of many detrital samples tends to be large, as such samples commonly contain a mixture of grains derived from different sediment source rocks (e.g. Bernet and Garver 2005; Galbraith 2005; Malusà and Fitzgerald 2020). In the case of such large fission-track age dispersion (>>20%), the central and pooled ages may differ and do not have a direct geological significance as reflecting a specific thermal or tectonic event. However, the central age can still be used for obtaining the first-order estimate of average drainage-basin exhumation rates (Bernet et al. 2009). For detrital white mica $^{40}Ar/^{39}Ar$ data commonly with no mean age estimates are determined but the data are interpreted in terms of different peak ages. This is because, due to the higher precision of individual grain-age measurements in this technique, detrital mica $^{40}Ar/^{39}Ar$ age distributions are practically always very significantly overdispersed.

4.2.2. *Peak age, minimum age and youngest age*

As an example of how detrital thermochronology data is generally presented, the combined dataset of the Rhône delta detrital zircon fission-track grain-age distribution from Bernet et al. (2004a) is presented in Figure 4.3 in radial, KDE and PD plots. The figure shows that the grain ages tend to cluster and form more or less distinct age populations which can be visually identified. However, visual estimates of population ages hinge on the type of scale selected for the plot, for example, a linear versus logarithmic scale for the KDE plots in Figure 4.3(B), and are not commonly used. In order to obtain a more robust estimate of the different age populations that may be detectable, it is common practice in fission-track studies to decompose the observed grain-age distribution into major grain-age components or peaks. Gaussian and (more commonly) binomial peak fitting approaches have been widely used for this in the past (e.g. Galbraith and Green 1990; Brandon 1992, 1996; Stewart and Brandon 2004). If the age distribution of a detrital sample contains different populations, then the peaks are commonly numbered from the youngest peak (P1) to the older peaks (P2, P3, P4, etc.). Fundamental concerns in peak fitting include how many peaks should be fitted to obtain a best-fit solution and whether these age peaks are geologically meaningful. Commonly used software packages such as Binomfit of Brandon (see Ehlers et al. 2005) or RadialPlotter of Vermeesch (2009), provide best-fit solutions, based on F-test comparison between different peak-fit solutions for determining the optimal data fit. However, these programs also provide the possibility to manually fit a certain number of peaks to a dataset. In general, peak-fitting results have to be treated with caution, as they depend to some degree on the number of grains analyzed (Vermeesch 2019), and the detection of peak ages may be biased if peaks overlap and are not well separated (Naylor et al. 2015). This is shown in Figures 4.3(C) and (D), where two different peak-fit solutions for the Rhône delta zircon fission-track dataset of Bernet et al. (2004a) are presented, using the Binomfit program of Brandon (see Ehlers et al. 2005). For

Figure 4.3(C), the Binomfit program was forced to fit two peaks to the observed grain-age distribution. In this peak-fit solution, the two peaks are well separated and do not overlap. Figure 4.3(D) shows the preferred best-fit solution proposed by the Binomfit program for the Rhône delta dataset. For this solution, a slight overlap between different peaks can be observed. The interpretation of peak ages in terms of exhumation rates may be biased in cases where the individual peaks largely overlap, as such peaks will be very difficult to separate (Naylor et al. 2015). Low track counts also render the decomposition into age peaks less reliable, complicating the interpretation (e.g. Naylor et al. 2015; Govin et al. 2020; Huyghe et al. 2020). Given the generally higher U concentrations of zircon in comparison to apatite, and the resulting more precise track-counting statistics, fission-track ages, here usually given with a 2-sigma uncertainty, tend to be better defined in detrital zircon fission-track data than in detrital apatite fission-track data. In the end, it is at the discretion of the analyst to decide if the peak-fit results are geologically meaningful and can be interpreted or not.

An alternative to the peak-fitting approach is to use the minimum-age method, developed by Galbraith and Laslett (1993) to determine the youngest age population present in an aged mixture. The advantage of the minimum-age approach is that it is independent of the number of analyzed grains (Vermeesch 2019). As can be seen from the Alps detrital zircon fission-track dataset of Bernet et al. (2001, 2009) in Figure 4.4, the P1 age estimates from best-fit solutions and the minimum age estimates overlap in their 2-sigma error ranges and are with the exception of outlier sample AP54 basically identical. It is therefore not too important which approach is used if the objective is to determine the age of the youngest age population. Many fission-track studies focus on the P1 and P2 age peaks (e.g. Bernet et al. 2001, 2004a, 2004b, 2009; Bermúdez et al. 2013; Jourdan et al. 2013) or the minimum age (e.g. Jourdan et al. 2018; Govin et al. 2020; Huyghe et al. 2020) for determining exhumation rates of the most rapidly exhuming sediment source areas in a drainage basin. In contrast, in $^{40}Ar/^{39}Ar$ studies commonly only the youngest single-grain age is used, which does not require any statistical data treatment. However, a potential problem with white mica $^{40}Ar/^{39}Ar$ ages is the possibility that they reflect metamorphic crystallization instead of cooling (Malusà and Fitzgerald 2020), which cannot be used for obtaining exhumation rates.

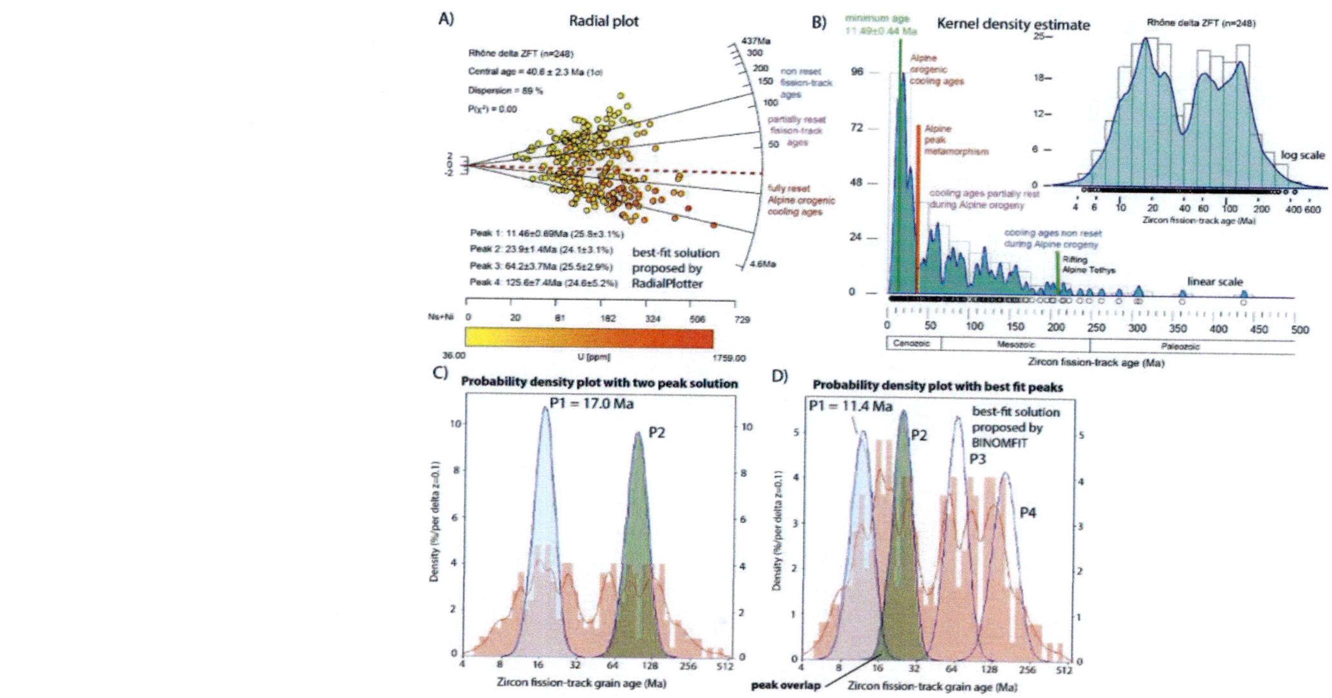

Figure 4.3. *Example of (A) radial plot and (B) Kernel density estimate of the combined Rhône river delta detrital zircon fission-track dataset from Bernet et al. (2004a). The radial plot shows also the minimum age estimate. Both types of graphs were made with the RadialPlotter program of Vermeesch (2009). (C) and (D) show the same Rhône river delta dataset, plotted using the Binomfit program (Brandon in Ehlers et al. 2005), once with a 2-peak solution (C) and once with the best-fit 4-peak solution (D). Fitting more peaks commonly results in a younger P1 age estimate and may cause peak overlap*

4.2.3. *Estimating exhumation rates from detrital thermochronology data*

Applying the lag-time concept outlined above, fission-track peak ages can be used for determining exhumation rates, if the time of deposition of the detrital apatite or zircon grains is reasonably well constrained (about ±1 Myr). Furthermore, fission-track peak ages can be used for provenance analysis, if they can be compared to a sufficiently dense bedrock dataset (e.g. Bernet et al. 2004a, 2004b).

Different approaches exist for obtaining estimates of exhumation rates and magnitude from peak/minimum/central or youngest ages. The simplest approach (Garver et al. 1999) assumes a constant geothermal gradient and effective closure temperature for the different analyzed thermochronological systems (Figure 4.2(B)); dividing the closure temperature (minus the surface temperature) by the geothermal gradient yields a first-order estimate of the amount of exhumation; dividing this by the thermochronological age yields a simple average exhumation rate. However, this back-of-the-envelope type of calculation does not consider the thermal advection effect of exhumation, the kinetics of annealing/diffusion (i.e. the dependence of closure temperature on cooling rate, grain size, chemistry, radiation damage, etc.), or any transient effects. The 1D model Age2edot was developed by Brandon (see Ehlers et al. 2005; Reiners and Brandon 2006; Willett and Brandon 2013) allows incorporating simplified annealing kinetics and thermal advection but supposes a thermal steady state. Transient effects and more complete annealing/diffusion models are incorporated in numerical codes such as QTQT (Gallagher and Parra 2020), or Pecube (Braun 2003; Braun et al. 2012), which have been used to predict detrital thermochronology ages both as a 1D model (e.g. Glotzbach et al. 2011a; Govin et al. 2020) and as a full 3D model considering laterally varying exhumation rates and topographic effects on the cooling-age distribution in source areas (Whipp et al. 2009). For a more extensive treatment of low-temperature thermochronology in general and detrital thermochronology and its applications to studying mountain building, exhumation and basin evolution in particular, the interested reader is referred to Bernet and Spiegel (2004), Reiners and Ehlers (2005), Reiners and Brandon (2006) and Malusà and Fitzgerald (2019, 2020).

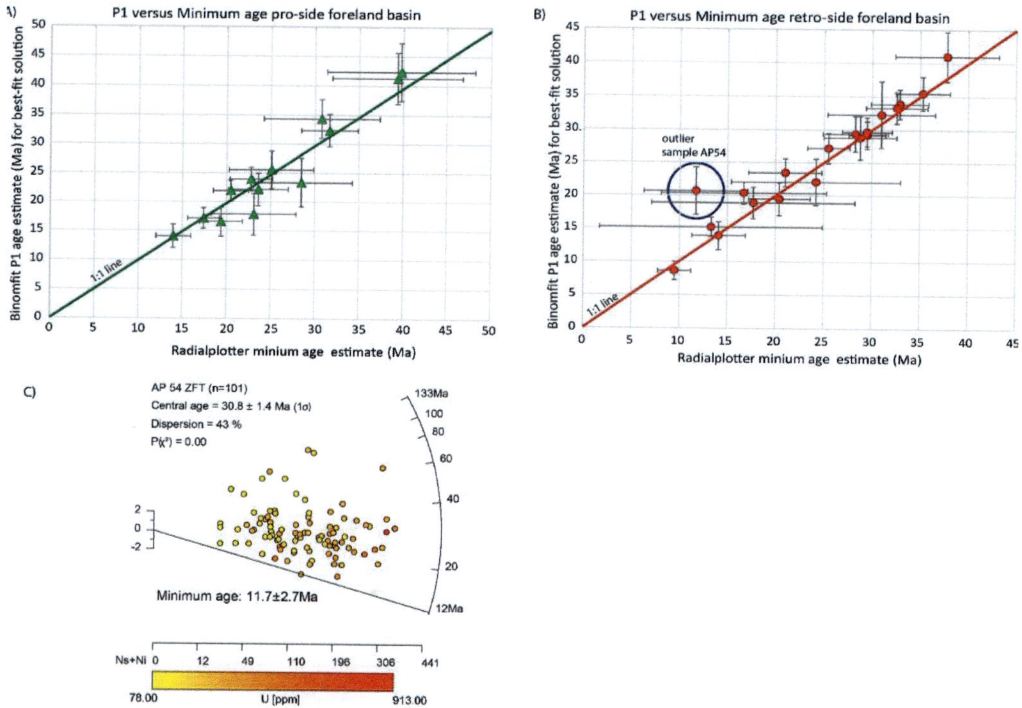

Figure 4.4. *Detrital zircon fission-track P1 ages from Binomfit best-fit solutions vs. RadialPlotter minimum age estimates of the Bernet et al. (2001, 2009) dataset from the Alps pro- (A) and retro-side (B) foreland basins. (C) shows the details of sample AP54 which appears as an outlier in (B); having the minimum age estimate based on one single grain*

4.3. The modern-river signal of Alpine exhumation

4.3.1. *Detrital white mica $^{40}Ar/^{39}Ar$ data*

$^{40}Ar/^{39}Ar$ ages of detrital white mica from modern river sands were published by Carrapa et al. (2003) for rivers draining the southern Western Alpine arc towards the Po basin. Considering this $^{40}Ar/^{39}Ar$ age as cooling ages and not as metamorphic crystallization ages, the youngest single-grain white mica $^{40}Ar/^{39}Ar$ ages observed in these river samples were 36–40 Ma, with visually estimated young peaks at 38 Ma (Carrapa et al. 2003). A 38 Ma white mica $^{40}Ar/^{39}Ar$ age from a modern river deposit is compatible with long-term average exhumation rates of ~0.3 km/Myr. The 38 Ma micas were possibly derived from crystalline basement rocks of the internal Western Alps, which were exhumed during the Late Eocene–Oligocene (Carrapa et al. 2003, 2004, 2016; Beucher et al. 2012; Malusà et al. 2015). Older white mica cooling ages of ~260–350 Ma determined for some of these rivers indicate <0.1 km/Myr long-term exhumation rates in the southernmost internal Western Alps.

For the Eastern Alps, Gemignani et al. (2017) published a study with 19 modern-river samples, using white mica and biotite $^{40}Ar/^{39}Ar$ dating. Their results showed age clusters at ~0.5–50 Ma, ~60–120 Ma, and ~250–350 Ma. For the Tauern Window, they determined exhumation rates on the order of 0.3–0.8 km/Myr since the Miocene, with the more rapid exhumation of the western than the eastern Tauern Window (Figure 4.1).

4.3.2. *Detrital zircon fission-track data*

The majority of the available modern-river detrital thermochronology data for the Alps are zircon fission-track data that were obtained from rivers draining both the pro- and retro-side of the mountain belt (Figure 4.5; data from Bernet et al. 2001, 2004a, 2004b). The single-grain age distribution of each river sample shows a large range, overall from <7 Ma to >400 Ma, which reflects both the various sediment source areas and levels of exhumation in the different parts of the Alps. Figure 4.6 shows radial plots for three of the main rivers draining the Alps, the Rhône, the Rhine and the Po. The Rhône delta dataset is a composite of three samples collected along the beachfront of the Rhône delta plain in southern France. The Rhine river data are from a single sample collected in the Rhine river estuary near Rotterdam (the Netherlands) and the Italian Po river dataset comprises the combined data of the Alpine retro-side river drainages shown in Figure 4.5. Bernet (2013) showed with the example of the Isère river drainage that such a combination of detrital zircon fission-track data from different tributaries is largely comparable to the age distribution observed further downstream of the same drainage system, despite potential lithological, fertility, erosion rate and sediment transport biases, which may all affect zircon yield in any given river drainage. The three radial plots show that the signal of Alpine exhumation with <35 Ma zircon fission-track cooling ages remains detectable in both the pro- and retro-side drainages even 500 to >1000 km downstream from the source areas, even though the signal may become diluted further from the source (Bernet et al. 2004a). Gemignani et al. (2018) reached similar conclusions from a study of the Brahmaputra river draining the eastern Himalayas. Exhumation-rate estimates calculated from the minimum and central ages of the Alpine river samples, using the simple approach of Garver et al. (1999), are summarized in Table 4.1. The most rapid exhumation rates, based on P1/minimum-age estimates, are on the order of 0.4–0.9 km/Myr and are detected in rivers draining the Lepontine dome in the Central Alps (Rhône, Rhine, Ticino, Adda rivers, Figure 4.5 and Table 4.1). Long-term average exhumation rates for the Alps are between 0.1 and 0.4 km/Myr, based on the central ages of all available modern river detrital zircon data (Figure 4.6(D)).

4.3.3. *Detrital apatite fission-track data*

For the pro-side of the Alps, Glotzbach et al. (2011a) published data for the Isère and Drac rivers, near Grenoble in the Western Alps. The P1 ages of these two rivers are respectively 6.4 ± 1.0 Ma (66% of dated grains) and 6.7 ± 1.4 Ma (28% of dated grains). Again, using the simple approach of Garver et al. (1999) and assuming an apatite fission-track closure temperature of 120°C (Figure 4.2(B)), a surface temperature of 10°C, and a geothermal gradient of 30°C/km, such P1 ages correspond roughly to exhumation rates of 0.5–0.6 km/Myr for the most rapidly exhuming rocks within these drainage areas in the Western Alps. Central ages of the Isère and Drac river samples are 12.1 ± 2.2 and 10.7 ± 1.0 Ma respectively, which correspond to average long-term exhumation rates of 0.3–0.4 km/Myr (Glotzbach et al. 2011a).

Reiter et al. (2013) published detrital apatite fission-track data for the Aare and Reuss rivers that drain the Central Alps toward the pro-side. For the Aare river, the P1 age is equal to the central age of 10.9 ± 1.6 Ma, as 100% of the dated grains fall into the P1. This corresponds to an exhumation rate of 0.3 km/Myr. In contrast, for the Reuss river, the P1 age is 4.5 ± 4.8 Ma (5% of dated grains), which indicates exhumation at a rate of 0.8 km/Myr. The central age of the Reuss river sample is 10.9 ± 1.8 and reflects a drainage basin average exhumation rate of ~0.3 km/Myr.

For the Rhine river, Glotzbach et al. (2017) provided detrital apatite fission-track ages from samples collected far from the Alps in the Upper Rhine Graben and downstream of Cologne, but which were nonetheless dominated by Alpine P1 cooling ages of 7.7 ± 2.7 Ma (60% of dated grains) and 8.9 ± 1.4 Ma (67% of dated grains), respectively. Such P1 ages correspond to exhumation rates of about 0.4–0.5 km/Myr. No central ages were provided.

River	Lithology	Sample No.	Deposition (Ma)	Minimum age (Ma)	2 SE (Myr)	Exhumation rate (km/Myr)	Central age (Ma)	2 SE (Myr)	Exhumation rate (km/Myr)
Rhine River drainage									
Aare River	river sediments	99MB36	0	16.28	1.2	0.47	20.1	3.4	0.38
Reuss River	river sediments	00MB44	0	17.2	1.12	0.45	20	2.4	0.38
Rhine River near Buchs	river sediments	99MB37	0	19.4	2.4	0.4	25.3	3.6	0.3
Rhine River near Worms	river sediments	99MB40	0	16.9	2.4	0.46	37.7	8.4	0.2
Rhine River estuary	river sediments	99MB41	0	24.5	3.6	0.31	87.9	16.8	0.09
Inn River	river sediments	99MB39	0	23.1	4.8	0.33	43.5	5.4	0.18
Rhône River drainage									
Rhône River	river sediments	99MB11	0	14.63	1.12	0.52	26.4	3.8	0.29
Isère River	river sediments	99MB12	0	14	2	0.55	35.7	5	0.21
Ardeche River	river sediments	99MB15	0	34.8	5.4	0.22	94.3	11	0.08
Durance River	river sediments	99MB20	0	37.7	5.6	0.2	77.3	7.2	0.1
Rhône Delta	beach sediments	99MB22+23+24	0	11.49	0.88	0.64	40.6	4.6	0.19
Rhône drainage	river sediments	merged data	0	16.59	0.94	0.46	50.6	4	0.15
Po River drainage									
Dora Riparia River	river sediments	98MB01	0	34	6	0.23	71	24	0.11
Orco River	river sediments	99MB33	0	31.6	2.6	0.24	31.7	2.4	0.24
Dora Baltea River	river sediments	99MB34	0	17.2	2.6	0.45	27.1	4.4	0.28
Sesia River	river sediments	99MB35	0	34	2	0.23	35.65	1.9	0.22
Ticino River	river sediments	98MB02	0	9.5	1.72	0.81	20.4	3.2	0.38
Adda River	river sediments	98MB03	0	16.8	2	0.46	36.1	4.4	0.21
Adige River	river sediments	98MB04	0	13.73	1.88	0.56	32	3.6	0.24
Brenta River	river sediments	98MB05	0	33	3.8	0.23	83.1	8.8	0.09
Piave River	river sediments	98MB06	0	54.4	9.6	0.14	93.4	8.6	0.08
Po River drainage	river sediments	merged data	0	14.02	0.62	0.55	41.3	2.4	0.18

Note: Data from Bernet et al. 2004a and 2004b. Minimum and central ages calculated with the RadialPlotter program of Vermeesch (2009).

Table 4.1. *Modern river detrital zircon fission-track minimum and central ages*

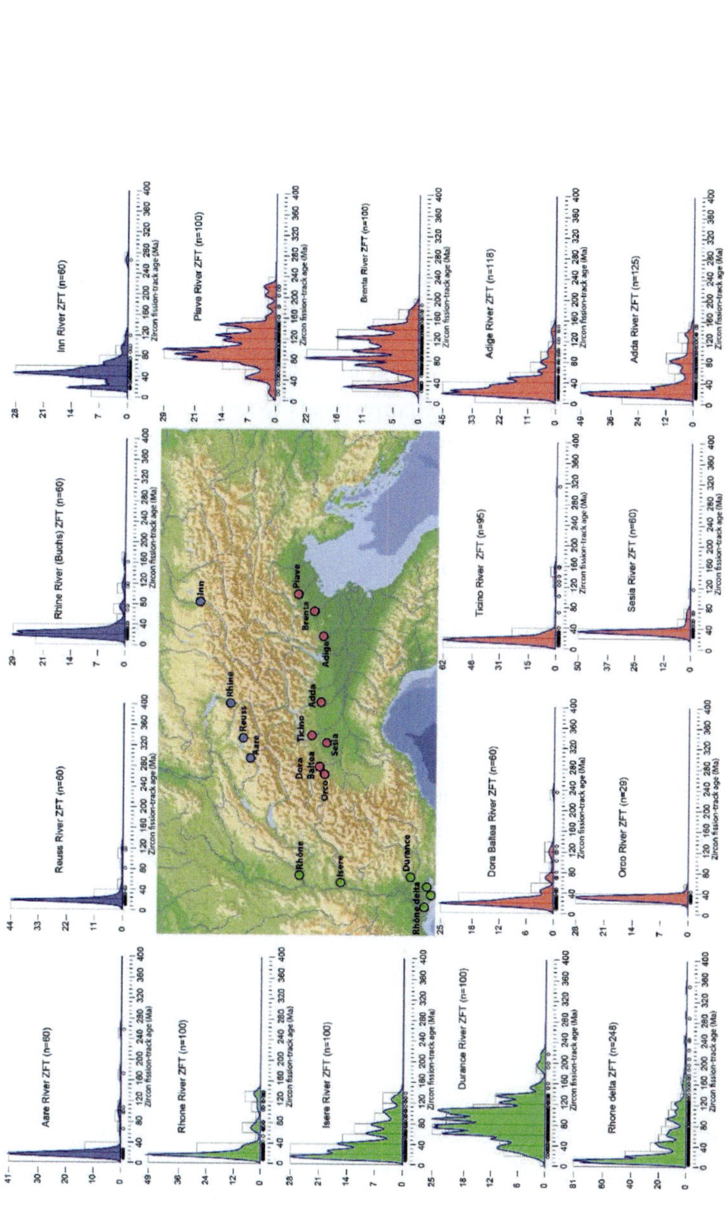

Figure 4.5. *Map of the present-day Alpine drainage system with detrital zircon fission-track grain-age distributions (KDE plots made with RadialPlotter; Vermeesch 2009). The plots were made using the modern river data of Bernet et al. (2004a, 2004b)*

For the retro-side Po river drainage system, the detrital apatite fission-track dataset is more complete. The first set of such data was published by Garver et al. (1999) for the small drainage of the Cordera river in the Bergell massif (Central Alps). The P1 age of 14.2 ± 1.4 Ma corresponds to an exhumation rate of ~0.3 km/Myr (Garver et al. 1999). The authors did not provide a central age for this sample.

Malusà et al. (2017) summarized the apatite fission-track data from Malusà et al. (2009), Garzanti et al. (2012), Malusà and Balestrieri (2012) and Resentini and Malusà (2012) which cover the entire Po river drainage system. The detrital apatite fission-track ages observed in the Po river delta show a wide age range with a P1 age of about 8 Ma (Malusà et al. 2009). Based on ε_{Nd} analyses of the same apatite grains, the Lepontine dome in the Central Alps was identified as the principal source area for the P1 age apatites. An 8 Ma P1 age corresponds to an exhumation rate of about 0.5 km/Myr. Unfortunately, no central-age data were provided by Malusà et al. (2009, 2017) for estimating average source-area exhumation rates.

4.4. The pro-and retro-side foreland basin record of Alpine exhumation

We now turn to the detrital thermochronological data of ancient deposits in the pro- and retro-side foreland basins of the Alps. The Northern Alpine foreland basin (Switzerland, Germany and Austria) is still largely preserved, even though parts of the Late Eocene to Late Miocene deposits were incorporated into the propagating orogenic wedge by thrusting. Thus, basin inversion provides ample access for sampling foreland basin deposits for detrital thermochronology studies (e.g. von Eynatten et al. 1999; Spiegel et al. 2000, 2001; Cederbom et al. 2004, 2011; von Hagke et al. 2012). The depositional ages of the foreland basin deposits in Switzerland are well constrained by bio- and magneto-stratigraphic data (e.g. Schlunegger 1999; Schlunegger et al. 1998, 2001). The Eocene to Oligocene foreland basin deposits in southeastern France, in contrast, are only preserved in remnants such as the Barrême basin (e.g. Ford et al. 1999). On the retro-side, Early Oligocene to Miocene basin sediments are preserved in the Tertiary Piedmont basin, Saluzzo basin and the Torino Hills to the east of the Western Alpine arc (Figure 4.1, e.g. Carrapa et al. 2003, 2016; Jourdan et al. 2013). Along the northern edge of the Po basin, between the southern ends of Lakes Maggiore and Como, outcrops of the Gonfolite Lombarda Group allow access to Oligocene–Miocene sedimentary rocks (e.g. Spiegel et al. 2001; Zattin et al. 2003). Finally, turbidite deposits exposed in the Apennines also provide access to the Oligocene to Pliocene exhumation record of the Central Alps (Bernet et al. 2001, 2009).

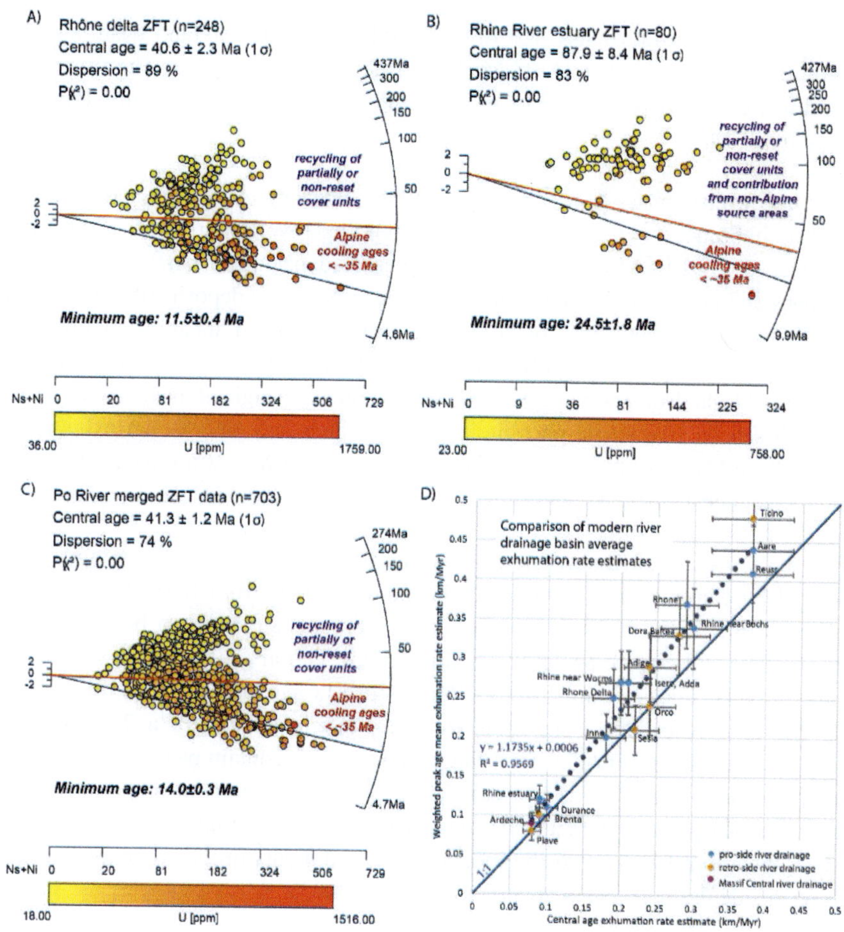

Figure 4.6. *Radial plots with minimum age estimates and central ages of A) the Rhône river delta, B) the Rhine estuary, and C) the merged Po river drainage zircon fission-track dataset (Bernet et al. 2004a, 2004b). D) Comparison of drainage-basin average exhumation rates based on detrital zircon fission-track central ages, with weighted-average exhumation rates, based on individual peak ages weighted by the percentage of grains belonging to each peak (data from Bernet et al. 2004a, 2004b) for the available modern-river sediment data*

4.4.1. *Detrital white mica $^{40}Ar/^{39}Ar$ data*

On the pro-side, the first detrital white mica $^{40}Ar/^{39}Ar$ data from the Northern Alpine foreland basin were published by von Eynatten et al. (1996, 1999). Samples collected from Oligocene (31 Ma) to Miocene (14 Ma) sedimentary rocks of the

Honneg-Napf fan in the Swiss part of the Northern Alpine foreland basin showed that white micas deposited at 31–25 Ma have Paleozoic cooling ages. The first micas with Alpine cooling ages started to appear at about 21 Ma, becoming the dominant fraction by 14 Ma.

Detrital white mica $^{40}Ar/^{39}Ar$ data for the Western Alps foreland basin in southeastern France are limited to the data reported by Morag et al. (2008) and Carrapa et al. (2016). Morag et al. (2008) presented phengite $^{40}Ar/^{39}Ar$ ages from sandstones and pebbles collected from Oligocene to Miocene deposits of the Barrême basin (Figure 4.1). About 95% of all analyzed micas showed apparent cooling ages between ~224 and 361 Ma. Only one blueschist-garnet pebble from the Clumanc Formation (29 Ma deposition age; Callec 2001) yielded a weighted mean age of 34.1 ± 3.4 Ma, based on five single-phengite analyses, which may indicate an exhumation rate of >2 km/Myr, given the rather short lag time of ~5 Myr. Carrapa et al. (2016) found the youngest $^{40}Ar/^{39}Ar$ ages of 51–251 Ma for the Barrême basin deposits.

On the retro-side of the Western Alps, detrital white mica $^{40}Ar/^{39}Ar$ ages were presented by Carrapa et al. (2003, 2004a, 2004b, 2016) from Early Oligocene to Miocene sedimentary rocks of the Tertiary Piedmont and Saluzzo basins (Figure 4.1). The Early Oligocene Molare Formation in the Tertiary Piedmont basin yielded among 39 dated micas a single mica with a 31.6 Ma age, and the Late Oligocene–Early Miocene Rocchetta Formation yielded a single mica with a 22.7 Ma age among 16 dated micas. Furthermore, these samples also contain peaks at 37 ± 1 Ma, based on visual estimation, as no peak fitting was performed (Carrapa et al. 2003). The 37 ± 1 Ma peak is in fact persistent throughout the stratigraphic record of the Tertiary Piedmont basin up to the modern river sediments in the southern Western Alps retro-side drainage basin area (Carrapa et al. 2003).

4.4.2. *Detrital zircon fission-track data*

The detrital zircon fission-track sample locations from the pro-side basins are regionally more dispersed than for the retro-side basins (Tables 4.2 and 4.3). For example, in southeastern France, detrital zircon fission-track data are available from the Annot, Barrême and Saint Antonin basin remnants and the Miocene Valence and Chambaran basin deposits (Bernet et al. 2009; Jourdan et al. 2013, 2018). However, the Annot and Saint Antonin basin sedimentary rocks were not derived from the Western Alps, but rather from erosion of the Maures–Esterel massif, Sardinia and Corsica located to the south, as inferred from fission-track and U-Pb double dating, sediment petrography and sedimentology (see Jourdan et al. 2018 and references therein).

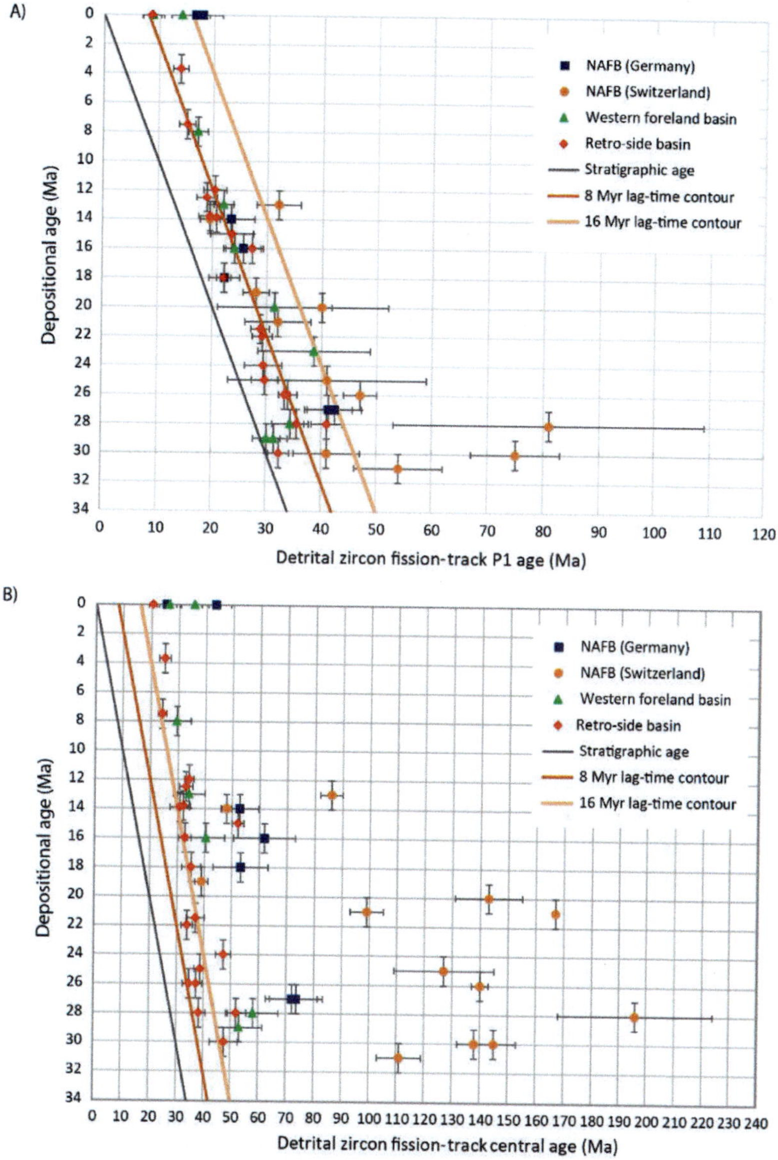

Figure 4.7. *A) Detrital zircon fission-track P1 and B) central age lag-time plots of pro- and retro-side foreland samples from the Western foreland basin (Bernet et al. 2009; Jourdan et al. 2013), Swiss Northern Alpine foreland basin (Spiegel et al. 2000), German Northern Alpine foreland basin and the Apennines (both from Bernet et al. 2009)*

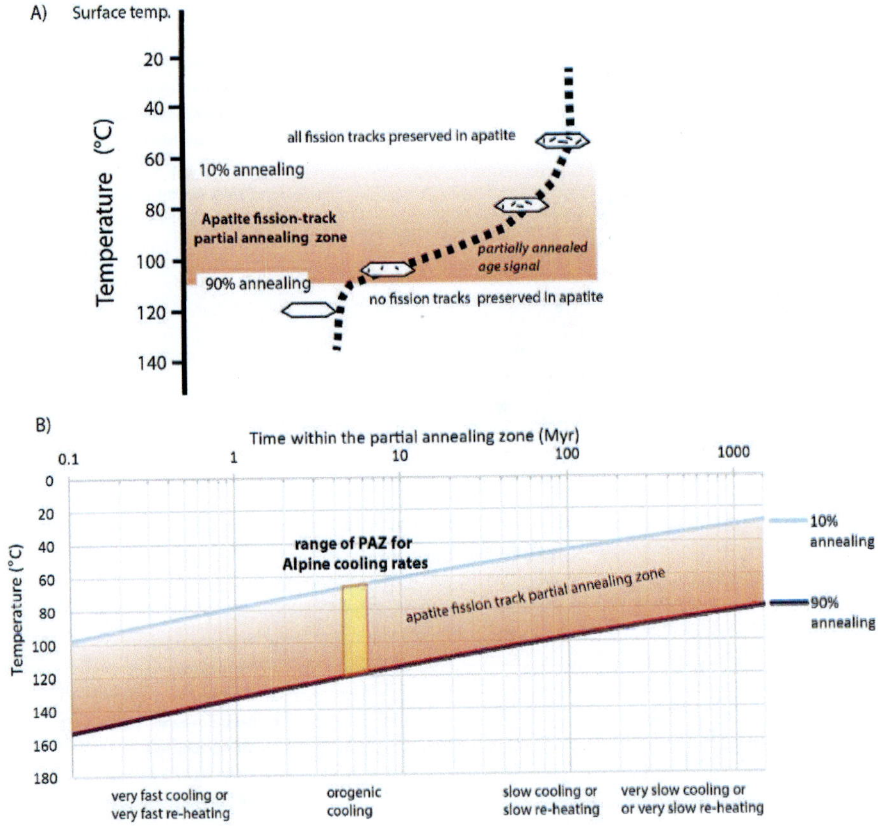

Figure 4.8. *A) Schematic diagram of the apatite fission-track partial annealing zone, defined by the 10% and 90% annealing limits. B) The apatite partial annealing zone for apatite of average chemical composition and in dependence on time spent with partial annealing temperature range (determined with the Age2edot program of Brandon (see Ehlers et al. 2005). For orogenic cooling rates of 15°C/Myr, the apatite fission-track partial annealing zone is about 60–120°C*

Zircon fission-track data from the Northern Alpine foreland basins have been published for the Lower Oligocene (Lu et al. 2020) and Upper Oligocene to Mid-Miocene deposits of the Swiss Northern Alpine foreland basin (Spiegel et al. 2000, 2001). A limited dataset obtained from outcrop and borehole samples of Upper Oligocene to Mid-Miocene sedimentary rocks from the German Northern Alpine basin was published by Bernet et al. (2009). Detrital zircon fission-track data from the Eastern Alps are very limited and have only been presented for Late Cretaceous and Palaeocene–Eocene sedimentary rocks of the Main flysch/Greifenstein nappe by Trautwein et al. (2002). As the depositional ages of these rocks are not very well defined, these data are not included in the discussion below.

The retro-side basin record is more complete and sample locations are regionally less dispersed. Late Oligocene to Mid-Miocene sedimentary rocks of Western Alps provenance is exposed in the Torino hills (Jourdan et al. 2013). Clastic sedimentary rocks of Central Alps provenance are preserved in the turbidite sequences of Oligocene to Pliocene sedimentary units in the Apennines (Bernet et al. 2001, 2009). All published zircon P1 and central ages of the Western Alps foreland basin, Northern Alpine foreland basin, and the Apennine turbidite sequences are plotted against the sediment depositional ages in Figure 4.7. The minimum, central and P1 ages are summarized in Tables 4.2 and 4.3. Figure 4.7(A) shows a fairly constant lag-time trend since about 27 Ma for the retro-side basin. The pro-side basin record shows more variability, particularly in the central age lag-time plot (Figure 4.7(B)). The Barrême basin samples from the Clumanc (La Poste) and St. Lions formations, with a depositional age of ~29 Ma contain zircons in the P1 fraction with very short lag times (Figure 4.7(A) and Table 4.2).

4.4.3. *Detrital apatite fission-track data*

The use of detrital apatite fission-track data from foreland basins for quantifying long-term exhumation may be complicated by the fact that fission tracks in apatite are susceptible to partial or full post-depositional annealing due to burial heating over temperature ranges of about 60–120°C (Figures 4.2 and 4.8), depending on how long the sediments resided within the apatite fission-track partial annealing zone (e.g. Reiners and Brandon 2006). Cederbom et al. (2004, 2011) used the post-depositional partial annealing of apatite fission tracks from Northern Alpine foreland basin deposits to demonstrate basin inversion at around 5 to 4 Ma, whereas von Hagke et al. (2012) used both apatite fission-track and (U-Th)/He data to propose thrusting in the Central Subalpine Molasse basin already during the Late Miocene, with distinct events at 10, 8, and 6–5 Ma. Mazurek et al. (2006) used the same partial annealing effect in combination with vitrinite reflectance and biomarker data from drill holes to constrain a Miocene heating and cooling phase in the Northern Alpine foreland basin. In the Western pro-side foreland basin, Schwartz et al. (2017) used apatite fission-track and (U-Th)/He data to determine the burial depth of Paleogene/Neogene foreland basin deposits beneath the Digne thrust sheet in southeastern France and subsequent exhumation starting at ~6 Ma.

Basin/location	Unit	Sample No.	Deposition (Ma)	2 SE (Myr)	Minimum age (Ma)	2 SE (Myr)	Lag time (Myr)	Central age (Ma)	2 SE (Myr)	Lag time (Myr)	P1 (Ma)	2 SE (Myr)
Northern Alpine foreland basin												
Swiss Molasse basin	Sternberg/Hörnli	ZC12	13.0	1	n/a	n/a	n/a	86.0	16.0	73.0	32.0	4.0
Swiss Molasse basin	Hengst/ H.-Napf	ZC3	14.0	1	n/a	n/a	n/a	48.0	8.0	34.0	19.5	1.8
Swiss Molasse basin	Eschholzmatt/H.-Napf	ZC4	19.0	1	n/a	n/a	n/a	39.0	6.0	20.0	28.0	2.4
Swiss Molasse basin	Goldinger Tobel/Hörnli	ZC10	20.0	1	n/a	n/a	n/a	143.0	20.0	123.0	40.0	12.0
Swiss Molasse basin	Fischenbach/H.-Napf	ZC6	21.0	1	n/a	n/a	n/a	99.0	18.0	78.0	32.0	6.0
Swiss Molasse basin	Spicher/Kronberg-G.	ZC18	21.0	1	n/a	n/a	n/a	167.0	12.0	146.0	n/a	n/a
Swiss Molasse basin	Reust/H.-Napf	ZC1	25.0	1	n/a	n/a	n/a	127.0	18.0	102.0	41.0	18.0
Swiss Molasse basin	Felsenweg/Rigi-H.	ZC22	26.0	1	n/a	n/a	n/a	140.0	22.0	114.0	47.0	3.0
Swiss Molasse basin	Walensee/Speer	ZC23	28.0	1	n/a	n/a	n/a	196.0	22.0	168.0	81.0	28.0
Swiss Molasse basin	Zuger See/Rigi-H.	ZC5	30.0	1	n/a	n/a	n/a	138.0	22.0	108.0	41.0	6.0
Swiss Molasse basin	Waldemme/H.-Napf	ZC24	30.0	1	n/a	n/a	n/a	145.0	20.0	115.0	75.0	8.0
Swiss Molasse basin	Steintal/Speer	ZC15	31.0	1	n/a	n/a	n/a	111.0	20.0	80.0	54.0	8.0
German Molasse basin	Glimmersande	00MB57	14.0	1	28.5	5.8	14.5	52.8	6.8	38.8	23.4	4.2
German Molasse basin	Molasse sandstone	99MB38	16.0	1	25.1	4.8	11.1	61.8	11.2	45.8	25.6	3.2
German Molasse basin	Molasse sandstone	00MB56	18.0	1	23.6	3.4	5.6	53.2	10	35.2	22.2	2.8
German Molasse basin	Bausteinschichten	99MB42	27.0	1	39.5	5.6	12.5	72.9	6.8	45.9	41.3	4.3
German Molasse basin	Bausteinschichten	99MB43	27.0	1	39.8	8.4	12.8	72.1	9.4	45.1	42.4	4.9
Western pro-side foreland basin (Western Alps sediment provenance)												
Sable de Montvendre	Molasse sandstone	99MB13	8.0	1	17.4	1.8	9.4	29.5	5.2	21.5	17.2	1.8
Sable de Valreas	Molasse sandstone	99MB16	13.0	1	20.5	2.0	7.5	34.2	5.8	21.2	22.0	1.8
Etang du Lavalduc	Molasse sandstone	99MB22	16.0	1	22.8	2.4	6.8	40.5	6.8	24.5	24.0	2.0
Mountmaur	Congl. de Montmaur	SJ05	20.0	1								
Barrême basin	Gres Verts	SJ01+00MB53	23.0	1								
Barrême basin	Molasse Rouge	00MB54	28.0	1	30.8	6.6	2.8	58.1	9.2	30.1	34.4	3.3
Barrême basin	Congl. de Clumanc	SJ03+00MB55	29.0	1								
Barrême basin	Congl. de St Lions	SJ02	29.0	1								

Note: Swiss Molasse basin (Spiegel et al. 2000); German Molasse basin and Western Alps foreland basin (Bernet et al. 2009 and Jourdan et al. 2013). P1 age are published ages. Minimum and central ages were calculated with the RadialPlotter program of Vermeesch (2009), except for the Swiss Molasse basin samples. Western foreland samples with Maures–Esterel, Sardinia and Corsica provenance of Bernet et al. (2009) and Jourdan et al. (2013, 2018) are not shown here.

Table 4.2. *Pro-side foreland basin detrital zircon fission-track minimum and central age lag time and P1 ages*

Basin/location	Unit	Sample No.	Deposition (Ma)	1 SE (Myr)	Minimum age (Ma)	2 SE (Myr)	Lag time (Myr)	Central age (Ma)	2 SE (Myr)	Lag time (Myr)	P1 (Ma)	2 SE (Myr)
Apennines	Borello sandstone	00MB50	3.7	1	14.1	2.8	10.4	25	2.8	21.3	13.9	2.1
Apennines	Marnoso arenacea	AP36	7.5	1	13.3	11.6	5.8	24.2	3.0	16.7	15.2	1.5
Apennines	Marnoso arenacea	AP5	12.0	1	16.7	8.6	4.7	34.2	4.2	22.2	20.4	1.7
Apennines	Marnoso arenacea	AP9	12.5	1	17.7	10.6	5.2	33.1	3.8	20.6	18.9	2.4
Apennines	Marnoso arenacea	AP17	13.8	1	20.4	3.2	6.6	32.1	4.2	18.3	19.5	2.5
Apennines	Marnoso arenacea	AP54	13.9	1	11.7	10.8	*-1.3*	30.8	2.8	16.9	20.7	3.6
Apennines	Marnoso arenacea	AP38-33	15.0	1	21.0	3.8	6.0	52.2	7.8	37.2	23.5	2.1
Apennines	Cervarola sandstone	C1	16.0	1	25.5	2.2	9.5	32.8	4.0	16.8	27.2	2.3
Apennines	Cervarola sandstone	C51	18.0	1	24.2	8.8	6.2	35.2	2.6	17.2	22.1	3.5
Apennines	Cervarola sandstone	C2	21.5	1	28.8	3.8	7.3	36.9	3.4	15.4	28.8	3.3
Apennines	Cervarola sandstone	VALD12	22.0	1	29.5	1.9	8.0	33.8	3.6	11.8	29.2	2.1
Apennines	Falterona sandstone	VALD2	24.0	1	28.3	3.4	4.3	47.2	6.8	23.2	29.4	2.7
Apennines	Falterona sandstone	VALD6	25.0	1	29.5	2.6	4.5	38.8	4.8	13.8	29.7	2.2
Apennines	Macigno Fm	GOM-1	26.0	1	32.9	3.0	6.9	37.3	3.4	11.3	33.9	2.2
Apennines	Macigno Fm	GOM-3	26.0	1	32.6	3.2	6.6	34.7	2.0	8.7	33.3	2.3
Apennines	Macigno Fm	CAS-3	28.0	1	35.3	2.8	7.3	38.3	2.8	10.3	35.5	2.5
Apennines	Macigno Fm	CAS-2	28.0	1	37.8	5.6	9.8	52.0	5.6	24.0	41.0	3.7
Apennines	M. Senario sandstone	C49	30.0	1	31.0	5.4	1.0	47.6	3.8	17.6	32.3	5.0
Torino hills	Termofoura Formation	TO52	17.2	1							47.2	29.8
Torino hills	Antognola Formation	TO03	23.5	4							25.5	4.0
Torino hills	Cardona Formation	TO02	29.0	5							77.7	43.0
Torino hills	Cardona Formation	TO53	33.0	5							43.3	17.2
Torino hills	Cardona Formation	TO54	33.0	5							38.9	11.1

Note : Data from Bernet et al. (2001, 2009) and Jourdan et al. (2013). P1 ages are published ages. Minimum and central ages were calculated with the RadialPlotter program of Vermeesch (2009).

Table 4.3. *Retro-side foreland basin detrital zircon fission-track minimum and central age lag times and P1 ages*

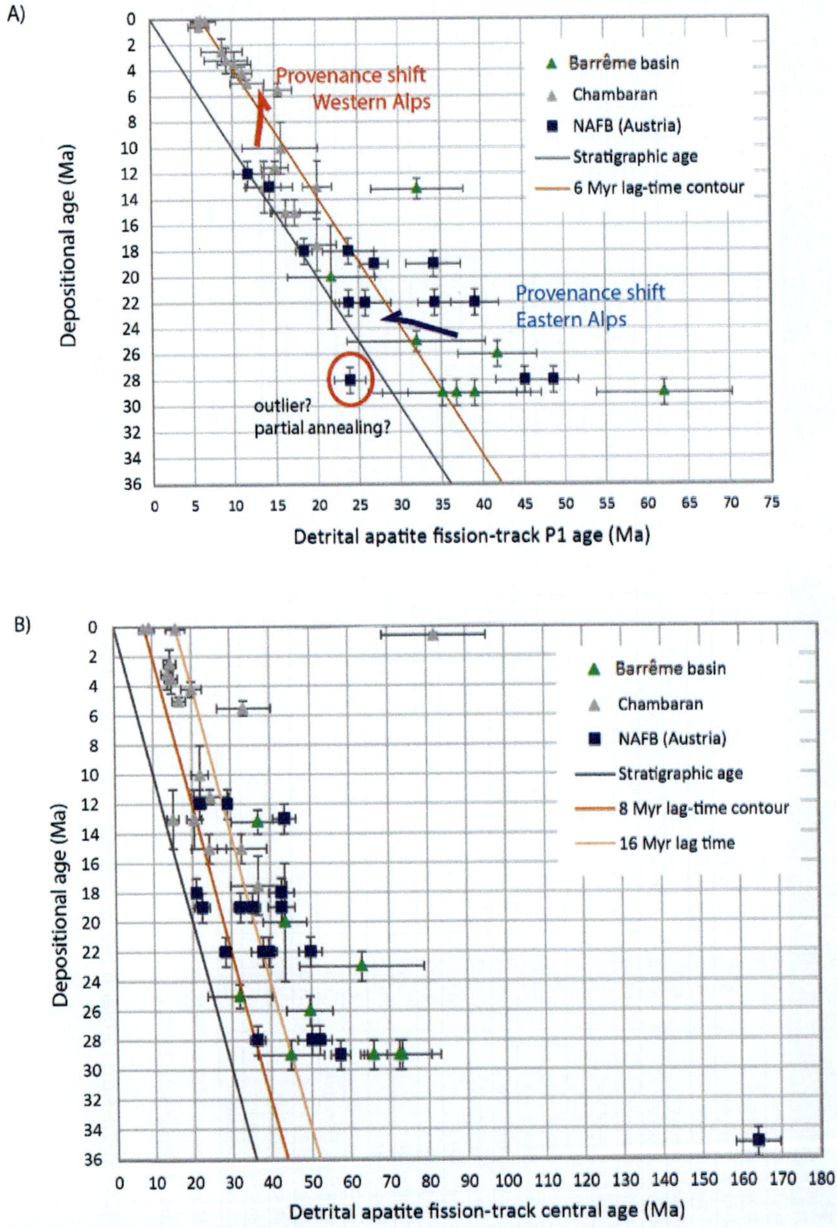

Figure 4.9. *(A) Detrital apatite fission-track P1 and (B) central age lag-time plots of pro-side foreland basin samples from the Chambaran basin and Barrême basin of the Western Alps foreland (Glotzbach et al. 2011a; Jourdan et al. 2013; Mark et al. 2016), and the Northern Alpine foreland basin in Austria (Kuhlemann et al. 2006)*

Here we focus on detrital apatite fission-track data that reflect bedrock exhumation in the source areas and were not affected by post-depositional heating leading to partial annealing. A pre-depositional exhumation signal can confidently be inferred from both the P1 and central ages in case they are older than the depositional age and increase with depositional age (Table 4.3). Such data are available in the pro-side foreland basin from Lower Oligocene to Upper Miocene and Pliocene sedimentary rocks of the Barrême basin (Jourdan et al. 2013; Mark et al. 2016), the Chambaran basin (Glotzbach et al. 2011a), and the Northern Alpine foreland basin from east of Lake Constance to the east of Salzburg (see Kuhlemann et al. 2006 for details). In the retro-side foreland basin, detrital apatite fission-track ages have been determined for the Venetian foreland basin (Zattin et al. 2003; Stefani et al. 2007), and from some of the Oligocene to Mid-Miocene sedimentary rocks of the Torino hills (Jourdan et al. 2013).

The apatite fission-track P1 and central ages of the pro-side foreland basin are shown in Figure 4.9(A). Data from the Chambaran basin show a shift in P1 ages between about 10 and 6 Ma, with relatively constant lag times of about 6 Myr since 6 Ma (Figure 4.9(A)). The data from the Barrême basins show a more dispersed signal without a clear trend (Figure 4.9(A)). Apatite fission-track and U-Pb double dating of apatite from the same Barrême basin units by Mark et al. (2016) did not detect any volcanically-derived apatite.

Kuhlemann et al. (2006) presented a detrital apatite fission-track dataset of 18 samples collected over a >300 km east–west profile along the northern edge of the Eastern Alps (NAFB, Austria; Figure 4.9(A)), which shows a decrease in lag times between 28 and 18 Ma and more constant lag times between 18 and 12 Ma. The P1 ages of the Kuhlemann et al. (2006) study were determined with the Popshare program of Dunkl and Székely (2002).

4.5. Discussion

4.5.1. *Exhumation signal from modern-river deposits*

Given the limited available modern-river white mica $^{40}Ar/^{39}Ar$ data, we focus the discussion on the present-day exhumation signal obtained from detrital apatite and, in particular, the detrital zircon fission-track data. The currently available bedrock apatite and zircon fission-track data are shown in two maps in Figure 4.10, as a reference for the following discussion. Both the apatite and zircon fission-track bedrock ages show three general trends: (A) Similar to the metamorphic grade, the

Western Alps show a west to east trend with younger bedrock cooling ages in the west; (B) the Central Alps in the area of the Lepontine dome show a nested pattern with orogenic cooling ages in the core of the mountain belt and older fission-track ages on the flanks; and (C) the Eastern Alps tend to have somewhat younger zircon fission-track ages on the southern flank and older ages on the northern flank, whereas the apatite fission-track ages show a more nested pattern in the core of the Eastern Alps (around the Tauern Window), but with cooling ages older than the ones observed in the Central Alps. The detrital zircon fission-track KDE plots shown in Figure 4.5 reflect how the detrital exhumation signal changes along the mountain belt in relation to differences in metamorphic and erosion history. The rivers draining the Central Alps show mainly narrow age distributions clustering to the most part at <35 Ma. In contrast, the Isère and Durance rivers draining the Western Alps to the pro-side show a wider age spectrum with a higher percentage of >35 Ma grains. This same trend to a higher percentage of grains with fission-track ages >35 Ma is also visible in the rivers draining the Eastern Alps and is coherent with the bedrock age distributions shown in Figure 4.10.

With the exhumation rate estimates mentioned above, both the apatite and zircon fission-track data of the present-day pro- and retro-side drainage systems indicate long-term average exhumation rates on the order of 0.3±0.1 km/Myr (Figure 4.6(D)). Based on the detrital zircon fission-track P1 ages, rapid exhumation of up to 0.7–0.9 km/Myr has occurred locally in the Lepontine dome of the Central Alps (Bernet et al. 2001, 2004a, 2004b). According to Bernet et al. (2004b), the P1 zircons of the Ticino river sample, which indicate such rapid exhumation, were derived from the western Lepontine dome, where exhumation was driven by erosion and normal faulting along the Simplon fault (Mancktelow 1992; Grasemann and Mancktelow 1993; Bernet et al. 2001; Campani et al. 2010). The eastern Lepontine dome, with zircon fission-track ages on the order of 16 Ma, corresponding to the P2 of the Ticino river sample (Bernet et al. 2001, 2004b), was exhumed by erosion alone at rates of 0.5 km/Myr. This interpretation is fully compatible with the detrital apatite fission-track data of Malusà et al. (2017) from the Maggia river draining the central-eastern part of the Lepontine dome. The Maggia river shows a P1 of 7 Ma, corresponding to an exhumation rate of 0.5 km/Myr, indicating steady long-term exhumation rates, at least between the times that the Lepontine dome cooled below the zircon and apatite fission-track closure temperatures, respectively. Detrital quartz ^{10}Be analyses from the western and eastern Lepontine dome indicate short-term erosion rates of 1.5 and 0.7 km/Myr respectively (Malusà et al. 2017). Whether these more rapid short-term erosion rates are related to deep-seated tectonic processes, climate change or human impact (Napoli et al. 2019), needs to be determined.

4.5.2. *Long-term pro- and retro-side foreland basin exhumation signal*

4.5.2.1. *Western Alps $^{40}Ar/^{39}Ar$ exhumation signal*

The detrital white mica $^{40}Ar/^{39}Ar$ age record comprises the Tertiary piedmont basin on the retro-side, with youngest ages of 31 Ma (Early Oligocene Molare Formation) and 22 Ma (Late Oligocene–Early Miocene Rocchetta Formation), as well as the static 37±1 Ma peak throughout the stratigraphic record of the Tertiary Piedmont basin. These age peaks were interpreted as reflecting rapid, episodic exhumation of the internal Western Alps (Carrapa et al. 2003, 2004). Malusà and Fitzgerald (2020) are critical of that interpretation, arguing that the 37 ± 1 Ma white mica $^{40}Ar/^{39}Ar$ ages are metamorphic crystallization rather than cooling ages. Nonetheless, the two youngest single grain ages of 31 and 22 Ma are close to their respective depositional ages (Figure 4.11), which would still indicate rapid exhumation after crystallization (if these were also crystallization ages). Furthermore, the white mica $^{40}Ar/^{39}Ar$ ages of 33–32 Ma of a blueschist pebble deposited in the pro-side Barrême basin in the Western Alps at around 29 Ma also indicate rapid exhumation, at least during the Early Oligocene (Morag et al. 2008). Substantial exhumation on the order of 3 ± 1 km/Myr would at least be necessary to exhume the blueschist rocks and deposit them in the foreland basin. This would be compatible with the detrital zircon fission-track data from the same units in the Barrême basin, which also show very short lag times (Figures 4.8 and 4.11; Bernet et al. 2009; Jourdan et al. 2013).

4.5.2.2. *Central Alps $^{40}Ar/^{39}Ar$ exhumation signal*

The detrital white mica $^{40}Ar/^{39}Ar$ record from the Swiss Northern Alpine foreland basin was interpreted as recording erosion of the Austroalpine cover of the Lepontine dome, with recycling of white micas from Cretaceous–Paleogene flysch deposits (Figure 4.11; von Eynatten et al. 1996, 1999). The samples collected from Early and Mid-Miocene strata show lag times between about 10 and 20 Myr, reflecting exhumation at rates of ~1.0–0.5 km/Myr. The micas with the youngest cooling ages were inferred to have been sourced from the footwall of the Simplon normal fault, which contributed to the exhumation of the Lepontine dome since at least the Early Miocene (von Eynatten et al. 1996, 1999). The trend of younger white mica $^{40}Ar/^{39}Ar$ cooling ages up-section reflects a typical unroofing signal (Figure 4.11).

4.5.2.3. *Detrital fission-track exhumation signal*

The detrital zircon fission-track data of the pro- and retro-side foreland basin samples are the most extensive dataset for tracing the exhumation history of the European Alps (Figures 4.7 and 4.11). The P1 values of samples collected from the

Early Oligocene sedimentary rocks show an unroofing trend with upward decreasing lag times. Zircons deposited at 29 ± 1 Ma in the Barrême basin show P1 ages of 30 ± 1 Ma. As such cooling ages are in the same range as the last volcanic activity in the Alps (i.e. 35–30 Ma; e.g. Boyet et al. 2001), the question arose if zircons with such short lag times reflect very rapid exhumation cooling or rather a volcanic contribution. To resolve this question, Jourdan et al. (2013) performed fission-track and U-Pb double dating on detrital zircons. They found that about 30% of the zircons with 30 Ma cooling ages are of volcanic origin, but even when removing these grains from the dataset, the 30 Ma cooling signal of rapidly exhumed zircons persists. This thermochronological signal may indicate the end of a rapid exhumation event during the Late Eocene-Early Oligocene, contemporary with the collision and building of topography (Fauquette et al. 2015) and the reorganization of the Western Alps convergence at that time (Dumont et al. 2012). Detrital zircons deposited since about 28 Ma in the Western foreland basin, the German Northern Alpine foreland basin, and the retro-side deposits preserved in the Apennines show relatively constant P1 lag times of about 8 Myr (Figure 4.7(A)). An 8 Myr lag time translates into an exhumation rate of about ~1 km/Myr for the fastest exhumation areas, using the simple calculation of Garver et al. (1999). The P1 values of the Swiss Northern Alpine foreland basin are more variable but also show an up-section younging trend with a relatively constant 16 Myr lag time, which corresponds to exhumation rates of ~0.5 km/Myr. The constant lag-time trend was interpreted as indicating an exhumation steady state of the European Alps (Bernet et al. 2001, 2009).

Malusà and Fitzgerald (2020) recently presented an interesting alternative interpretation of the Bernet et al. (2001, 2009) zircon fission-track data, arguing that the observed peak-age trends shown in Figure 4.7 do not reflect steady exhumation, but rather a shift in sediment provenance over time-related to moving the focus of exhumation from the internal to the external European Alps. These authors used the peak age and size (percentage of single grains per peak) to assign the shift in sediment provenance, arguing that the P1 zircons were not always derived from the same source areas. Given that the data shown in Figure 4.7 for the pro-side foreland basin is compiled from different parts of the basin (Western foreland basin remnants in France, the Swiss Northern Alpine foreland basin and the German Northern Alpine foreland basin) the arguments of Malusà and Fitzgerald (2020) seem reasonable. For example, Kuhlemann et al. (2006) argued, based on their dataset of apatite fission-track P1 ages from the Austrian Northern Alpine foreland basin, that the P1 age trend may indicate a shift in sediment provenance at around 22 Ma. Pre-22 Ma deposits supposedly recorded the erosion of the Austroalpine cover rocks of the Central Alps and/or Tauern Window, and post-22 Ma deposits received apatite from Penninic units (Kuhlemann et al. 2006). Instead of a shift in provenance, which would imply changes in the drainage system to provide sediments from spatially different source areas, this apatite

P1 trend from the Austrian Northern Alpine foreland basin could simply reflect an unroofing signal, with the removal of partially to non-reset cover units and subsequent exposure of rocks with orogenic cooling ages. In general, in the Eastern Alps, rocks with Alpine (i.e. <35 Ma) apatite and zircon fission-track cooling ages are mainly exposed in the Tauern Window (Figure 4.10; e.g. Foeken et al. 2007; Keil and Neubauer 2015; Wölfler et al. 2015; Heberer et al. 2017), and in the Dolomites (Zattin et al. 2006). Furthermore, Glotzbach et al. (2011a) interpreted the shift in the apatite fission-track P1 age trend of the Chambaran basin deposits as a change in sediment provenance rather than a decrease in exhumation rates for the Western Alps (Figure 4.9(A)). Not only do lag times change up-section, but the peak ages themselves increase, which is inconsistent with the continuous exhumation of the same source area. This interpretation is consistent with paleogeographic reconstructions of the Northern Alpine foreland basin drainage system by Kuhlemann and Kemp (2002). Therefore, the P1 grains deposited between 18 and ~12 Ma were interpreted as being derived from the Central Alps via long-shore drift in a Mid-Miocene seaway (Kuhlemann and Kemp 2002), whereas P1 grains deposited since 6 Ma were derived locally from the Western Alps.

For these reasons, the discussions on unroofing, steady or non-steady exhumation, changing provenance, etc. are important for synthesizing the information retained in the detrital exhumation record. One aspect that is important in this discussion is if the sediments or sedimentary rocks were collected near the source areas (proximal) or far from the source areas (distal) as shown in Figure 4.2(A). As discussed above, the proximal or distal sampling issue is well reflected in the modern river detrital zircon fission-track dataset (Figure 4.5; Bernet et al. 2004b). The more distal the sampling location, the more mixing of the exhumation signal from the orogenic source areas with the increasing size of the drainage area (e.g. Rhône river vs. Rhône delta KDE plots in Figure 4.5). Whereas the pro-side foreland basin zircon fission-track samples provide a much more locally influenced signal, the retro-side basin samples were collected from turbidite sequences in the Apennines (Bernet et al. 2001, 2009). These turbidite sediments reflect a very distal sampling location, which was continuously sourced from the Central Alps (Bernet et al. 2001, 2009), with no important shift in sediment provenance over time. Whereas the discussion about changes in sediment provenance has been mainly based on peak size, as suggested by Malusà and Fitzgerald (2020), and less on peak age, it has to be considered that: (A) only bedrock lithologies with present-day cooling ages equal or younger than the peak ages defined from detrital zircon samples may have been potential source areas (Bernet et al. 2004a, 2004b); (B) it has been shown within the detrital system that P1 ages tend to be more stable than peak sizes, given that peak size primarily depends on (short-term) sediment yield (i.e. erosion rate per surface area) for a given source area (Bernet et al. 2004a, 2004b), rather than mineral fertility alone; and (C) the number of peaks and peak sizes of a sedimentary sample may not only depend on how many grains were dated (Vermeesch 2019) and how well the peaks are separated (Naylor et al. 2015), but also on which program or algorithm was used to determine peak ages.

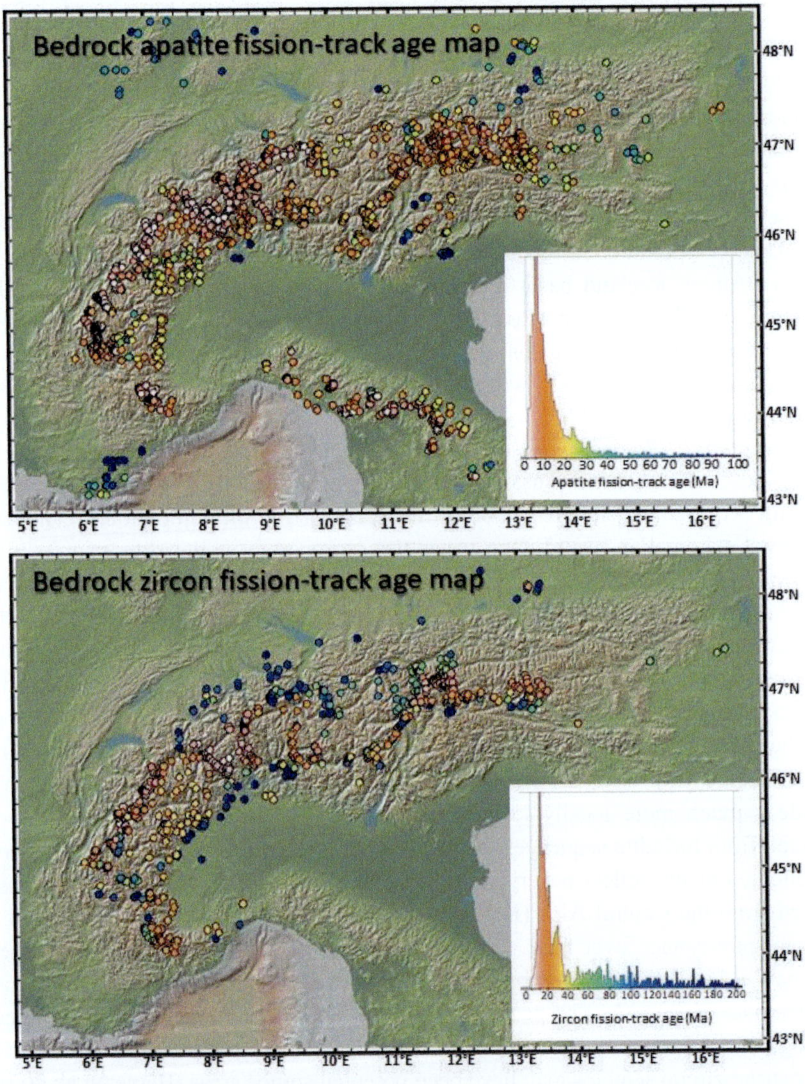

Figure 4.10. *Maps of A) apatite and B) zircon fission-track bedrock ages based on the fission-track data compilation of Herman et al. (2013), with additional data from Dunkl et al. (1998), Herberer et al. (2017), Beucher et al. (2012), and Girault (2020)*

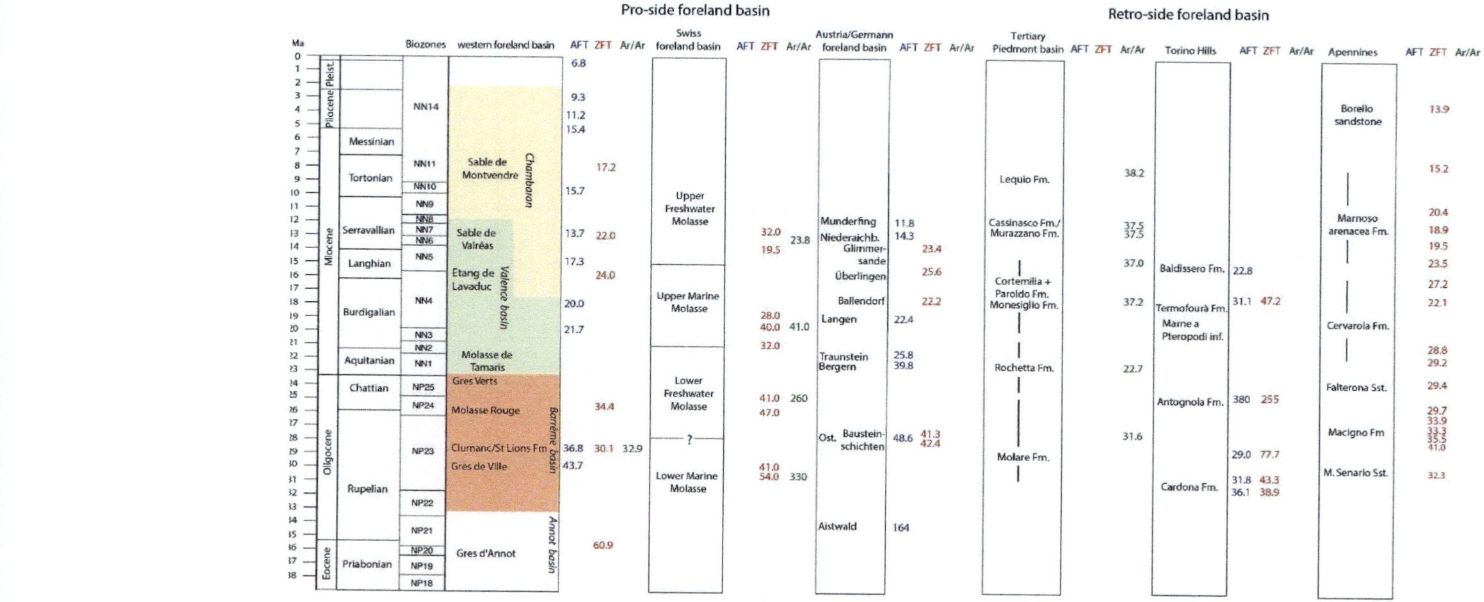

Figure 4.11. *Stratigraphy and detrital thermochronology data of pro- and retro-side basin samples, compiled from von Eynatten et al. (1999), Spiegel et al. (2000), Bernet et al. (2001, 2009), Carrapa et al. (2003), Kuhlemann et al. (2006), Morag et al. (2008), Jourdan et al. (2013) and Mark et al. (2016). For apatite fission-track (AFT) and zircon fission-track (ZFT), data P1 ages are shown. For 40Ar–39Ar results are the youngest grains shown. The Western Alps pro-side foreland basin data were obtained from basin remnants, which are indicated by different colors. For the Apennines, no apatite fission-track data are given, as apatite fission-track were widely reset during the tectonic evolution of the Apennines and no longer reflect Alpine exhumation (e.g. Thomson et al. 2010)*

Sample	Deposition	n	Binomfit P1	P2	P3	P4	RadialPlotter P1	P2	P3	P4
No.	(Ma)		(Ma)	(Ma)	(Ma)	(Ma)	(Ma)	(Ma)	(Ma)	(Ma)
00MB50	3.7	90	13.9±2.1 (30.0%)	24.2±3.1 (42.0%)	38.5±7.9 (21.0%)	72.9±22.1 (7.0%)	15.4±2 (41.9%)	29.1±3.4 (46.3%)	62.9±12 (12%)	-
AP36	7.5	102	15.2±1.5 (39.8%)	24.7±2.2 (51.1%)	106.7±13.7 (9.1%)	-	15.7±1.6 (39.6%)	24.5±2.2 (51.3%)	106.6±14 (9%)	-
AP5	12.0	102	20.4±1.7 (49.2%)	34.2±3.6 (25.4%)	69.5±8.1 (16.8%)	131.2±20.6 (8.6%)	20.5±1.7 (49.1%)	34.1±3.6 (25.4%)	69.9±8.4 (16.8%)	131±22 (8.7%)
AP9	12.5	105	18.9±2.4 (29.3%)	28.8±3.2 (44.3%)	51.7±6.0 (14.0%)	115.6±14.8 (12.4%)	19.0±2.4 (30.4%)	29.1±3.2 (43.4%)	51.8±6.0 (13.9%)	116±7.5 (12.0%)
AP17	13.8	78	19.5±2.5 (5.7%)	27.7±4.2 (13.9%)	51.7±12.4 (8.7%)	94.9±14.6 (15.9%)	20.1±2.2 (41.0%)	29.7±3.8 (39.0%)	82.3±10.8 (20.0%)	-
AP54	13.9	101	20.7±3.6 (33.5%)	29.9±4.9 (50.0%)	49.9±8.8 (36.0%)	114.1±20.6 (6.7%)	21.1±3.2 (38.0%)	30.6±5.0 (47.0%)	51.4±9.8 (9%)	113±10 (7%)
AP38-33	15.0	85	23.5±2.1 (35.1%)	43.9±4.5 (28.9%)	110.6±9.6 (40.6%)	-	17.8±3.8 (8.3%)	26.5±3.2 (30.0%)	45.7±2.5 (21.3%)	112±10 (40.0%)
C1	16.0	60	27.2±2.3 (83.0%)	82.1±12.5 (17.0%)	-	-	26.1±2.4 (76.1%)	48.3±11.4 (13.2%)	95.3±16.6 (10.7%)	-
C51	18.0	100	22.1±3.5 (15.0%)	33.2±3.1 (69.0%)	72.9±10.3 (16.0%)	-	22.1±3.6 (15.1%)	33.4±3.0 (69.2%)	73.0±10.6 (15.7%)	-
C2	21.5	60	28.8±3.3 (52.0%)	43.0±6.2 (40.0%)	80.2±17.2 (8.0%)	-	28.9±3.4 (53.0%)	43.2±6.4 (39.0%)	80.2±17.8 (9.0%)	-
VALD12	22.0	49	29.2±2.1 (77.0%)	39.6± 6.5 (12.0%)	65.0±16.4 (5.0%)	106. ±23.5 (6.0%)	31.2±2.0 (88.6%)	84.0±14.0 (11.4%)	-	-

Sample	Deposition	n	Binomfit P1	P2	P3	P4	RadialPlotter P1	P2	P3	P4
VALD2	24.0	46	29.4±2.7 (49.0%)	75.1±7.1 (51.0%)	-	-	29.4±2.0 (48.6%)	75.3±7.2 (51.4%)	-	-
VALD6	25.0	51	29.7±2.2 (70%)	57.4±9.0 (20%)	103.9±22.1 (10.0%)	-	31.7±1.1 (76.9%)	78.2±4.7 (23.1%)	-	-
GOM-1	26.0	46	33.9±2.2 (83%)	57.2±8.1 (14%)	100.7±38.6 (3.0%)	-	34.0±2.2 (83.9%)	62.0±7.8 (16.1%)	-	-
GOM-3	26.0	50	33.3±2.3 (92.0%)	50.9±19.5 (8.0%)	-	-	34.5±2 (100%)	-	-	-
CAS-3	28.0	49	35.5±2.5 (87.0%)	50.0±10.9 (11.0%)	111.2±80.5 (2.0%)	-	35.5±2.5 (87.7%)	61.1±10.6 (12.3%)	-	-
CAS-2	28.0	49	41.0±3.7 (65.0%)	83.1±11.7 (35.0%)	-	-	41.0±3.8 (65.0%)	83.1±12.0 (35.0%)	-	-
C49	30.0	100	32.3±5.0 (36.0%)	50.8±9.8 (43.0%)	83.9±14.6 (21.0%)	-	33.1±4.8 (40.0%)	52.7±10.4 (41.3%)	85.7±15.2 (19.0%)	-

Note: n = number of dated grains. Published best-fit Binomfit (Brandon, see Ehlers et al. 2005) peak age data from Bernet et al. (2001, 2009). RadialPlotter (Vermeesch 2009) peak ages calculated with "auto" option of the program. Peak ages are given with 2-sigma error estimates. Peak size is given with the percentage of grains belonging to each peak. Samples marked with grey background show similar Binomfit and RadialPlotter peak-fit results.

Table 4.4. *Retro-side foreland basin detrital zircon fission-track peak age comparison between Binomfit and RadialPlotter results*

This last point is demonstrated by the comparison of 18 samples from the retro-side foreland basin data published by Bernet et al. (2001, 2009), with the "best-fit" Binomfit peak ages compared to the "auto" fit RadialPlotter peak ages of the same data. As shown in Table 4.4, only eight of the samples show similar Binomfit and RadialPlotter peak ages and peak sizes, whereas ten samples show differences in the number and/or the size of the peaks determined with Binomfit or RadialPlotter. We also note that there is no systematic difference between the two programs. For certain samples, Binomfit detects more peaks and for other samples, RadialPlotter does so. Also, the number of dated grains (100 or much less) cannot explain these differences. Supposedly both programs use the same peak-fitting algorithm, but Binomfit assigns asymmetric grain-age errors, whereas RadialPlotter uses symmetric errors. This may explain part of the observed variations in the peak age/size results. These differences between the two programs in which peak ages and sizes are detected highlight one of the shortcomings of the peak-fitting approach and the interpretation of multiple peaks per sample. Depending on which peak-age estimates are used, the provenance shift interpretations of Malusà and Fitzgerald (2020) may change considerably. Malusà and Fitzgerald (2020) rightly emphasized the importance of sediment mixing and the potential spatio-temporal evolution of the exhumation signal, which has also been shown by Bernet et al. (2004a) for the Rhône and Rhine river systems and by Gemigani et al. (2018) for the Brahmaputra system. However, the Rhine river dataset of Bernet et al. (2004a) showed that the signal of Alpine exhumation is not significantly altered over 1300 km of fluvial transport, despite intermediate sediment storage in floodplains and the addition of zircons from a very large non-Alpine source area. As a consequence, we think that the steady detrital zircon fission-track exhumation signal is a valid observation, at least for the erosional exhumation of the Central Alps towards the retro-side foreland basin, and reflects relatively steady long-term exhumation conditions in this central part of the Alps. This is compatible with the rather constant plate convergence rates between the Eurasian and Adriatic plates of 0.45–0.5 cm/yr between about 35 and at least 7 Ma (Schmid et al. 1997). In contrast, various model interpretations of the bedrock thermochronological data in the Western Alps suggest that exhumation rates possibly were locally more variable over time (e.g. Vernon et al. 2008; Fox et al. 2015, 2016), and it is well accepted that the locus of exhumation shifted from the internal to the external zones, at least in the Western Alps during the Late Oligocene/Early Miocene (e.g. Fügenschuh and Schmid 2003; Tricart et al. 2007; Beucher et al. 2012).

Several authors have argued for a significant increase in exhumation rates in particularly the Western Alps over the last ~2 Myr, which has been linked to the possible effects of slab detachment, glaciation, or both (e.g. Vernon et al. 2008; Baran et al. 2014; Fox et al. 2015, 2016). However, it has recently been shown that the inferred increase in exhumation rates is in fact an artifact of the modeling techniques used in these studies (Schildgen et al. 2018), which translate spatial variations in exhumation rates into temporal increases. The detrital apatite fission-track record does not show any evidence for such an increase, although it has been shown by Glotzbach et al. (2011a) that it should have the resolution to track a two-fold increase in source-area erosion rates since 2.5 Ma. The effect of glaciation has been to carve out deep glacial valleys, which locally expose very young thermochronological ages at their valley bottoms (e.g. Glotzbach et al. 2011b; Valla et al. 2011, 2012). Not considering this localized valley incision also leads to biased predictions of a regional increase in exhumation rates, as shown by Glotzbach et al. (2011b).

4.5.3. *Topographic evolution of the Alps since the Late Eocene*

When discussing the exhumation history of the European Alps, the question of how the topography of this mountain belt evolved over time naturally arises. Chapter 3 in Volume 1 of this book, by Gradwohl et al., provides an excellent overview of the topographic evolution of the Alps. Here we will touch only on certain aspects in relation to what was learned from combining exhumation studies based on detrital thermochronology with sediment petrological or paleo-elevation (i.e. palynology, stable-isotope geochemistry) studies, and mainly with respect to the Western Alps. In general, the rate of elevation increase of orogenic topography is primarily driven by the rate of plate convergence, causing surface uplift by crustal thickening, if erosion rates are less than rock uplift rates (e.g. Molnar 2018). The growth of topography has a local and regional impact on climate, resulting in effects such as changes in mean annual temperatures with respect to elevation and orographic precipitation (e.g. Willett 1999; Roe 2005; Napoli et al. 2019). Precipitation in the form of rain and snow is important for erosion and sediment transport. Erosion will wear down the mountainous topography by removing rock at the surface (van der Beek 2018); this removal of mass will to some degree be compensated by the isostatic response of the over-thickened crust (Molnar 2018).

Continent–continent collision in the Western Alps started during the Late Eocene (e.g. Dumont et al. 2012), and pollen analysis of Oligocene–Miocene foreland basin sediments from the Western Alps has shown that mean elevation similar to today were reached during the Early Oligocene (Fauquette et al. 2015). The early stages of topographic growth during the Late Eocene–Early Oligocene were therefore contemporaneous with the phase of rapid exhumation documented in the 31–32 Ma detrital white mica $^{40}Ar/^{39}Ar$ ages on both sides of the Western Alps, as proposed by Carrapa et al. (2003) and Morag et al. (2008), and the zircon fission-track/U-Pb double dating results of Jourdan et al. (2013). Once the topography of the Western Alps had reached mean elevations similar to today during the Early Oligocene, lag times increased and long-term average exhumation rates decreased to ~0.3 ± 0.1 km/Myr, based on detrital zircon central ages. These long-term average exhumation rates have remained fairly constant since then (Tables 4.1 and 4.2), although the position of the drainage divide may have changed over time. For example, a change in the position of the drainage divide of the Western Alps from Oligocene to Miocene times has been traced with sediment petrological studies using basalt pebble geochemistry and Raman spectroscopy of serpentinite pebbles (e.g. Jourdan et al. 2012; Schwartz et al. 2012, Fauquette et al. 2015).

Schlunegger and Kissling (2015) proposed a model for the Central Alps in which topography started to grow at about 32 Ma and reached mean elevations similar to present-day elevations by about 30 Ma. These authors concluded that slab rollback and buoyancy-driven surface uplift were the driving forces to create and maintain orogenic topography since the Early Oligocene. Sinclair (1997a) and Kuhlemann and Kempf (2002) argued that slab breakoff and the associated dynamic and isostatic response had an important impact on the topographic evolution of the Central Alps, resulting in significant surface uplift followed by an increase in erosion, relief and sediment flux towards the foreland basins. Campani et al. (2012) proposed that high topography in the Central Alps existed at least since the Mid-Miocene, inferred from stable-isotope data on hydrothermal minerals. A change in the position of the drainage divide in the Central Alps has also been traced with detrital thermochronology and sediment petrography studies (e.g. von Eynatten et al. 1999; Spiegel et al. 2002; von Eynatten 2003), which apparently migrated to the south between 30 and 20 Ma (von Eynatten et al. 1999; von Eynatten and Gaupp 1999; Spiegel et al. 2000, 2001). In contrast to the Central Alps pro-side foreland basin, where white micas and zircons with Alpine (<35 Ma) cooling ages appear only in sediments younger than ~21 Ma (Figure 4.11), zircons with Alpine cooling ages can be detected in sedimentary rocks with depositional ages as early as 28 Ma in the retro-side basin as discussed above (Bernet et al. 2009).

No such record is available for the Eastern Alps, as Eocene to Miocene pro- and retro-side foreland basins with exclusively Eastern Alps provenance are not well exposed, with exception of the Venetian foreland basin (Zattin et al. 2003). Nonetheless, on the basis of palinspastic reconstructions, Frisch et al. (1998) proposed that mountainous topography in the Eastern Alps started to develop between 29 and 22 Ma, with an eastward propagation of topographic growth. At least from the Frisch et al. (1998), Schlunegger and Kissling (2015) and Fauquette et al. (2015) studies emerge the picture that there is a general west to east trend in the initiation of orogenic topographic growth and that the Western and Central Alps reached a mean topography similar to today by about 30 Ma, whereas the Eastern Alps followed this trend during the Miocene (Figure 4.11).

Today, tectonic shortening driven by plate convergence seems to have come to an end in the Western and Central Alps, given the low seismic activity and the prevalence of extension over-compression in these parts of the Alps (e.g. Calais et al. 2002; Sue et al. 2007; Nocquet et al. 2016), but may still continue in the Eastern Alps (Zattin et al. 2006; Heberer et al. 2017; Robl et al. 2017). In the absence of significant convergence, the mountainous topography of the Alps should decay over millions of years until the crustal root of the mountain belt has been consumed by erosion and isostatic compensation (e.g. Whipple 2009). Nonetheless, recent GPS measurements show that in the apparent absence of convergence, isostatic uplift in response to erosion and ice removal since the last glacial maximum (LGM) can only account for 50%–60% of the surface uplift observed in the Western Alps today (Mey et al. 2016; Nocquet et al. 2016). In the Central and Eastern Alps, in contrast, post-LGM erosion and ice removal can explain the full surface uplift pattern (Mey et al. 2016). This difference between the Western and Central Alps led Nocquet et al. (2016) and Sternai et al. (2019) to the conclusion that deep-seated processes (e.g. slab breakoff or mantle flow) in the Alps still have an important impact on the Alpine topography and the organization of the drainage system during the evolution of the mountain belt (e.g. Schlunegger and Hinderer 2001, 2003; Robl et al. 2017).

Given the present-day topography of the Western, Central and Eastern Alps and the relatively similar climatic conditions in terms of temperature and precipitation distribution, at least the present-day climate cannot explain the different levels of exhumation observed in the different parts of the Alps (Fox et al. 2015). It seems that the underlying tectonic processes controlled almost exclusively the evolution of the Alps. Variable erosion rates in relation to Quaternary climate change and successive glacial/interglacial periods are not reflected in the detrital apatite and zircon fission-track datasets, even though at least apatite fission-track dating has the temporal resolution for detecting changes in erosion rates with the Pliocene (Glotzbach et al. 2011a). The Quaternary topographic evolution of the Alps, particularly with respect to glaciations is treated in detail in Chapter 3 of Volume 1 (by Gradwohl et al.).

4.6. Conclusion

The existing thermochronological data of the present-day river and pro- and retro-side foreland basin deposits surrounding the Alps reflect the exhumation history and magnitude for the Western, Central, Southern, and Eastern Alps. The most rapid exhumation rates have locally been on the order of 1–3 km/Myr in the internal Western Alps and the Lepontine dome of the Central Alps, but overall the mountain belt has experienced long-term average exhumation rates of about 0.3 ± 0.1 km/Myr on the orogen scale since the Late Oligocene. The topographic evolution of the European Alps was contemporaneous with its exhumation history, showing an apparent west to east trend in the onset of topographic growth. Despite the extensive datasets available, the record of exhumation remains fragmentary and more information is needed, particularly for the Eastern Alps. The application of thermochronometers with even lower closure temperatures to detrital sediments, such as apatite (U-Th)/He, is needed to trace changes in exhumation rates on shorter time scales to better understand the impact of climate change on surface erosion and topographic evolution in the European Alps. Detrital thermochronology data definitely allow for more than one interpretation, and further research, completing the existing datasets, will allow the continuation of this discussion.

4.7. Acknowledgments

We thank the editors Claudio L. Rosenberg and Nicolas Bellahsen for the invitation to contribute to this chapter on detrital thermochronology. Constructive and detailed reviews by Jocelyn Barbarand and an anonymous reviewer helped us to improve the manuscript.

4.8. References

Baran, R., Friedrich, A.M., Schlunegger, F. (2014). The late Miocene to Holocene erosion pattern of the Alpine foreland basin reflects Eurasian slab unloading beneath the western Alps rather than global climate change. *Lithosphere*, 6, 124–131.

van der Beek, P.A. (2018). Mountain relief, climate and surface processes. In *Mountains, Climate and Biodiversity*, Hoorn, C., Perrigo, A., Antonelli, A. (eds). John Wiley & Sons Ltd., Hoboken.

van der Beek, P.A., Valla, P.G., Herman, F., Braun, J., Persano, C., Dobson, K., Labrin, E. (2010). Inversion of thermochronological age–elevation profiles to extract independent estimates of denudation and relief history II: Application to the French Western Alps. *Earth Planet. Sci. Lett.*, 296, 9–22.

Bermúdez, M.A., van der Beek, P., Bernet, M. (2013). Strong tectonic and weak climatic control on exhumation rates in the Venezuelan Andes. *Lithosphere*, 5, 3–16.

Bernet, M. (2013). Detrital zircon fission track thermochronology of the present-day Isère River drainage system in the Western Alps: No evidence for increasing erosion rates at 5 Ma. *Geosciences*, 3, 528–542.

Bernet, M. and Spiegel, C. (2004). Detrital thermochronology, provenance analysis, exhumation and landscape evolution of mountain belts. In *Detrital Thermochronology – Provenance Analysis, Exhumation and Landscape Evolution of Mountain Belts*, Bernet, M. and Spiegel, C. (eds). Geological Society of America Special Publication.

Bernet, M., Zattin, M., Garver, J.I., Brandon, M.T., Vance, J.A. (2001). Steady-state exhumation of the European Alps. *Geology*, 29, 35–38.

Bernet, M., Brandon, M.T., Garver, J.I., Molitor, B. (2004a). Downstream changes of Alpine zircon fission-track ages in the Rhône and Rhine rivers. *Journal of Sedimentary Research*, 74, 82–94.

Bernet, M., Brandon, M.T., Garver, J.I., Molitor, B. (2004b). Fundamentals of detrital zircon fission-track analysis for provenance and exhumation studies. In *Detrital Thermochronology – Provenance Analysis, Exhumation and Landscape Evolution of Mountain Belts*, Bernet, M. and Spiegel, C. (eds). Geological Society of America Special Publication.

Bernet, M., Brandon, M.T., Garver, J.I., Balestrieri, M.L., Ventura, B., Zattin, M. (2009). Exhuming the Alps through time: Clues from detrital zircon fission-track ages. *Basin Research*, 21, 781–798.

Beucher, R., van der Beek, P., Braun, J., Batt, G.E. (2012). Exhumation and relief development in the Pelvoux and Dora-Maira massifs (western Alps) assessed by spectral analysis and inversion of thermochronological age transects. *Journal of Geophysical Research – Earth Surface*, 117, F03030.

Bertrand, A., Rosenberg, C.L., Rabaute, A., Herman, F., Fügenschuh, B. (2017). Exhumation mechanisms of the Tauern Window (Eastern Alps) inferred from apatite and zircon fission track thermochronology. *Tectonics*, 36, 207–228.

Bousquet, R., Oberhänsli, R., Schmid, S.M., Berger, A., Wiederkehr, M., Robert, C., Möller, A., Rosenberg, C.L., Zeilinger, G., Molli, G. et al. (2012). Metamorphic framework of the Alps. Map, 1:1,000,000, CCGM CCMV.

Boyet, M., Lapierre, H., Tardy, M., Bosch, D., Maury, R. (2001). Sources of the andesitic components in the Taveyannaz Sandstones and Champsaur Sandstones: Implications for the Paleogene geodynamic evolution of the Alps. *Bulletin de la Société géologique de France*, 172, 487–501.

Brandon, M.T. (1996). Probability density plot for fission-track grain-age samples. *Radiation Measurements*, 26, 663–676.

Brandon, M.T., Roden-Tice, M.K., Garver, J.I. (1998). Late Cenozoic exhumation of the Cascadia accretionary wedge in the Olympic Mountains, northwest Washington State. *Geological Society of America Bulletin*, 110, 985–1009.

Braun, J. (2003). Pecube: A new finite-element code to solve the 3D heat transport equation including the effects of time-varying, finite amplitude surface topography. *Computers and Geosciences*, 29, 787–794.

Calais, E., Nocquet, J.-M., Jouanne, F., Tardy, M. (2002). Current strain regime in the Western Alps from continuous Global Positioning System measurements, 1996–2001. *Geology*, 30, 651–654.

Callec, Y. (2001). La déformation synsédimentaire des bassins paléogènes de l'Arc de Castellane (Annot, Barrême, St-Antonin). PhD Thesis, École des Mines de Paris.

Campani, M., Mancktelow, N., Seward, D., Rolland, Y., Muller, W., Guerra, I. (2010). Geochronological evidence for continuous exhumation through the ductile-brittle transition along a crustal-scale low-angle normal fault: Simplon Fault Zone, Central Alps. *Tectonics*, 29, TC3002, 1–25.

Campani, M., Mulch, A., Kempf, O., Schlunegger, F., Mancktelow, N. (2012). Miocene paleotopography of the central Alps. *Earth Planet. Sci. Lett.*, 337–338, 174–185.

Carrapa, B., Wijbrans, J., Bertotti, G. (2003). Episodic exhumation in the Western Alps. *Geology*, 31, 601–604.

Carrapa, B., Wijbrans, J., Bertotti, G. (2004a). Detecting provenance variations and cooling patterns within the western Alpine orogen through $^{40}Ar/^{39}Ar$ geochronology on detrital sediments: The Tertiary Piedmont Basin, northwest Italy. In *Detrital Thermochronology — Provenance Analysis, Exhumation, and Landscape Evolution of Mountain Belts*, Bernet, M. and Spiegel, C. (eds). Geological Society of America Special Paper.

Carrapa, B., Di Giulio, A., Wijbrans, J. (2004b). The early stages of the Alpine collision: An image derived from the upper Eocene-lower Oligocene record in the Alps-Apennines junction area. *Sedimentary Geology*, 171, 181–203.

Carrapa, B., DeCelles, P.G., Reiners, P.W., Gehrels, G.E., Sudo, M. (2009). Apatite triple dating and white mica $^{40}Ar/^{39}Ar$ thermochronology of syntectonic detritus in the Central Andes: A multiphase tectonothermal history. *Geology*, 37, 407–410.

Carrapa, B., Di Giulio, A., Nicoletta, M., Stockli, D., Fantoni, R., Hughes, A., Gupta, S. (2016). Tectonic significance of Cenozoic exhumation and foreland basin evolution of the Western Alps. *Tectonics*, 35, 1892–1912.

Cederbom, C.E., Sinclair, H.D., Schlunegger, F., Rahn, M.K. (2004) Climate-induced rebound and exhumation of the European Alps. *Geology*, 32, 709–712.

Cederbom, C.E., van der Beek, P., Schlunegger, F., Sinclair, H., Oncken, O. (2011). Rapid, extensive erosion of the North Alpine foreland basin at 5–4 Ma: Climatic, tectonic and geodynamic forcing on the European Alps. *Basin Research*, 23, 528–550.

Dumont, T., Schwartz, S., Guillot, S., Simon-Labric, T., Tricart, P., Jourdan, S. (2012). Structural and sedimentary records of the Oligocene revolution in the western Alpine arc. *Journal of Geodynamics*, 56–57, 18–38.

Dunkl, I. and Székely, B. (2002). Component analysis with visualization of fitting PopShare, a Windows program for data analysis, Goldschmidt Conference Abstracts 2002. *Geochim. Cosmochim. Acta*, 66/15A, 201.

Ehlers, T.A., Chaudhri, T., Kumar, S., Fuller, C.S., Willett, S.D., Ketcham, R.A., Brandon, M.T., Belton, D.X., Kohn, B.P., Gleadow, A.J.W. et al. (2005). Computational tools for low-temperature thermochronometer interpretation. In *Low-temperature Thermochronology: Techniques, Interpretations, and Applications*, Reiners, P.W. and Ehlers, T.A. (eds). De Gruyter, Berlin.

Fauquette, S., Bernet, M., Suc, J.P., Grosjean, A.S., Guillot, S., van der Beek, P., Jourdan, S., Popescu, S.M., Jiménez-Moreno, G., Bertini, A. et al. (2015). Quantifying the Eocene to Pleistocene topographic evolution of the southwestern Alps, France and Italy. *Earth Planet. Sci. Lett.*, 412, 220–234.

Foeken, J.P.T., Persano, C., Stuart, F.M., ter Voorde, M. (2007). Role of topography in isotherm perturbation: Apatite (-Th)/He and fission-track results from the Malta tunnel, Tauern Window, Austria. *Tectonics*, 26, TC3006.

Ford, M., Lickorish, W.H., Kusznir, N.J. (1999). Tertiary foreland sedimentation in the Southern Subalpine Chains, SE France: A geodynamic appraisal. *Basin Research*, 11, 315–336.

Fox, M., Herman, F., Kissling, E., Willett, S.D. (2015). Rapid exhumation in the Western Alps driven by slab detachment and glacial erosion. *Geology*, 43, 379–382.

Fox, M., Herman, F., Willett, S.D., Schmid, S.M. (2016). The exhumation history of the European Alps inferred from linear inversion of thermochronological data. *American Journal of Science*, 316, 505–541.

Frey, M., Desmons, J., Neubauer, F. (1999). The new metamorphic map of the Alps: Introduction. *Schweizerische mineralogische und petrographische Mitteilungen*, 79, 1–4.

Frisch, W., Kuhlemann, J., Dunkl, I., Brügel, A. (1998). Palinspastic reconstruction and topographic evolution of the Eastern Alps during late Tertiary tectonic extrusion. *Tectonophysics*, 297, 1–15.

Fügenschuh, B. and Schmid, S.M. (2003). Late stages of deformation and exhumation of an orogen constrained by fission-track data: A case study in the Western Alps. *Geol. Soc. Am. Bull.*, 115, 1425–1440.

Galbraith, R.F. (2005). *Statistics for Fission Track Analysis*. CRC Press, Boca Raton.

Galbraith, R.F. and Laslett, G.M. (1993) Statistical models for mixed fission track ages. *Nuclear Tracks and Radiation Measurements*, 21, 459–470.

Gallagher, K. and Parra, M. (2020). A new approach to thermal history modelling with detrital low temperature thermochronological data. *Earth Planet. Sci. Lett.*, 529, 115872.

Garver, J.I., Brandon, M.T., Roden-Tice, M., Kamp, P.J.J. (1999). Exhumation history of orogenic highlands determined by detrital fission-track thermochronology. In *Exhumation Processes: Normal Faulting, Ductile Flow and Erosion*, Ring, U., Brandon, M.T., Willett, S.D., Lister, G.S., (eds). Geol. Soc. Lond. Spec. Publ., London.

Garzanti, E., Resentini, A., Vezzoli, G., Andò, S., Malusà, M.G., Padoan, M. (2012). Forward compositional modelling of Alpine orogenic sediments. *Sedimentary Geology*, 280, 149–164.

Gemignani, L., Sun, X., Braun, J., van Gerve, T.D., Wijbrans, J.R. (2017). A new detrital mica $^{40}Ar/^{39}Ar$ dating approach for provenance and exhumation of the Eastern Alps. *Tectonics*, 36, 1521–1537.

Gemignani, L., van der Beek, P.A., Braun, J., Najman, Y., Bernet, M., Garzanti, E., Wijbrans, J.R. (2018). Downstream evolution of the thermochronologic age signal in the Brahmaputra catchment (eastern Himalaya: Implications for the detrital record of erosion. *Earth Planet. Sci. Lett.*, 499, 48–61.

Glotzbach, C., Bernet, M., van der Beek, P. (2011a). Detrital thermochronology records changing source areas and steady exhumation in the Western and Central European Alps. *Geology*, 39, 239–242.

Glotzbach, C., van der Beek, P.A., Spiegel, C. (2011b). Episodic exhumation and relief growth in the Mont Blanc massif, Western Alps from numerical modelling of thermochronology data. *Earth Planet. Sci. Lett.*, 304, 417–430.

Glotzbach, C., Busschers, F.S., Winsemann, J. (2017). Detrital thermochronology of Rhine, Elbe and Meuse river sediment (Central Europe): Implications for provenance, erosion and mineral fertility. *Int. J. Earth Sci.*, 107, 459–479.

Govin, G., van der Beek, P., Najman, Y., Millar, I., Gemignani, L., Huyghe, P., Dupont-Nivct, G., Bernet, M., Mark, C., Wijbrans, J. (2020). Early-onset and late acceleration of rapid exhumation in the Namche Barwa syntaxis, eastern Himalaya. *Geology*, 48, 1139–1143.

Gradwohl, G., Stüwe, K., Robl, J., Liebl, M. (2021). Topography and landscape evolution of the Alps. In *Geodynamics of the Alps 1*, Rosenberg, C.L. and Bellahsen, N. (eds). ISTE Ltd, London, and John Wiley & Sons, New York.

Grasemann, B. and Mancktelow, N.S. (1993). Two-dimensional thermal modelling of normal faulting: The Simplon fault zone, Central Alps, Switzerland. *Tectonophysics*, 225, 155–165.

von Hagke, C., Cederbom, C.E., Oncken, O., Stöckli, D.F., Rahn, M.K. Schlunegger, F. (2012). Linking the northern Alps with their foreland: The latest exhumation history resolved by low-temperature thermochronology. *Tectonics*, 31, 10TC5010.

Heberer, B., Reverman, R.L., Fellin, M.G., Neubauer, F., Dunkl, I., Zattin, M., Seward, D., Genser, J., Brack, P. (2017). Postcollisional cooling history of the Eastern and Southern Alps and its linkage to Adria indentation. *Int. J. Earth Sci.*, 106, 1557–1580.

Huyghe, P., Bernet, M., Galy, A., Naylor, M., Cruz, J., Gyawali, B.R., Gemignani, L., Mugnier, J.L. (2020). Rapid exhumation since at least 13 Ma in the Himalayas recorded by detrital apatite fission-track dating of Bengal fan (IODP Expedition 354) and modern Himalayan river sediments. *Earth Planet. Sci. Lett.*, 534, 116078.

Jacob, J.B., Guillot, S., Thiéblemont, D., Vanardois, J., Trap, P., Faure, M., Marquer, D. (2021). Structuration of the Paleozoic basement and Variscan inheritance in the Alps. In *Geodynamics of the Alps 2*, Rosenberg, C.L., and Bellahsen, N. (eds). ISTE Ltd, London, and John Wiley & Sons, New York.

Jourdan, S., Bernet, M., Schwartz, S., Guillot, S., Tricart, P., Chauvel, C., Dumont, T., Montagnac, G., Bureau, S. (2012). Tracing the Oligocene–Miocene evolution of the Western Alps drainage divide with pebble petrology, geochemistry and Raman spectroscopy of foreland basin deposits. *The Journal of Geology*, 120, 603–624.

Jourdan, S., Bernet, M., Tricart, P., Hardwick, E., Paquette, J.-L., Guillot, S., Dumont, T., Schwartz, S. (2013). Short-lived, fast erosional exhumation of the internal Western Alps during the late early Oligocene: Constraints from geothermochronology of pro- and retro-side foreland basin sediments. *Lithosphere*, 5, 211–225.

Jourdan, S., Bernet, M., Hardwick, E., Paquette, J.L., Tricart, P., Senebier, F., Coeur, F. (2018). Geo-thermochronology of the Saint Antonin basin, south-eastern France. *BSGF – Earth Sciences Bulletin*, 189, 12.

Keil, M. and Neubauer, F. (2015). Orogen-parallel extension and topographic gradients east of the Tauern Window: A possible indication of intra-orogenic raft tectonics? *Austrian Journal of Earth Sciences*, 108, 4–15.

Ketcham, R.A., Donelick, R.A., Carlson, W.D. (1999). Variability of apatite fission-track annealing kinetics III: Extrapolation to geological time scales. *American Mineralogist*, 84, 1235–1255.

Kuhlemann, J. and Kempf, O. (2002). Post-Eocene evolution of the North Alpine Foreland Basin and its response to Alpine tectonics. *Sedimentary Geology*, 152, 45–78.

Kuhlemann, J., Frisch, W., Székely, B., Dunkl, I., Kázmér, M. (2002), Post-collisional sediment budget history of the Alps: Tectonic versus climatic control. *Int. J. Earth Sci.*, 91, 818–837.

Kuhlemann, J., Dunkl, I., Brügel, A., Spiegel, C., Frisch, W. (2006). From source terrains of the Eastern Alps to the Molasse basin: Detrital record of non-steady-state exhumation. *Tectonophysics*, 413, 301–316.

Lu, G., Fellin, M.G., Winkler, W., Rahn, M., Guillong, M., von Quadt, A., Willett, S.D. (2020). Revealing exhumation of the central Alps during the Early Oligocene by detrital zircon U–Pb age and fission-track double dating in the Taveyannaz Formation. *Int. J. Earth Sci.*, 109, 2425–2446.

Malusà, M.G. and Balestrieri, M.L. (2012). Burial and exhumation across the Alps–Apennines junction zone constrained by fission-track analysis on modern river sands. *Terra Nova*, 24, 221–226.

Malusà, M.G. and Fitzgerald, P.G. (2019). *Fission-track Thermochronology and its Application to Geology*. Springer, New York.

Malusà, M.G. and Fitzgerald, P.G. (2020). The geologic interpretation of the detrital thermochronology record within a stratigraphic framework, with examples from the European Alps, Taiwan and the Himalayas. *Earth-Science Reviews*, 201, 103074.

Malusà, M.G., Zattin, M., Andò, S., Garzanti, E., Vezzoli, G. (2009). Focused erosion in the Alps constrained by fission-track ages on detrital apatites. *Geological Society, London, Special Publications*, 324, 141–152.

Malusà, M.G., Carter, A., Limoncelli, M., Villa, I.M., Garzanti, E. (2013). Bias in detrital zircon geochronology and thermochronology. *Chemical Geology*, 359, 90–107.

Malusà, M.G., Faccenna, C., Baldwin, S.L., Fitzgerald, P.G., Rossetti, F., Balestrieri, M.L., Danišík, M., Ellero, A., Ottria, G., Piromallo, C. (2015). Contrasting styles of (U)HP rock exhumation along the Cenozoic Adria-Europe plate boundary (Western Alps, Calabria, Corsica). *Geochemistry, Geophysics, Geosystems*, 16, 1786–1824.

Malusà, M.G., Wang, J., Garzanti, E., Liu, Z.C., Villa, I.M., Wittmann, H. (2017). Trace-element and Nd-isotope systematics in detrital apatite of the Po river catchment: Implications for provenance discrimination and the lag-time approach to detrital thermochronology. *Lithos*, 290–291, 48–59.

Mancktelow, N.S. (1992). Neogene lateral extension during convergence in the Central Alps: Evidence from interrelated faulting and backfolding around the Simplonpass (Switzerland). *Tectonophysics*, 215, 295–317.

Mark, C., Cogné, N., Chew, D. (2016). Tracking exhumation and drainage divide migration of the Western Alps: A test of the apatite U-Pb thermochronometer as a detrital provenance tool. *Geological Society of America Bulletin*, 128, 1439–1460.

Mazurek, M., Hurford, A.J., Leu, W. (2006). Unravelling the multi-stage burial history of the Swiss Molasse Basin: Integration of apatite fission track, vitrinite reflectance and biomarker isomerisation analysis. *Basin Research*, 18, 27–50.

McDougall, I. and Harrison, T. (1988). *Geochronology and Thermochronology by the $^{40}Ar/^{39}Ar$ Method*. Oxford University Press, New York.

McPhillips, D. and Brandon, M.T. (2010). Using tracer thermochronology to measure modern relief change in the Sierra Nevada, California. *Earth Planet. Sci. Lett.*, 296, 373–383.

Mey, J., Scherler, D., Wickert, A.D., Egholm, D.L., Tesauro, M., Schildgen, T.F., Strecker, M.R. (2016). Glacial isostatic uplift of the European Alps. *Nature Communications*, 7, 13382.

Molnar, P. (2018). Simple concepts underlying the structure, support and growth of mountain ranges, high plateaus and other high terrain. In *Mountains, Climate and Biodiversity*, Hoorn, C., Perrigo, A., Antonelli, A. (eds). John Wiley & Sons Ltd, Hoboken.

Morag, N., Avigad, D., Harlvan, Y., McWilliams, M.O., Michard, A. (2008). Rapid exhumation and mountain building in the Western Alps: Petrology and $^{40}Ar/^{39}Ar$ geochronology of detritus from Tertiary basins of southeastern France. *Tectonics*, 27, TC2004.

Napoli, A., Crespi, A., Ragone, F., Maugeri, M., Pasquero, C. (2019). Variability of orographic enhancement of precipitation in the Alpine region. *Scientific Reports*, 9, 13352.

Naylor, M., Sinclair, H.D., Bernet, M., van der Beek, P., Kirstein, L.A. (2015). Bias in detrital fission track grain-age populations: Implications for reconstructing changing erosion rates. *Earth Planet. Sci. Lett.*, 422, 94–104.

Nocquet, J.M., Sue, C., Walpersdorf, A., Tran, T., Lenôtre, N., Vernant, P., Cushing, M., Jouanne, F., Masson, F., Baize, S. et al. (2016). Present-day uplift of the western Alps. *Scientific Reports*, 6, 28404.

Pfiffner, A.O. (2016). Basement-involved thin-skinned and thick-skinned tectonics in the Alps. *Geological Magazine*, 153, 1085–1109.

Rahn, M.K., Brandon, M.T., Batt, G.E., Garver, J.I. (2004). A zero-damage model for fission-track annealing in zircon. *American Mineralogist*, 89, 473–484.

Reiners, P.W. and Brandon, M.T. (2006). Using thermochronology to understand orogenic erosion. *Ann. Rev. Earth Planet. Sci.*, 34, 419–466.

Reiners, P.W. and Ehlers, T.A. (2005). Low-temperature thermochronology: Techniques, interpretations, and applications. *Reviews in Mineralogy and Geochemistry*, 58, 662.

Reiter, W., Elfert, S., Glotzbach, C., Bernet, M., Spiegel, C. (2013). Relations between denudation, glaciation, and sediment deposition: Implications from the Plio-Pleistocene Central Alps. *Basin Research*, 25, 659–674.

Reiter, W., Elfert, S., Glotzbach, C., Spiegel, C. (2015). Plio-Pleistocene evolution of the north Alpine drainage system: New constraints from detrital thermochronology of foreland deposits. *Int. J. Earth Sci.*, 104, 891–907.

Resentini, A. and Malusà, M.G. (2012). Sediment budgets by detrital apatite fission-track dating (Rivers Dora Baltea and Arc, Western Alps). *Geological Society of America Special Papers*, 487, 125–140.

Ring, U., Brandon, M.T., Willett, S.D., Lister, G.S. (1999). Exhumation processes. In *Exhumation Processes: Normal Faulting, Ductile Flow and Erosion*, Ring, U., Brandon, M.T., Lister, G.S., Willett, S.D. (eds). Geological Society London Special Publications, London.

Roe, G.H. (2005). Orographic Precipitation. *Ann. Rev. Earth Planet. Sci.*, 33, 645–671.

Schegg, R. and Leu, W. (1998). Analysis of erosion events and palaeogeothermal gradients in the North Alpine Foreland Basin of Switzerland. In *Basin Modelling: Practice and Progress*, Duppenbecker, S.J. and Iliffe, J.E. (eds). Geological Society of London, Special Publications, London.

Schildgen, T.F., van der Beek, P.A., Sinclair, H.D., Thiede, R.C. (2018). Spatial correlation bias in late-Cenozoic erosion histories derived from thermochronology. *Nature*, 559, 89–93.

Schlunegger, F. (1999). Controls of surface erosion on the evolution of the Alps: Constraints from the stratigraphies of the adjacent foreland basins. *Int. J. Earth Sci.*, 88, 285–304.

Schlunegger, F. and Hinderer, M. (2001). Crustal uplift in the Alps: Why the drainage pattern matters. *Terra Nova*, 13, 425–432.

Schlunegger, F. and Hinderer, M. (2003). Pleistocene/Holocene climate change, re-establishment of fluvial drainage network and increase in relief in the Swiss Alps. *Terra Nova*, 15, 88–95.

Schlunegger, F. and Kissling, E. (2015). Slab rollback orogeny in the Alps and evolution of the Swiss Molasse basin. *Nature Communications*, 6, 8605.

Schlunegger, F., Slingerland, R., Matter, A. (1998). Crustal thickening and crustal extension as controls on the evolution of the drainage network of the central Swiss Alps between 30 Ma and the present: Constraints from the stratigraphy of the North Alpine Foreland Basin and the structural evolution of the Alps. *Basin Research*, 10, 197–212.

Schlunegger, F., Melzer, J., Tucker, G.E. (2001). Climate, exposed source-rock lithologies, crustal uplift and surface erosion: A theoretical analysis calibrated with data from the Alps/North Alpine Foreland Basin system. *Int. J. Earth Sci.*, 90, 484–499.

Schmid, S.M., Pfiffner, O.A., Schoenborn, G., Froitzheim, N., Kissling, E. (1997). Integrated cross section and tectonic evolution of the Alps along the eastern traverse. In *Results of NRP 20: Deep Structure of the Swiss Alps*, Pffifner, O.A., Lehner, P., Heitzman, P., Müller, S., Steck A., (eds). Birkhäuser Verlag, Basel.

Schwartz, S., Guillot, S., Tricart, P., Bernet, M., Jourdan, S., Dumont, T., Montagnac, G. (2012). Source tracing of detrital serpentinite in the Oligocene molasse deposits from the western Alps (Barrême basin): Implications for relief formation in the internal zone. *Geological Magazine*, 149, 841–856.

Schwartz, S., Gautheron, C., Audin, L., Dumont, T., Nomade, J., Barbarand, J., Pinna-Jamme, R., van der Beek, P. (2017). Foreland exhumation controlled by crustal thickening in the Western Alps. *Geology*, 45, 139–142.

Simon-Labric, T., Rolland, Y., Dumont, T., Heymes, T., Authemayou, C., Corsini, M., Fornari, M. (2009). $^{40}Ar/^{39}Ar$ dating of Penninic Front tectonic displacement (W Alps) during the Lower Oligocene (31–34 Ma). *Terra Nova*, 21, 127–136.

Sinclair, H.D. (1997a). Flysch to molasse transition in peripheral foreland basins: The role of the passive margin versus slab breakoff. *Geology*, 25, 1123–1126.

Sinclair, H.D. (1997b). Tectonostratigraphic model for underfilled peripheral foreland basins: An Alpine perspective. *Geological Society of America Bulletin*, 109, 324–346.

Spiegel, C., Kuhlemann, J., Dunkl, I., Frisch, W., von Eynatten, H., Balogh, K. (2000). The erosion history of the Central Alps: Evidence from zircon fission track data of foreland basin sediments. *Terra Nova*, 12, 163–170.

Spiegel, C., Kuhlemann, J., Dunkl, I., Frisch, W. (2001). Paleogeography and catchment evolution in a mobile orogenic belt: The Central Alps in Oligo-Miocene times. *Tectonophysics*, 341, 33–47.

Spiegel, C., Siebel, W., Frisch, W., Berber, Z. (2002). Nd and Sr isotopic ratios and trace element geochemistry of epidote from the Swiss Molasse Basin as provenance indicators: Implications for the reconstruction of the exhumation history of the Central Alps. *Chemical Geology*, 189, 231–250.

Spiegel, C., Kuhlemann, J., Frisch, W. (2007). Tracing sediment pathways by zircon fission-track analysis: Oligocene marine connections in the Central Europe. *Int. J. Earth Sci.*, 96, 363–374.

Sternai, P., Sue, C., Husson, L., Serpelloni, E., Becker, T.W., Willett, S.D., Faccenna, C., Di Giulio, A., Spada, G., Jolivet, L. et al. (2019). Present-day uplift of the European Alps: Evaluating mechanisms and models of their relative contributions. *Earth-Science Reviews*, 190, 589–604.

Sue, C., Delacou, B., Champagnac, J.-D., Allanic, C., Tricart, P., Burkhard, M. (2007). Extensional neotectonics around the bend of the Western/Central Alps: An overview. *Int. J. Earth Sci.*, 96, 1101–1129.

Tagami, T., Galbraith, R.F., Yamada, R., Laslett, G.M. (1998). Revised annealing kinetics of fission tracks in zircon and geological implications. In *Advances in Fission-track Geochronology*, van den Haute, P. and De Corte, F. (eds). Kluwer, Dordrecht.

Thomson, S., Brandon, M.T., Reiners, P.W., Zattin, M., Isaacson, P.J., Balestrieri, M.L., (2010). Thermochronologic evidence for orogen-parallel variability in wedge kinematics during extending convergent orogenesis of the northern Apennines, Italy. *Geological Society of America Bulletin*, 122, 1160–1179.

Trautwein, B., Dunkl, I., Kuhlemann, J., Frisch, W. (2002). Cretaceous-Tertiary Rhenodanubian flysch wedge (Eastern Alps): Clues to sediment supply and basin configuration from zircon fission-track data. *Terra Nova*, 13, 382–393.

Tricart, P., van der Beek, P., Schwartz, S., Labrin, E. (2007). Diachronous late-stage exhumation across the western Alpine arc: Constraints from apatite fission-track thermochronology between the Pelvoux and Dora-Maira Massifs. *J. Geol. Soc. London*, 164, 163–174.

Vermeesch, P. (2009). RadialPlotter: A Java application for fission track, luminescence and other radial plots. *Radiation Measurements*, 44, 409–410.

Vermeesch, P. (2019). Statistics for fission-track thermochronology. In *Fission-track Thermochronology and its Application to Geology,* Malusà, M. and Fitzgerald, P. (eds). Springer, New York.

Vernon, A.J., van der Beek, P.A., Sinclair, H.D., Rahn, M.K. (2008). Increase in late Neogene denudation of the European Alps confirmed by analysis of a fission track thermochronology database. *Earth Planet. Sci. Lett.*, 270, 316–329.

Von Blanckenburg, F. and Davies, J.H. (1995). Slab breakoff: A model for syncollisional magmatism and tectonics in the Alps. *Tectonics*, 14, 120–131.

Von Eynatten, H. (2003). Petrography and chemistry of sandstones from the Swiss Molasse Basin: An archive of the Oligocene to Miocene evolution of the Central Alps. *Sedimentology*, 50, 703–724.

Von Eynatten, H. and Gaupp, R. (1999). Provenance of Cretaceous synorogenic sandstone in the Eastern Alps: Constraints from framework petrography, heavy minerals analysis and mineral chemistry. *Sedimentary Geology*, 124, 81–111.

Von Eynatten, Gaupp, R., Wijbrans, J.R. (1996). $^{40}Ar/^{39}Ar$ laser-probe dating of detrital white micas from Cretaceous sedimentary rocks of the Eastern Alps: Evidence for Variscan high-pressure metamorphism and implications for Alpine orogeny. *Geology*, 24, 691–694.

Von Eynatten, H., Schlunegger, F., Gaupp, R., Wijbrans, J.R. (1999). Exhumation of the Central Alps: Evidence from $^{40}Ar/^{39}Ar$ laserprobe dating of detrital white micas from the Swiss Molasse Basin. *Terra Nova*, 11, 284–289.

Whipple, K.X. (2009). The influence of climate on the tectonic evolution of mountain belts. *Nature Geoscience*, 2, 97–104.

Willett, S.D. (1999). Orogeny and orography: The effects of erosion on the structure of mountain belts. *Journal of Geophysical Research*, 104, 28975–28981.

Willett, S.D. (2010). Late Neogene erosion of the Alps: A climate driver? *Ann. Rev. Earth Planet. Sci.*, 38, 409–435.

Willett, S.D. and Brandon, M.T. (2013). Some analytical methods for converting thermochronometric age to erosion rate. *Geochemistry Geophysics Geosystems*, 14, 209–222.

Whipp Jr., D.M., Ehlers, T.A., Braun, J., Spath, C.D. (2009). Effects of exhumation kinematics and topographic evolution on detrital thermochronometer data. *Journal of Geophysical Research*, 114, F04021.

Wölfler, A., Dekant, C., Frisch, W., Danisik, M., Frank, W. (2015). Cretaceous to Miocene cooling of the Austroalpine units southeast of the Tauern Window (Eastern Alps) constrained by multi-system thermochronometry. *Austrian Journal of Earth Sciences*, 108, 16–33.

Zattin, M., Stefani, C., Martin, S. (2003). Detrital fission-track analysis and sedimentary petrofacies as keys of alpine exhumation: The example of the Venetian foreland (European Southern Alps, Italy). *Journal of Sedimentary Research*, 73, 1051–1061.

Zattin, M., Cuman, A., Fantoni, R., Martin, S., Scottu, P., Stefani, C. (2006). From Middle Jurassic heating to Neogene cooling: The thermochronological evolution of the southern Alps. *Tectonophysics*, 414, 191–202.

5

The Northern Deformation Front of the European Alps

Hugo ORTNER[1], Christoph VON HAGKE[2], Anna SOMMARUGA[3], Samuel MOCK[4,5], Jon MOSAR[3], Ralph HINSCH[6] and Andreas BEIDINGER[6]

[1]*Institute of Geology, University of Innsbruck, Austria*
[2]*Department of Geography and Geology, University of Salzburg, Austria*
[3]*Department of Geosciences, University of Fribourg, Switzerland,*
[4]*Institute of Geological Sciences, University of Bern, Switzerland*
[5]*Fachstelle Gefahrenprävention Kantonales Laboratorium Basel-Stadt, Switzerland*
[6]*OMV Exploration & Production GmbH, Vienna, Austria*

The northern deformation front of the European Alps, i.e. the northern limit of Alpine thrusting, is characterized by structural variability along strike. It can be subdivided into four regions: the deformation front in (i) France and Switzerland, (ii) western Austria and Bavaria, (iii) eastern Bavaria, Salzburg and Upper Austria and (iv) Lower Austria. In France and Switzerland, the deformation front can be subdivided into the external Jura Mountains, the Plateau Molasse and the Subalpine Molasse, i.e. the part of the foreland that has been included in the orogenic wedge after deposition. The deformation front in France and Switzerland is directly linked to the basal décollement below the External Crystalline Massifs. The external position of the Jura fold-thrust belt is associated with the occurrence of Triassic salt deposits in the European margin sediments. The Subalpine Molasse can be considered the frontal-most part of the Helvetic thrust sheets. South of the Jura fold-thrust belt, it is characterized by a continuous triangle zone extending from Lake Geneva to SW of

Geodynamics of the Alps 3,
coordinated by Claudio L. ROSENBERG and Nicolas BELLAHSEN.

Munich. South of Lake Constance lies the transition between the Central and Eastern Alps. While this transition is pronounced within the Alpine orogen, it is less evident at the deformation front. East of Lake Constance, structural styles as well as mechanical stratigraphy are comparable to observations in Switzerland and France. Further to the east, however, the Subalpine Molasse tapers off and completely disappears east of Munich. This coincides with a gradual change of structures exposed that the surface, as deeper sections of the Subalpine Molasse are exposed in the west. East of Munich in eastern Bavaria, Salzburg and Upper Austria, the location of the basal décollement of the deformation front varies between Cretaceous and Rupelian units. No frontal triangle zone developed; instead, duplexes buried by foreland deposits dominate the structural style. On the eastern side of the Bohemian Spur, there is a marked change in the strike of the Alpine deformation front, paralleling pre-existing geometries of the underlying European Basement units. This is also reflected in the structural styles and timing of thrusting along the deformation front.

This large variability of the geometries of the Alps' northern deformation front is associated with the variable mechanical stratigraphy along strike, the different links to more internal units of the Alps, as well as links to the deep structure of the Alps, in particular, the geometry and overall mechanical properties of the down-going European Plate. Similar to different geometries, the timing and sequence of thrusting of the deformation front are highly variable along strike. This difference is caused by a link of the deformation front to the deep and shallow dynamics of the Alps. This stresses the need for a comprehensive approach to resolving the dynamics of the deformation front, including sedimentological and structural studies from outcrop to mountain belt scale. At the same time, the complexity of the deformation front is a unique record of the dynamics of the orogen through time.

5.1. The deformation front of the Alps – an overview

The northern deformation front of the Alps at the orogen scale shows major lateral changes that partly coincide with changes in orogen architecture as mapped at the surface. West of Lake Constance, the Western and Central Alps largely consist of units which can be associated with the European Plate. Large structures such as the External Crystalline Massifs (ECM, such as Aar, Aiguilles Rouges, Mont Blanc and Belledonne; Figure 5.1) appear to be related to the structural architecture of the foreland. There, the deformation front corresponds to the frontal Jura deformational zone. Associated with the formation of the external Jura Mountains, the whole foreland in the west is detached above an evaporitic décollement level and the Molasse Basin is in a wedge top position (Plateau Molasse; PM of Figure 5.1). East of Lake Constance, remnants of the Apulian plate dominate and units related to the

European Plate are exposed only in isolated windows (e.g. Tauern Window; Figure 5.1). There, the deformation front separates the undeformed Foreland Molasse (FM of Figure 5.1) from the deformed parts of the foreland, the Subalpine Molasse, (Figure 5.1).

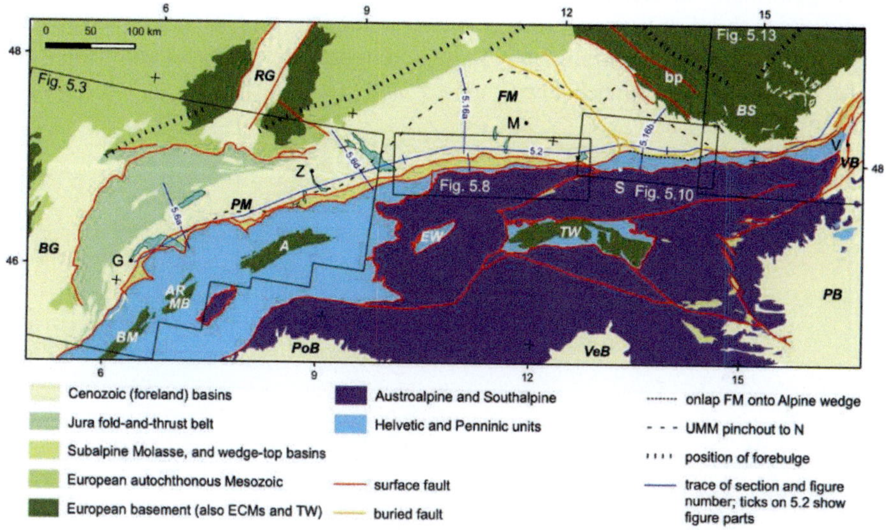

Figure 5.1. *Overview map of the peripheral Alpine foreland basin and the Alpine front at its southern margin. The positions of Figures 5.3, 5.8, 5.10 and 5.13 are shown*

COMMENTS ON FIGURE 5.1.– *Cenozoic foreland basins: FM = Foreland Molasse, PM = Plateau Molasse, PoB = Po basin, VeB = Venetian basin. Other Cenozoic basins: RG = Rhine Graben, BG = Bresse Graben, PB = Pannonian basin, VB = Vienna basin. bp = Pfahl fault ("Bayrischer Pfahl"). ECMs = External Massifs: A = Aar Massif, AR = Aiguille Rouge Massif, BM = Belledonne Massif, MB = Mont Blanc Massif. Major tectonic windows: EW = Engadine Window, TW = Tauern Window. BS = Bohemian Spur. G = Geneva, M = Munich, S = Salzburg, V = Vienna, Z = Zurich. The northern pinchout of the UMM after Freudenberger and Schwerd (1996) and Berger et al. (2005a). The position of the present-day forebulge has been drawn along topographic culminations in a smoothed digital elevation model with 10 km resolution and roughly coincides with the position given by Burkhard and Sommaruga (1998).*

The Subalpine Molasse forms the frontal part of a foreland fold-thrust belt. In this frontal portion, the décollement steps up from beneath the Mesozoic sedimentary cover to the base of the Cenozoic sediments (Figure 5.2). The

Subalpine Molasse has a rather uniform size in map view between Lake Geneva and Munich. East of Munich, it narrows, until it completely disappears west of Salzburg at the surface (Figure 5.1). Despite the apparent uniformity of the Subalpine Molasse along strike, structural differences can be observed, such as the number of tectonic slices, deformational style, sequence of deformation and amount of exhumation. Furthermore, the décollement horizon is not continuous and its location depends largely on the mechanical stratigraphy preserved in the foreland basin sediments and the Mesozoic platform series below (Figure 5.2).

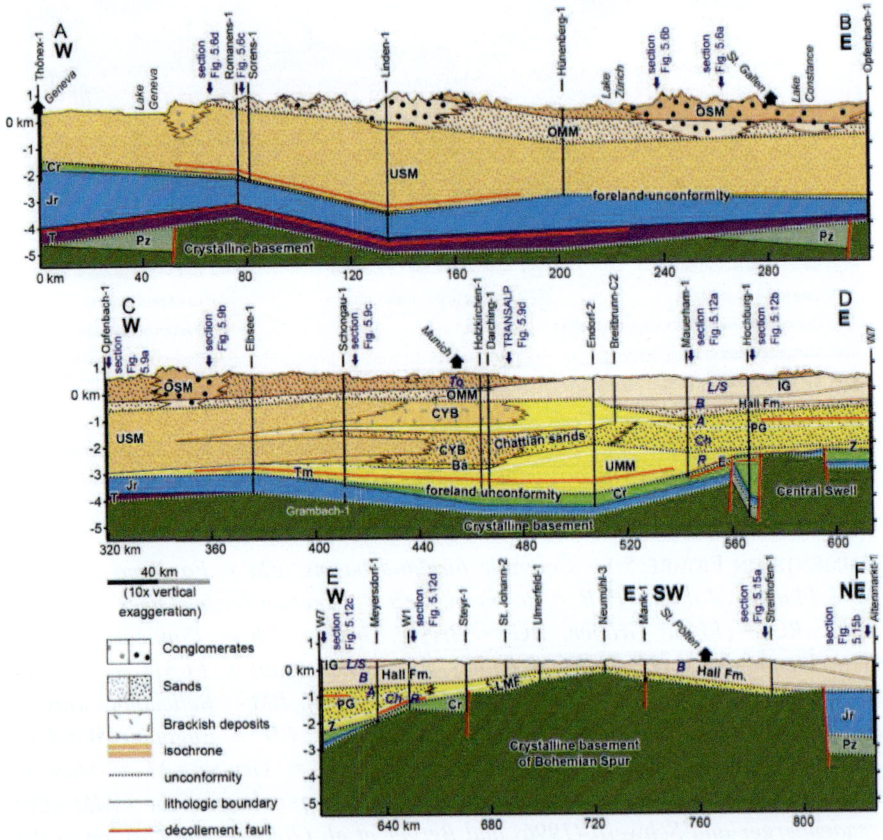

Figure 5.2. *Cross-section from Lake Geneva to Vienna in the foreland close to the Subalpine Molasse thrust front (see Figure 5.1 for location) based on the information about wells (see Table 5.1)*

COMMENTS ON FIGURE 5.2.– *The middle part (320–600 km) is inspired by Lemcke (1988), the 600–720 km part follows Rupp (2008). Informal and formal lithostratigraphic units (according to Piller et al. (2004) in the Austrian part of the foreland basin): OSM = Upper Freshwater Molasse, OMM = Upper Marine Molasse, IG = Innviertel Group, USM = Lower Freshwater Molasse, CYB = Cyrena Beds, Ba = Baustein Beds, Tm = Tonmergel Beds, PG = Puchkirchen Group, LMF = Linz-Melk Fm., Z = Rupelian (Schöneck, Dynow, Eggerding, Zupfing Fms.), E = Eocene, Cr = Cretaceous, Jr = Jurassic, T = Triassic, Pz = Paleozoic. Abbreviations for ages (in blue): R = Rupelian, Ch = Chattian, A = Aquitanian, B = Burdigalian, L/S = Langhian/Serravallian, To = Tortonian and younger. The red line indicates the position of the décollement horizons within the sedimentary succession. See also Figure 5.4 for chronostratigraphy of the Swiss foreland (A and B), Figure 5.7 for southern Germany (C and D), Figure 5.10 for northern Austria west of the Bohemian Spur and Figure 5.14 for northern Austria east of the Bohemian Spur (E and F).*

East of the Subalpine Molasse and west of the Bohemian Massif (Figure 5.1), the deformation front of the Alps is located within foreland deposits that are not separated from the underlying sediments of the European margin. The deformation front is buried below Early Miocene foreland sediments (Figure 5.1).

Another major structural change within the European Plate is associated with the Bohemian Massif, a prominent structural element on top of which the base of the Alpine foreland rises to a shallow depth in a culmination before it sinks toward the Vienna basin (Figure 5.2). Along the SW margin of the Bohemian Massif a bundle of faults such as the Pfahl fault ("Bayerischer Pfahl"; bp of Figure 5.1; e.g. Hofmann 1962), the Franconian fault and parallel faults in the subsurface of the foreland basin can be observed (Figures 5.1 and 5.2; Bergerat and Geyssant 1983; Bergerat 1987; Lemcke 1988). These faults delimit a zone of thickened crust in the Bohemian Massif (Ziegler and Dèzes 2006; Tesauro et al. 2008; Grad et al. 2009). West of these fault sediments on the European epicontinental platform and slope have Germanic-/Helvetic and Ultrahelvetic facies (e.g. Freudenberger and Schwerd 1996), while east of the faults, a Jurassic Gresten-type facies (i.e. fluvial, limnic to coastal coal deposits overlain by open marine marls) on rifted European continental crust is present (Granado et al. 2016; Tari and Strauss 2018). There, km-scale basement half-grabens are filled by these Gresten-type syn-rift deposits and covered by carbonates (Wagner 1998; Picha et al. 2006). During Cenozoic inversion, slices detached from such horsts were incorporated into the overthrusting allochthons to form the so-called "Klippenzonen" (Prey 1980; Wessely et al. 2006; Ultrahelvetic Hauptklippenzone or Gresten Klippen Zone, but also the Waschberg Super Unit of the Subalpine Molasse; see section 5.2.4). These occur only east of Salzburg (Prey 1953; Plöchinger and Prey 1968; Schuster et al. 2007).

Name of well	Abbr.	Reference	Name of well	Abbr.	Reference
Altenmarkt-1	A1	Brix and Götzinger (1964); Beidinger and Decker (2014)	**Meyersdorf-1**	Me1	Kohl and Krenmayr (1997)
Breitbrunn-C2	B1	Ganss (1983)	**Neumühl-3**	N3	Grill and Waldmann (1949); Brix and Götzinger (1964)
Darching-1	D1	Lange (1985)	**Opfenbach-1**	O1	Lemcke (1988)
Elbsee-1	E1	Müller et al. (1988); Freitag (2003)	**Thônex-1**	T1	Jenny et al. (1995)
Endorf-2	Ed2	Ganss (1977, 1983)	**Romanens-1**	R1	Maurer and Nabholz (1980); Lahusen and Wyss (1995); Sommaruga et al. (2012)
Feilnbach-1	F1	Paulus and Brockert (1968)	**Schongau-1**	S1	Müller (2011)
Grambach-1	G1	Hiltmann et al. (1999)	**Sorens-1**	So1	Maurer (1983)
Hindelang-1	Hd1	Huber and Schwerd (1995)	**Steyr-1**	St1	Brix and Schultz (1993); Kröll and Wessely (2001)
Hochburg-1	Hb1	Grunert et al. (2013); Grunert et al. (2015)	**Streithofen-1**	Sh1	Wachtel and Wessely (1981); Beidinger and Decker (2014)
Holzkirchen-1	H1	Lange (1985)	**St. Johann-2**	SJ2	Grill and Waldmann (1949); Brix and Götzinger (1964); Brix and Schultz (1993)
Hünenberg-1	Hu1	Lemcke et al. (1968)	**Tuggen-1**	Tu1	Mühlberg (1943); Sommaruga et al. (2012)
Linden-1	Li1	Kern (1982); Choignard (1989); Sommaruga et al. (2012)	**Ulmerfeld-1**	U1	Brix and Schultz (1993); Wessely et al. (2006)
Mank-1	Ma1	Brix and Götzinger (1964)	**W1 – W11**	-	Hinsch (2013)
Mauerham-1	M1	Müller (1978)			

Table 5.1. *Alphabetic list of deep wells used in this paper and sources of information. Abbr. = Abbreviations for wells in Figures 5.3, 5.8, 5.10 and 5.13*

5.2. Regional description

5.2.1. *The deformation front in Switzerland and France*

The Molasse Basin s.str. in Switzerland and France shows differences in structural styles between western and eastern parts (Figure 5.3). In the western Swiss part, typical structures within the basin are broad anticlines oriented NE–SW and tear faults trending N–S, NW–SE and WNW–ESE in a conjugate system (Sommaruga et al. 2012). The eastern part comprises the thickest and most complete part of the Cenozoic infill and it is only weakly structured. The northern limit of the Plateau Molasse corresponds to an erosional limit along the most internal folds of the Jura Mountains, while the Jura Molasse represents the remnants of the Molasse Basin that have been passively involved in the Jura folding and thrusting. To the south, the so-called Folded Molasse forms the transition zone to the tectonic imbricates of the Subalpine Molasse. All these domains form together with the North Alpine Molasse Basin or Molasse Basin s.l. The deformation front in Switzerland and France takes a special position within the deformation front of the Alps, as strain is partitioned between the external Jura fold-thrust belt (JFTB) and the Subalpine Molasse.

5.2.1.1. *Sedimentary units of the Molasse Basin s.l. in Switzerland and France*

The western North Alpine foreland basin is classically subdivided into five lithostratigraphic units (Figure 5.4; Homewood et al. 1986) representing a succession of depositional cycles and erosional events linked to the development of the orogenic wedge: the North Helvetic Flysch (NHF) and the Lower Marine Molasse (UMM), the Lower Freshwater Molasse (USM), the Upper Marine Molasse (OMM) and the Upper Freshwater Molasse (OSM). These five units correspond to two shallowing-, coarsening- and thickening-upward megasequences: the first megasequence from deep marine NHF (Priabonian) to marine UMM (Rupelian) to fluvial USM (Rupelian to Aquitanian) and a second megasequence from marine OMM (Burdigalian) to fluvial OSM (Late Burdigalian to Serravallian) (Füchtbauer 1967; Kempf et al. 1999).

The basin started as a narrow, underfilled Flysch-type basin during Paleocene to Eocene times. Clastic sediments derived from the orogen and marls and limestones were deposited in today's Helvetic domain (Homewood 1983; Homewood and Lateltin 1988; Caron et al. 1989; Sinclair 1997). In Switzerland, Paleocene to Eocene deposits are very sparse and characterized by their reddish color. Karst pockets containing iron-rich conglomerates (Siderolithic), Bohnerz deposits and bauxitic paleosols represent the sedimentary record of the Eocene age (Hooker and Weidmann 2000; Berger et al. 2005b; Menkveld-Gfeller et al. 2016).

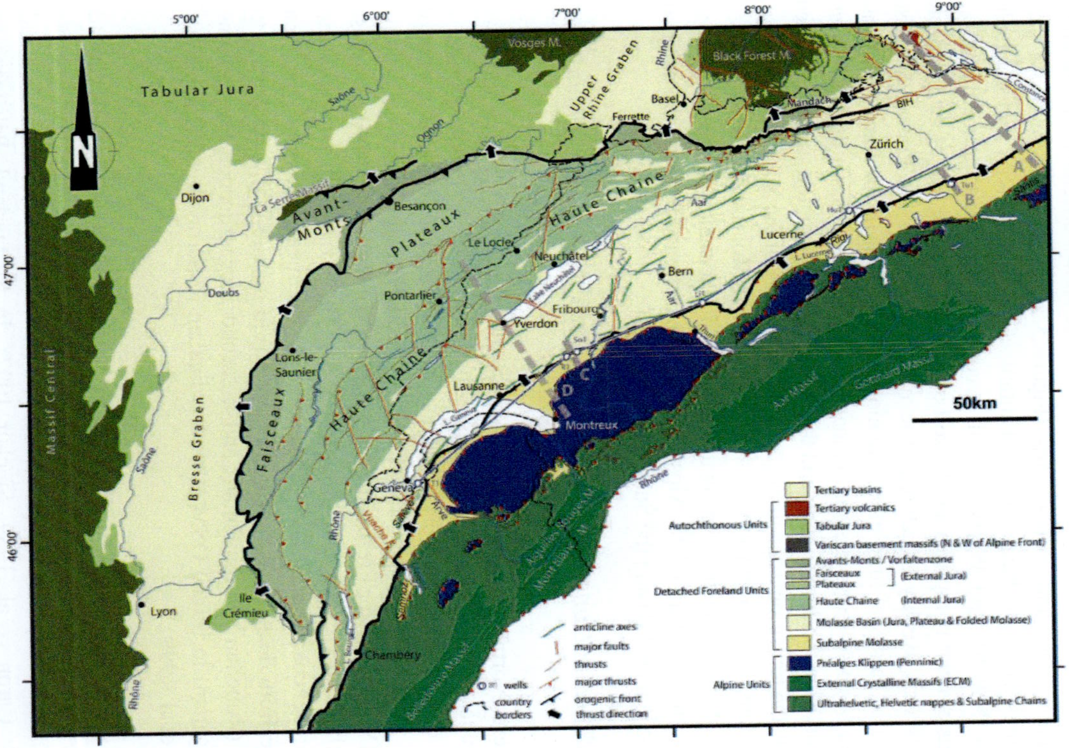

Figure 5.3. *Tectonic map of the Central and Western Alps including the Jura fold-thrust belt and the western Molasse Basin. Anticline axes are represented only in the Plateau and Folded Molasse. Grey dashed lines show the location of the cross-sections in Figure 5.6. The strike blue line represents sections A–B of Figure 5.2. Abbreviations: names of wells are given in Table 5.1; BIH: Baden-Irchel-Herdern lineament. Modified from Sommaruga et al. (2017)*

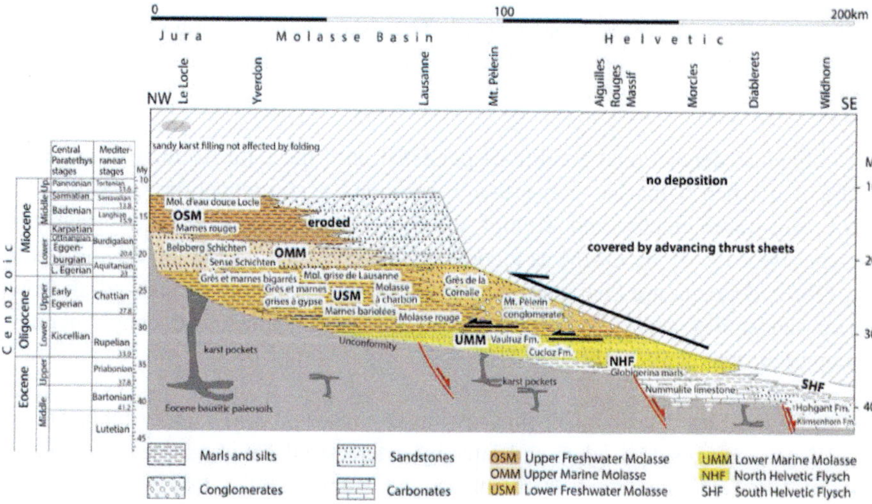

Figure 5.4. *Lithostratigraphic groups and facies across the western Molasse Basin (modified from Burkhard and Sommaruga, 1998; Strunck 2001; Gruber, 2017). NHF: North Helvetic Flysch; OMM: Upper Marine Molasse ("Obere Meeresmolasse"); OSM: Upper Freshwater Molasse ("Obere Süsswassermolasse"); SHF: South Helvetic Flysch; UMM: Lower Marine Molasse ("Untere Meeresmolasse"); USM: Lower Freshwater Molasse ("Untere Süsswassermolasse"). The correlation between Central Paratethys stages and Mediterranean stages follows Hilgen et al. (2012), absolute ages are from Cohen et al. (2013; updated)*

Following Flysch deposition, the foreland basin evolved into a filled stage, Molasse type basin. The onset of the filled stage is characterized by the deposition of the UMM unit during the Rupelian in the southern part of the basin. It is composed of a sequence of claystones and marls (lower part) and a more sandstone-dominated upper part. UMM layers exist in the allochthonous Subalpine Molasse (Weidmann et al. 1993; Strunck 2001) whereas autochthonous UMM layers up to 70 m thick are documented in different wells in the southern part of the Plateau Molasse (Sommaruga et al. 2012; Gruber 2017). Autochthonous UMM recorded deposits are spatially restricted suggesting a pinch out of UMM units toward the NW (Figures 5.1 and 5.4; Homewood et al. 1986; Strunck 2001).

During the Late Rupelian times, related to a global sea-level drop, the depositional environment evolved toward a continental fluvial and lacustrine environment, where USM sediments were deposited (Berger et al. 2005b). Large-scale alluvial fans supplied by clastic material from the Alpine orogeny (especially the Penninic Klippen in the west) formed. They show a coarsening and thickening upward trend due to radial progradation of the alluvial fans toward the foreland basin (Homewood et al. 1986).

Continental depositional conditions persisted until the Late Aquitanian or Early Burdigalian when a marine transgression provided accommodation space for the OMM unit (Kuhlemann and Kempf 2002). Allen et al. (1986) describe the transgressive base of the OMM as a conformable contact with the underlying USM unit. OMM deposits are characterized by thick-bedded, tidally influenced shallow marine sandstones and, less abundant, mudstones, siltstones and conglomerates (Keller 1989; Jost et al. 2016).

During the Late Burdigalian, this marine environment was rapidly replaced by isolated brackish residual lakes and later by a terrestrial environment defining the onset of OSM sedimentation (Kuhlemann and Kempf 2002). OSM deposits are, similar to USM units, characterized by radial alluvial fans (Homewood et al. 1986; Schlunegger et al. 1997). OSM sedimentation continued until the Serravallian. The stratigraphic evolution of the western North Alpine foreland basin after the Serravallian stage is difficult to reconstruct as corresponding younger sediments were eroded (Kuhlemann and Kempf 2002). Outcrops of OSM exist mainly in the eastern part of the basin. Different authors postulate ongoing sedimentation of OSM units during the Late Miocene based on critical taper models and vitrinite reflectance data (Schegg et al. 1997; Schegg and Leu 1998; Willett and Schlunegger 2010).

5.2.1.2. *Subalpine Molasse and Folded Molasse units*

The Subalpine Molasse unit in Switzerland and France can be described as a set of tectonic imbricates (from 4 km to 10 km in width). They are separated by south-dipping thrusts that root at the base of the Cenozoic layers in the main thrust, which continues southward beneath the Chaînes Subalpines in the west (France) and the Penninic Préalpes Klippen or the Helvetic nappes further east (Menkveld 1995; Schmid et al. 1996). Locally, folds develop inside individual imbricates and en-échelon relays can be observed between neighboring imbricates. The thrusts of the individual tectonic imbricate may be locally cut down to the main décollement level in the Triassic evaporites on top of the mechanical basement (Sommaruga et al. 2012). The thrust slices of the Subalpine Molasse extend beneath the most external Penninic Préalpes Klippen and/or the Helvetic nappes, reaching the northern edge of the External Crystalline Massifs (Figure 5.5). The displacement along these imbricates can be correlated with the exhumation and stacking of the External Crystalline Massifs (ECM) (Pfiffner 1986; Burkhard 1990; Burkhard and

Sommaruga 1998; Mosar 1999; Affolter and Gratier 2004; von Hagke et al. 2014b; see below also).

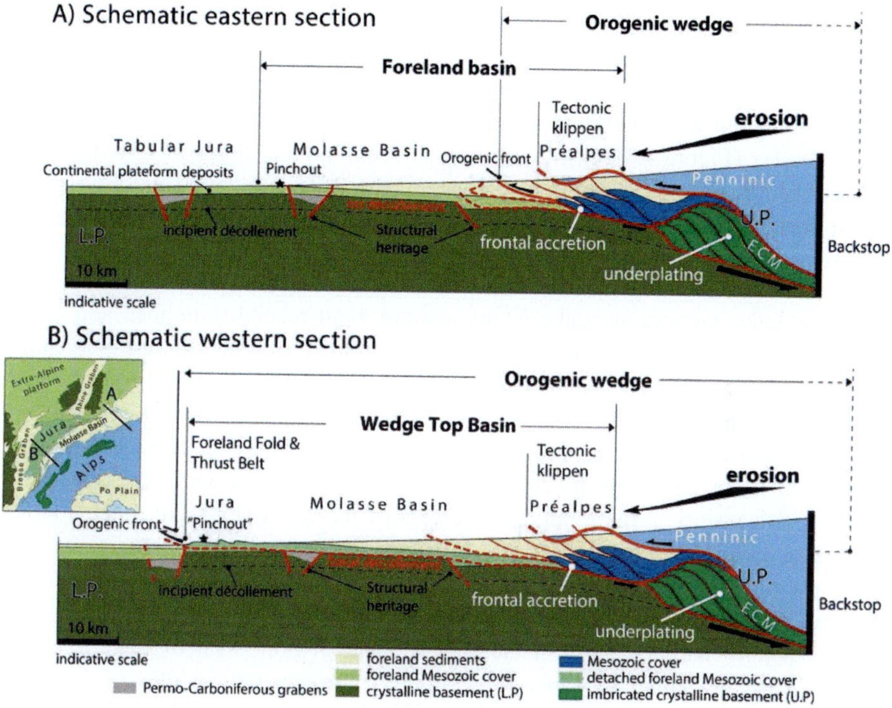

Figure 5.5. *Two conceptual cross-sections (eastern and western part) defining the North Alpine foreland basin with the following tectonic elements LP: lower plate; UP: upper plate; ECM: External Crystalline Massif. Modified from Bonnet et al. (2007)*

Overall, the Subalpine Molasse unit changes from an imbricated thrust system in western Switzerland to a triangle zone east of Lake Thun (Sommaruga et al. 2012). North-dipping layers along the northern edge of the Subalpine Molasse form the Folded Molasse. To the SW, in France, the Subalpine Molasse imbricates terminate against the SE extension of the Vuache fault system in the Lake Annecy area (Figure 5.3). The transition to the Subalpine Chains (south of the city of Geneva) leads to a tectonic style, which involves the folded and thrusted Mesozoic cover series carrying the Subalpine Molasse (Beck et al. 1998). Further south, the Subalpine Chains merge with the JFTB making a clear separation between Subalpine Chains and Subalpine Molasse difficult. For instance, the Semnoz anticline (Figure 5.3) is considered part of the Subalpine Molasse located within the Subalpine Chains. Similarly, entire units such as the Salève ramp fold could be considered as one single tectonic unit of the Subalpine Molasse.

Along the Penninic Préalpes Klippen, the Subalpine Molasse can be found not only in front but also below or behind this allochthonous klippen, such as in the Rhône Valley (Figure 5.3). USM units can be found on the shores of Lake Geneva, but also south of the Préalpes in a tributary of the Rhône Valley. This observation led Schardt (1893) in the late 19th century to propose nappe tectonics and helped define the Préalpes unit as an allochthonous klippen belt.

From Lake Geneva to Lake Thun, several fault-related anticlines and relay structures can be observed along the northern border of the Molasse Subalpine tectonic contact (Gruber 2017) (Figure 5.3). Between the cities of Thun and Lucerne, the Subalpine Molasse triangle structure replaces the thrust anticline structures and it is observed both at the surface and in seismic data (Sommaruga et al. 2012; Mock and Herwegh 2017). A lateral transition zone within the Subalpine Molasse unit occurs at the topographic recess of the Aar river–Lake Thun (Figure 5.3). Here the frontal most thrust of the Subalpine Molasse shows a lateral transition to local anticlines oriented ENE–WSW (the Wachseldorn and Balmegg anticlines; Gruber 2017). The western segment is offset to the NW by about 6–8 km to the front of the Préalpes Klippen lobe. In the central Swiss region, west of Lake Lucerne (Figure 5.3), the Subalpine Molasse is slightly narrower with widths of 4–5 km. Locally the outcrop width of the zone is broader due to the erosional incisions of the Aar, Sarine, Rhône and Arve rivers. This sector of the Subalpine Molasse units correlates with the appearance of the Penninic Klippen belts resting on top and in front of the Helvetic nappes (Figures 5.3 and 5.6). These large allochthonous klippen possibly inhibit the development of a broad and morphologically prominent Subalpine Molasse ridge as observed further east (Burkhard 1990; Burkhard and Sommaruga 1998).

In eastern Switzerland, between the city of Lucerne and Lake Constance, the Subalpine Molasse has an average width of 6–10 km. It is recognized in the topography due to its prominent summits (e.g. the Rigi, 1798 m altitude) (Figure 5.3). This south-dipping ridge of fluviatile conglomerates of the Chattian USM overthrusts the Folded Molasse unit, which is composed of north-dipping layers of OMM (Aquitanian-Burdigalian age) that belong to the frontal part of the triangle zone (Figure 5.6).

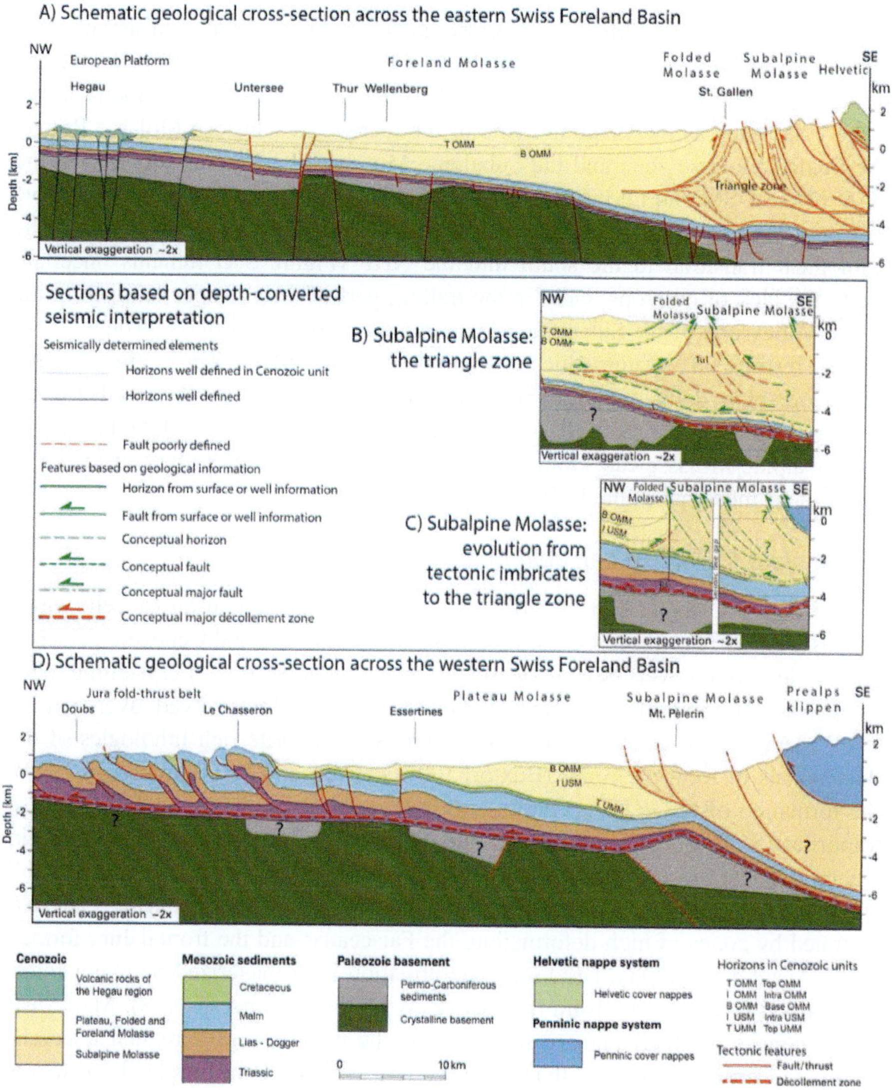

Figure 5.6. *Simplified cross-sections along the Swiss North Alpine foreland basin (modified from Sommaruga et al. 2012). A) The eastern Swiss foreland basin: the JFTB is not developed. D) The western Swiss foreland basin: from the JFTB to the Subalpine Molasse. In between B) and C) are sections based on seismic interpretation. Note that the décollement zone disappears in the eastern part. For location see Figure 5.3. Positions of wells are given in Figure 5.3 and full names of wells are in Table 5.1*

Thrust faults of the Subalpine Molasse units are mostly located in incompetent layers such as the marly series of the Vaulruz Fm. (Figure 5.4) of Rupelian age in the UMM (Sommaruga et al. 2016). Locally, thrusts are also found in marly layers in the USM such as the Chattian marls or the marly and schistose Molasse Rouge. The Folded Molasse zone and the Subalpine Molasse triangle zone develop in the series of the USM and OMM. Thus, the detachment is at the base of the Cenozoic series and the thrusts are stepping up progressively into younger Molasse units. The lithological transition to the south into the NHF (Figure 5.4) remains uncertain due to the lack of outcrops. Only in the trailing part of the Préalpes Klippen in the Rhône Valley one of the Subalpine Molasse units crops out (Molasse Rouge of the USM), resting in stratigraphic continuity on the NHF. Thus, the original paleogeographic domain of the Subalpine Molasse extends to the northernmost ECM, resting on top of the NHF of the Helvetic nappes and parautochthonous units. The Subalpine Molasse can, from a tectonic perspective, therefore be considered as the northern-most unit of the Helvetic nappe system.

5.2.1.3. *Jura fold-thrust belt deformation front*

The Jura fold-thrust belt (JFTB) is an arcuate foreland fold-thrust belt, which extends over 350 km (Figure 5.3). It consists of a Mesozoic and Cenozoic detached cover series, which overlies a pre-Mesozoic basement consisting of crystalline rocks and Permo-Carboniferous Continental Grabens. The JFTB evolved over a main décollement level/zone located within the Triassic evaporite rich lithologies of the Muschelkalk or Keuper Groups (Buxtorf 1907; Laubscher 1961; Sommaruga 1999). The formation of the JFTB is conditioned by the distribution and presence of these evaporites (Sommaruga et al. 2017). The JFTB can be subdivided into three main structural domains from SE to N–NW: the Internal or Haute Chaine Jura, with well-developed thrust-related folds; the External Jura, made of flat-lying Plateaux, separated by zones of high deformation, the Faisceaux; and the frontal Jura formed by the Avant-Monts. On a large scale, deformation is characterized by major folds, with a trend swinging 90° from the east (E–W) to the southwest (N–S). Major tear faults oriented at a high angle to fold axes cut the internal Jura at regular intervals. During the formation of the JFTB, the western North Alpine foreland basin was transported toward the NNW. During this event Triassic evaporite-cored low-amplitude folds and duplex structures formed, while strike-slip (tear) faults developed in the detached Mesozoic and Cenozoic cover. Locally, Mesozoic and Early Cenozoic normal faults were reactivated.

The outermost structures of the Avant-Monts, forming the deformation front, (the three dark green zones in Figure 5.3) change their structural style along the strike of the JFTB. At the SW tip of the JFTB, the frontal thrusts ramp onto the Ile Crémieu autochthonous promontory and have an NNW–SSE orientation. To the south, they merge with the frontal thrusts of the Subalpine Chains (Philippe et al. 1996). South of the city of Besançon, the deformation front is formed by the westernmost thrusts of the Faisceaux which have an overall N–S to NNE–SSW orientation. This Faisceaux overthrust the eastern edge of the Bresse Graben by some 5 km, as shown in several boreholes (Michel et al. 1953; Chauve et al. 1988). To the west of the city of Besançon, the La Serre Crystalline Massif forms a structural high just NW of the deformation front. The eastern frontal part of this basement massif is inverted near the Jura frontal thrust of the Avant-Monts (Madritsch et al. 2008; Madritsch et al. 2009). In this transition zone (Besançon zone) thin-skinned deformation above the main décollement interferes with thick-skinned reactivation of inherited normal basement faults (La Serre zone) (Madritsch et al. 2008). South of the Upper Rhine Graben, the Ferrette zone promontory forms the frontal thrust system (Ustaszewski and Schmid 2006, 2007a, 2007b). This zone presents thin-skinned and thick-skinned deformation of the Mesozoic layers and gentle folds north of the Ferrette thrust may represent the outermost deformation front (Ustaszewski and Schmid 2006). Further east between the zones of Ferrette and Mandach, the frontal deformation zone is formed by shallow thrusts and imbricates, blind thrusts and folds resting on the Tabular Jura (Bläsi et al. 2018). West of Lake Constance, along the southern slopes of the Black Forest basement, the Mandach thrust laterally grades into blind subsurface thrusts (Madritsch 2015; Madritsch et al. 2018). At the easternmost termination of the JFTB, we can observe the transition from extensional to compressional setting along the Baden-Irchel-Herdern (BIH) lineament (Malz et al. 2016; Egli et al. 2017b; Madritsch et al. 2018).

In summary, the structural style of the Jura deformation front is conditioned by the involvement of inherited paleo-topography linked to normal faults in the basement. In the east the inherited faults are associated with the Permo-Carboniferous Grabens of northern Switzerland, whereas to the west and south the inherited faults are linked to the formation of the Cenozoic Rhine–Bresse Graben system and the associated transfer zone. The main décollement steps up or down over these inherited ramps and basement highs.

5.2.2. *Subalpine Molasse of western Austria and southern Germany*

In eastern Switzerland and Bavaria, where the JFTB is absent, outcrops show an onlapping wedge of Cenozoic sediments on the Tabular Jura (e.g. Pfiffner 1986; Bachmann et al. 1987). The dominant deformation north of the Subalpine Molasse, which is also visible on seismic data, is characterized by small normal faults oriented WSW–ENE, parallel to the basin and affecting the Mesozoic and Cenozoic strata (Bachmann et al. 1982; Nachtmann and Wagner 1987; Diebold and Noack 1997; Lüschen et al. 2004; Pfiffner 2009; Ibele 2011; Moeck and Mraz 2015; Heuberger et al. 2016). The Subalpine Molasse of western Austria and Bavaria forms a 10–15 km wide fold-thrust belt that tapers off east of Munich before it disappears in map view near Lake Chiemsee between Munich and Salzburg (Figure 5.1). The along-strike geometry is variable, including changes in décollement depth. This variability in structural style is linked to the mechanical stratigraphy of the foreland basin.

5.2.2.1. *Sedimentary units involved in the Subalpine Molasse*

In its western part, the foreland basin of western Austria and southern Germany features the same sedimentary units as the Swiss and French parts of the foreland (see section 5.2.1). The oldest, Eocene sedimentary units of the Alpine foreland are found on top of units that are part of the Helvetic domain (Figure 5.7). The Eocene units are mostly limestones and sandstones containing nummulites (e.g. Stackelberg 1960; Schwerd 1983; Heyng 2012). The Eocene succession ends with pelagic globigerina marls, known under various local names such as Stadschiefer or Stockletten. The entire succession is overthrust by tectonically higher units.

Within the Subalpine Molasse, the succession starts with clayey marls (Tonmergel Beds). These represent the principal décollement that separates the Subalpine Molasse from the European margin (Figure 5.7; Ortner et al. 2015). At their base, the Tonmergel Beds are intercalated with the Deutenhausen Beds, which are turbiditic sandstones that have their source area in the growing orogen (Deutenhausen Beds; e.g. Maurer et al. 2002). The pelagic sedimentary environment progressively shallowed and the upper part of the succession witnesses influence by brackish conditions. This trend to shallower water depths culminates in the deposition of the littoral Baustein Beds that form the topmost unit of the UMM (Schwerd 1978). As the UMM is on top of the décollement, the preserved thickness varies, but generally increases to the east: east of the Rhine Valley, the UMM is 250 m thick (Wilhemine Alpe; Kempf and Pross 2005), in the central part of the thickness reaches 1200 m (Ammer; van der Boon et al. 2018) and in eastern southern Bavaria the thickness is more than 2500 m (well Feilnbach 1; Paulus and Brockert 1968). The age of the UMM succession is poorly constrained but ranges from the Latest Eocene to the end of the Early Oligocene (Ganss and Schmidt-Thomé 1955; Kempf and Pross 2005; van der Boon et al. 2018).

The Baustein Beds are overlain by the continental deposits of the USM. The USM is by far the thickest unit of the Alpine foreland. Within the western Subalpine Molasse of Bavaria, its thickness reaches up to 4.5 km (Scholz 2000; Ortner et al. 2015). These large thicknesses are related to the Pfänder and Hochgrat-Adelegg alluvial mega-fans entering the foreland basin from the orogen (pf and ha of Figure 5.8; Brügel 1998; Kuhlemann et al. 2006). South of Schongau, the USM starts to interfinger with brackish, coal-bearing marls and sands (Cyrena Beds, CYB; Gillitzer

1955; Geissler 1975), which replace the USM entirely east of Weilheim (Figure 5.8; Paulus 1963), thus the thickness is reduced to 1.5 km (Ortner et al. 2015). Within the brackish succession, marine ingressions appear near the Oligocene–Miocene boundary (Figure 5.2). The last occurrences of the Cyrena Beds are observed SE of Rosenheim; further to the east, they disappear and are replaced by deep marine marls and sands of the Puchkirchen Group (Figure 5.2). Paleogeographically, this transition can be envisioned as a sedimentary system consisting of a roughly N–S striking shelf at or close to sea level and a slope into the deep basin farther east, transected by submarine canyons incising both the Cyrena Beds and the mouth of the paleo-Inn system (Ortner and Stingl 2001; Kuhlemann and Kempf 2002; de Ruig and Hubbard 2006).

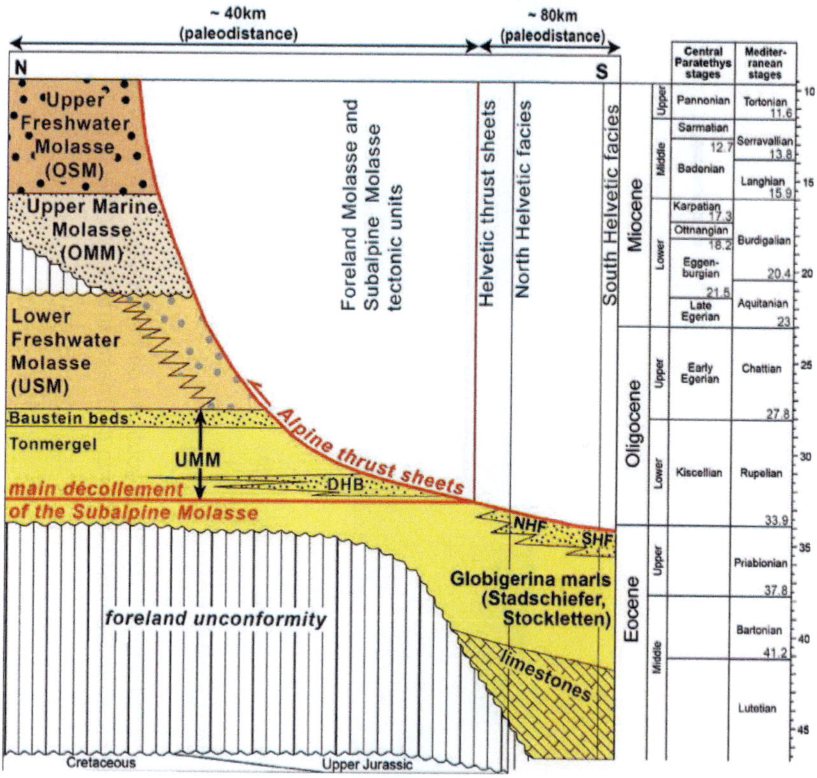

Figure 5.7. *Chronostratigraphy of the Alpine foreland of western Austria and southern Germany in an N–S transect. The correlation between Central Paratethys stages and Mediterranean stages follows Hilgen et al. (2012), absolute ages are from Cohen et al. (2013; updated). Note the changing of horizontal scale. Abbreviations: DHB = Deutenhausen Beds, NHF = North Helvetic Flysch, SHF = South Helvetic Flysch, UMM = Lower Marine Molasse*

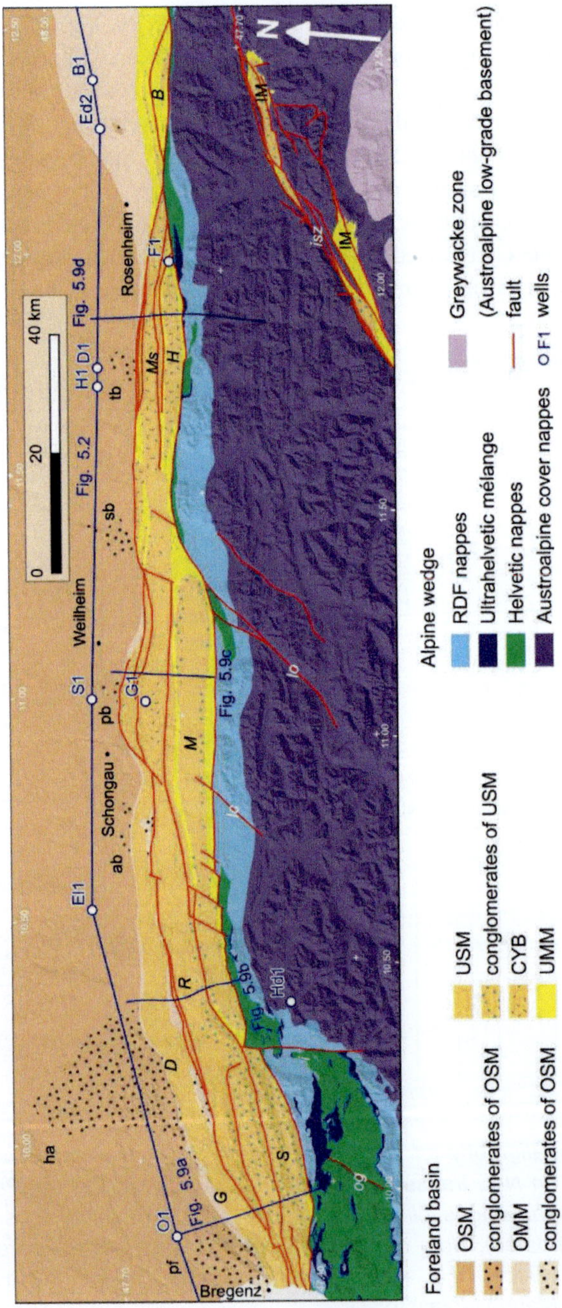

Figure 5.8. *Tectonic map of the southern part of the Alpine foreland basin of western Austria and southern Germany, showing the major structures (adapted from Ortner et al. 2015)*

COMMENTS ON FIGURE 5.8.– *Locations of cross-sections of Figure 5.9 are shown. ab = Auerberg fan, ha = Hochgrat-Adelegg fan, pb = Peissenberg fan, pf = Pfänder fan, sb = Schellenberg fan, tb = Taubenberg fan. ab, pb and sb are also known as Nesselburg fans. Slices and folds of the Subalpine Molasse: B = Bernau slice, D = Durach anticline, G = Gerholz anticline, H = Hausham slice/syncline, M = Murnau slice/syncline, Ms = Miesbach slice, S = Steineberg slice/syncline, R = Rottenbuch slice. Faults: i = Illertal fault, lc = Lech fault, lo = Loisach fault, isz = Inntal shear zone, og = Ostergunten fault. IM = Inneralpine Molasse. See Table 5.1 for full names and references for wells.*

The transition of the USM to the OMM is continuously only close to the Alpine wedge, whereas in the larger, northern part of the Alpine foreland an unconformity encompassing the Eggenburgian separates the Egerian from the Ottnangian (Figure 5.7; Lemcke 1988). This transition from the first to the second shallowing megasequence of the NAFB represents a major change in the dynamics of the Alpine wedge.

OMM sedimentation starts in the youngest Burdigalian to earliest Langhian and is only exposed north of the Subalpine Molasse within the tilted and locally folded belt in the southernmost part of the Foreland Molasse. In the western part of the Bavarian Molasse, the thickness of the OMM is relatively small and uniformly close to 400 m between Bregenz and Rosenheim (Ganss and Schmidt-Thomé 1955; Müller 1970; Schaad et al. 1992; Frieling et al. 2009). Toward the east, where the Cyrena Beds disappear, the OMM increases dramatically in thickness, reaching more than 2 km (Figure 5.2). Immediately north of the Subalpine Molasse, growth strata are observed within the OMM, which onlap the northernmost structure toward the south (Ortner et al. 2015).

In the Badenian, the entire NAFB returned to an overfilled state and the continental OSM was deposited. The fans east of the Rhine Valley persisted, the Hochgrat-Adelegg fan bringing coarse clastics 17 km into the foreland basin; there, 2 km thick sediments of the OSM are preserved (Zerlauth et al. 2014), whereas in most other areas less than 500 m are left (Ganss and Schmidt-Thomé 1955). However, more than 1 km of the OSM has been erosionally removed, as witnessed by (partially) reset low-temperature thermochronological data (e.g. von Hagke et al. 2014b). Similar to the OMM, the OSM also onlaps the northernmost structure, displaying growth strata. The youngest units (13–9 Ma) are still tilted toward the north (Eberhard 1986; Ortner et al. 2015).

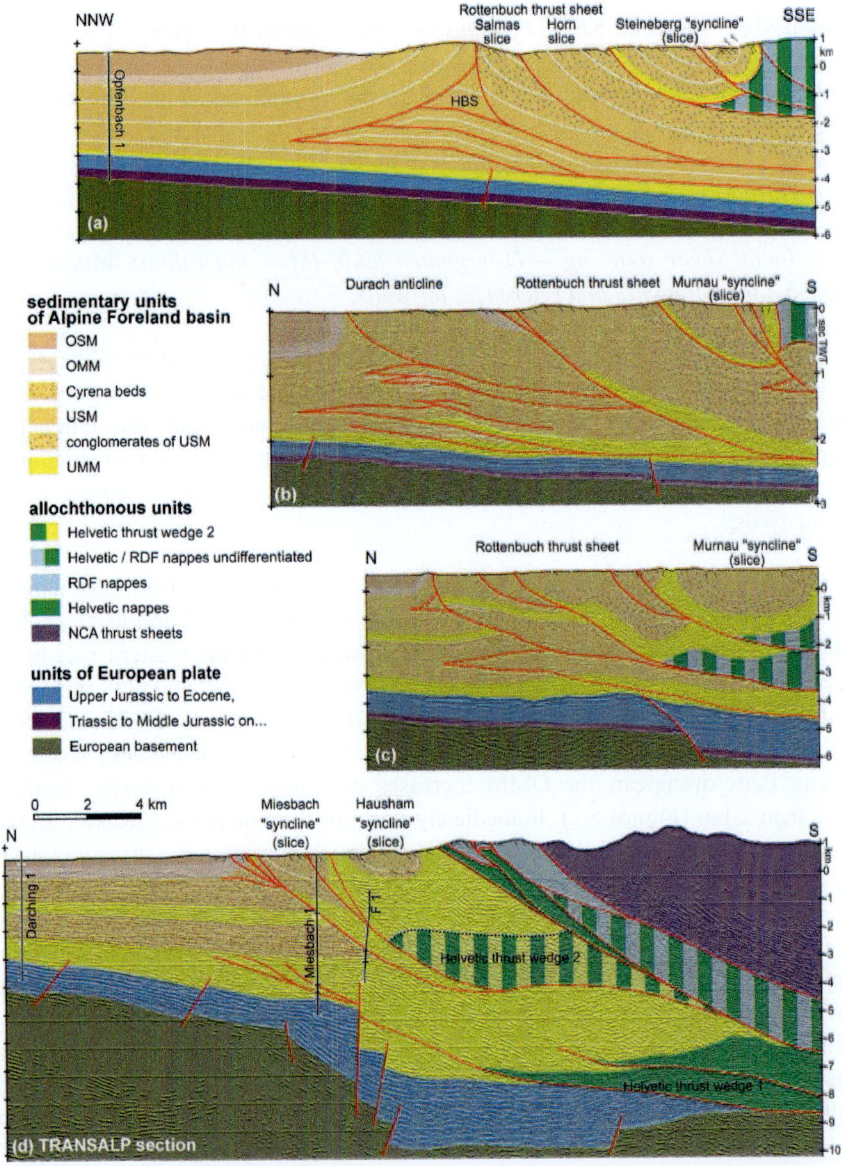

Figure 5.9. *Cross-sections of the Subalpine Molasse of western Austria and southern Germany between Bregenz in the west and Rosenheim in the east. For the location of cross-sections see Figure 5.8. All sections modified from Ortner et al. (2015). Seismogram of (B) was stretched to fit the depth of the deepest reflectors to the base of the autochthonous Mesozoic of the nearby Elbsee 1 well (Müller et al. 1988)*

5.2.2.2. *Geometry of tectonic units*

We will discuss the geometry of the Subalpine Molasse in Bavaria and western Austria using four cross-sections that illustrate different aspects of the geometry of this thrust belt as well as the change in structural style from west to east. Generally, the Subalpine Molasse narrows toward the east and the number of tectonic slices decreases. The thrust distances between individual slices are moderate (mostly <5 km) and the basal tectonic contacts disappear into anticlines laterally.

Some of the structural features are laterally persistent. For instance, the Murnau slice is cylindrical over a distance of 80 km. To the west, it continues into the Steinberg slice, which is a 40 km long cylindrical structure. Both are also known under the former German terms *Murnauer Mulde* and *Steinberg Mulde*, based on the observation of their synformal geometry exposed at the surface (Figures 5.9(A), (B) and (C)). At their base, these slices bring the basal décollement within the Tonmergel to the surface. To the north of both slices lies the Rottenbuch thrust sheet, which consistently dips to the south. In the area of the Hochgrat-Adelegg fan, the Rottenbuch slice displaces the complete 4.5 km thick USM succession *en block* (Figure 5.9(B)), whereas further west the Horn thrust splits it into two units, the structurally higher Horn slice and the structurally deeper Salmas slice (Figure 5.9(A)). East of the Hochgrat-Adelegg fan, where USM-Cyrena Beds succession is less than 3 km thick, the Rottenbuch slice starts to get folded and an anticline appears along the northern margin of the Murnau slice (Figure 5.9(C)).

The most external structures are the Gerholz and Durach anticlines in the footwall of the Rottenbuch thrust, which are cored by a triangle zone. The anticlines are in structural continuity with the undeformed Foreland Molasse. The Gerholz anticline is overturned to the south and formed together with the passive backthrust of the triangle zone, whereas the Durach anticline is overturned to the north and formed on top of a foreland-directed splay of the roof thrust (Ortner et al. 2015). In the core of the triangle zone, two 700 m thick slices are stacked, which are both detached at the basal Tonmergel décollement.

At Peissenberg, east of Schongau (Figure 5.8), this triangle zone disappears, at the expense of a triangle zone in the upper USM with a basal thrust at shallow depth. This reduces the width of the tilted part of the Foreland Molasse to less than 1 km. The Peissenberg thrust separates an additional slice from the Rottenbuch slice but dies out rapidly to the east.

East of Weilheim, where the USM is replaced by the Cyrena Beds, the Subalpine Molasse narrows and the individual slices decrease in width from 5–8 to 3–4 km. The northern margin of the Subalpine Molasse is a thrust system which evolved from a narrowly overturned anticline. Slices of this fault-related fold are preserved between the splays of the thrust system (Paulus 1963, 1981). The triangle zone below the frontal thrust has disappeared. However, the OSM of the Schellenberg and Taubenberg fans still shows growth strata that accumulated on top of the marginal anticline during its growth (Ganss and Schmidt-Thomé 1955; Pilser 2008; Zerlauth 2008).

Southeast of Rosenheim, the Bernau slice/syncline (B of Figure 5.8) is an easternmost and in map view isolated part of the Subalpine Molasse. Based on a section across the well Endorf 2 (Ed2 of Figure 5.8), Ganss (1977) interpreted a thrust at the northern margin of the slice. This thrust may also have initiated as an anticline, as can be judged from a more than 2 km wide zone of growth strata within the OMM (Ortner et al. 2015).

5.2.2.3. *Relationship of the Subalpine Molasse to the Helvetic and Flysch nappes*

As seen in the map of Figure 5.8, the southern limb of the most internal Subalpine Molasse synclines systematically exposes a north-younging succession where the oldest sedimentary units are in contact with the tectonically higher Helvetic or Rhenodanubian Flysch (RDF; see also above) nappes. Fold axes within the Helvetic and RDF nappes diverge from the contact and the fold axes of the Subalpine Molasse. However, in seismic sections (Zerlauth et al. 2014; Ortner et al. 2015; Pomella et al. 2015) and outcrops in eastern Switzerland (e.g. Trümpy 1969; Pfiffner 1993; Schmid et al. 1996), the Helvetic nappes are seen to be thrust onto the Subalpine Molasse. According to Ortner et al. (2015), both observations can be reconciled when the Helvetic and RDF nappe stack forms a triangle zone underlain and covered by foreland deposits. This triangle possibly evolved in the Early Oligocene when the Helvetic-RDF nappe stack was initially emplaced onto the Alpine foreland. Following emplacement, this nappe stack must then have been eroded at the top and covered by a foreland wedge-top that is still preserved in the Inneralpine Molasse of the Inn Valley in a more internal position of the Alpine wedge (IM of Figure 5.8; Ortner and Stingl 2001). This scenario is also necessary to explain the thermal history of the Helvetic-RDF nappe stack, where low-temperature thermochronological data indicate a reheating event in the Miocene (Zerlauth et al. 2016). Upon further shortening, the Helvetic-RDF nappe stack was transported onto younger foreland deposits and the contact to the wedge top was reactivated as a back thrust (Figure 5.9(A), (B) and (C)).

In all cross-sections east of Schongau, the trough line of any marker horizon in synclines, e.g. the base of the USM/CYB rises toward the south (see Figure 5.9(C) and (D) and cross-sections in Ganss and Schmidt-Thomé 1955; Paulus 1963, 1981;

Ortner et al. 2015), while the base of the autochthonous foreland deposits below the southernmost syncline sinks from 4.5 to 7 km (Figure 5.9(C) and (D); Ortner et al. 2015). In the TRANSALP section (Figure 5.9(D)), the space between the base Cyrena Beds and the base of the autochthonous foreland deposits cannot be easily filled by UMM deposits, even if duplication is considered, because of their limited thickness (see section 5.2.2.1). Therefore, we introduced a Helvetic thrust wedge to fill part of this space, based on the observation of slightly increased reflectivity at 2–4 km depth. Also in the other cross-sections, it is necessary to introduce such wedges, if the thickness of sedimentary units is kept in the range of values observed in outcrops or wells.

Changes in the structural architecture of the Subalpine Molasse slices occur across N- to NE-striking transfer faults (e.g. different number of slices, vertical offsets and new thrusts; see Figure 5.8). These root in the basal detachment of the Helvetic nappe stack, as shown by Ortner et al. (2006) for the Inntal shear zone (isz in Figure 5.8) in a more southern part of the TRANSALP seismic line and often separate structural domains (Ostergunten fault, og of Figure 5.8; Zerlauth 2014), that is entirely contained in the Helvetic nappe stack. Other slices terminate against thrusts within the Subalpine Molasse (Illertal fault, i in Figure 5.8), or dissect one limb of a syncline (Lech fault, lc of Figure 5.8). This demonstrates that similar to farther west, the Subalpine Molasse is the most external part of the Helvetic-RDF nappe stack.

5.2.3. *Imbricated European margin of eastern Bavaria, Salzburg and Upper Austria*

In eastern Bavaria, Salzburg and Upper Austria, (Figure 5.10) not only sediments of the Molasse are deformed, but partly also sediments of the former European shelf. Therefore, it is necessary to briefly review the sedimentary succession of the pre-Molasse sediments and after that, the characteristics of the Molasse sediments of this part of the basin.

5.2.3.1. *Non-Molasse sediments of the European margin*

The Molasse sediments in this part of the orogeny rest unconformably on older sediments of the European shelf/margin or directly on the crystalline basement (Nachtmann and Wagner 1987; Malzer et al. 1993), a result of the paleo-high of the southern spur of the Bohemian Massif (see Figure 5.2). Several episodes of pre-Molasse faulting and erosion resulted in a complex subcrop, with varying thicknesses of the Mesozoic deposits, angular unconformities and the presence of basement highs and sub-basins, which contain Mesozoic rocks (Figure 5.10; Nachtmann and Wagner 1987; Wagner 1998). From west to east these are the

Salzach Basin, the Central Swell, the Ried-Schwanenstadt Basin and the Bad Hall Basin (Nachtmann and Wagner 1987; Malzer et al. 1993). The Central Swell is a subsurface basement high (Nachtmann and Wagner 1987; Wagner 1996) that is the southeastern continuation of the Landshut-Neuöttinger-Hoch in Bavaria (Freudenberger and Schwerd 1996 and references therein). Consequently, Jurassic to Cretaceous siliciclastics, limestones and marls cover the basement with varying thicknesses, depending on the relative position to the crystalline highs (Figure 5.2; Nachtmann and Wagner 1987; Malzer et al. 1993).

The Jurassic sediments are mainly Upper Jurassic limestones, which show erosion and karstification in Early Cretaceous times (Nachtmann and Wagner 1987). Only a few Lower Cretaceous Beds are preserved in erosional pockets on top of the karstified Jurassic limestones. The Jurassic is covered by a series of Upper Cretaceous shallow marine clays, marls and some sandstone layers (Nachtmann and Wagner 1987; Wagner 1998). Uppermost Cretaceous sediments (Maastrichtian) are removed by erosion in the autochthonous foreland before being covered by the Eocene Molasse sediments (Nachtmann and Wagner 1987). The Cretaceous sediments can be considered to represent the northern shelf equivalent of the Helvetic slope sediments which are known only from allochthonous nappes (Prey 1980; Nachtmann and Wagner 1987) and in elongated windows within the RDF wedge ("Streifenfenster"; Prey 1980).

The Molasse sediments in this sector are the gradual continuation of the deposits in the west (Figure 5.2). Molasse sedimentation began in the Late Eocene with deposition of fluvial and shallow-marine sandstones, shales and carbonates (Figure 5.10; Malzer et al. 1993; Wagner 1998; Sachsenhofer and Schulz 2006). The main difference for this sector in comparison to the west is that deep marine conditions prevailed from Late Eocene (Kiscelian) to Early Miocene (Ottnangian/Middle Burdigalian; Rögl et al. 1979; Wagner 1998; Grunert et al. 2013; Grunert et al. 2015; Hülscher et al. 2019). The respective deep marine basin is referred to as Puchkirchen trough (Zweigel et al. 1998; Grunert et al. 2013). Its transition to shelf conditions is considered to be located east of Munich (Lemcke 1983; Zweigel et al. 1998; de Ruig 2003). The deepening of the Molasse Basin is recorded by a basin-wide unconformity (NSU, Northern Slope Unconformity, see Figure 5.11), possibly a result of loading of the lower plate by the growing Alpine wedge (Masalimova et al. 2015; Hülscher et al. 2019). A dense network of synthetic and antithetic tensional faults developed in the Latest Eocene to Early Oligocene (Nachtmann and Wagner 1987), which is attributed to the bending of the lower plate (Kittler and Neumayer 1983). Most of the fault-related morphology is leveled by the Early Oligocene (Chattian) deep marine shales of the Zupfing Fm. The shales are interbedded with distal turbiditic sands at the southern margin of the basin (Rögl et al. 1979; Wagner 1998).

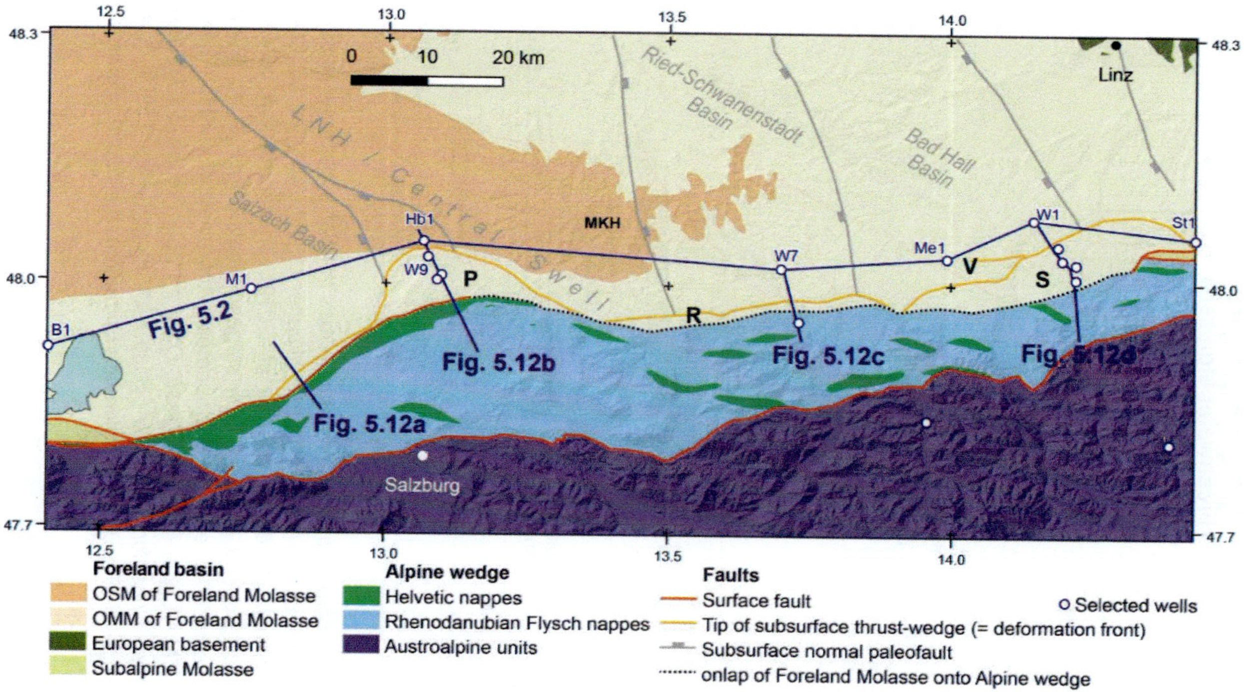

Figure 5.10. *Simplified geological map of the Northern Alpine front in the area of eastern Bavaria, Salzburg and Upper Austria modified from Schuster et al. (2015) and Schwerd (1996). Subsurface features adapted from Nachtmann and Wagner (1987), Kröll et al. (2006) and Hinsch (2013). The locations of the sections in Figure 5.12 are indicated. MKH = Munderfing-Kobernaus-Hausruck fan. P = Perwang imbricates, R = Regau imbricates, S = Sierning imbricates, LNH = Landshut Neuötting High, and V = Voitsdorf inversion*

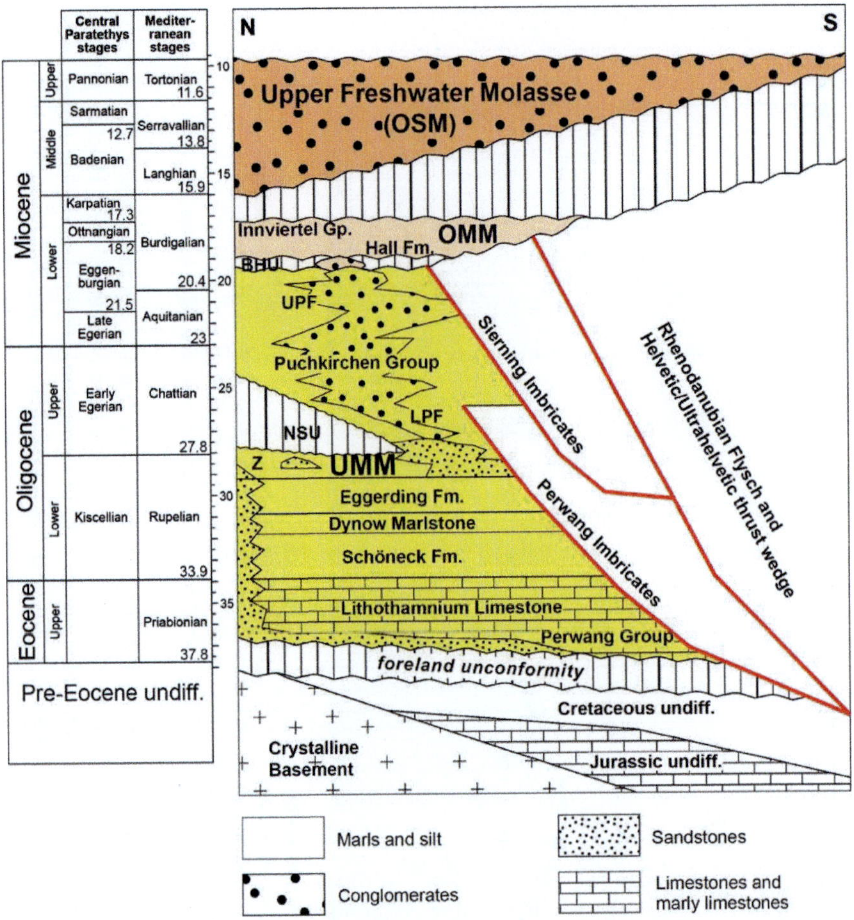

Figure 5.11. *Simplified chronostratigraphic chart across the Salzburg/Upper Austrian sector of the Molasse basin (modified after Wagner 1998; Piller et al. 2004; Sachsenhofer and Schulz 2006; Hülscher et al. 2019). The correlation between Central Paratethys stages and Mediterranean stages follows Hilgen et al. (2012), absolute ages are from Cohen et al. (2013; updated). Abbreviations: UMM = Lower Marine Molasse; OMM = Upper Marine Molasse; OSM = Upper Freshwater Molasse; NSU = Northern Slope Unconformity; BHU = Base Hall Unconformity; LPF = Lower Puchkirchen Formation; UPF = Upper Puchkirchen Formation; Z = Zupfing Formation*

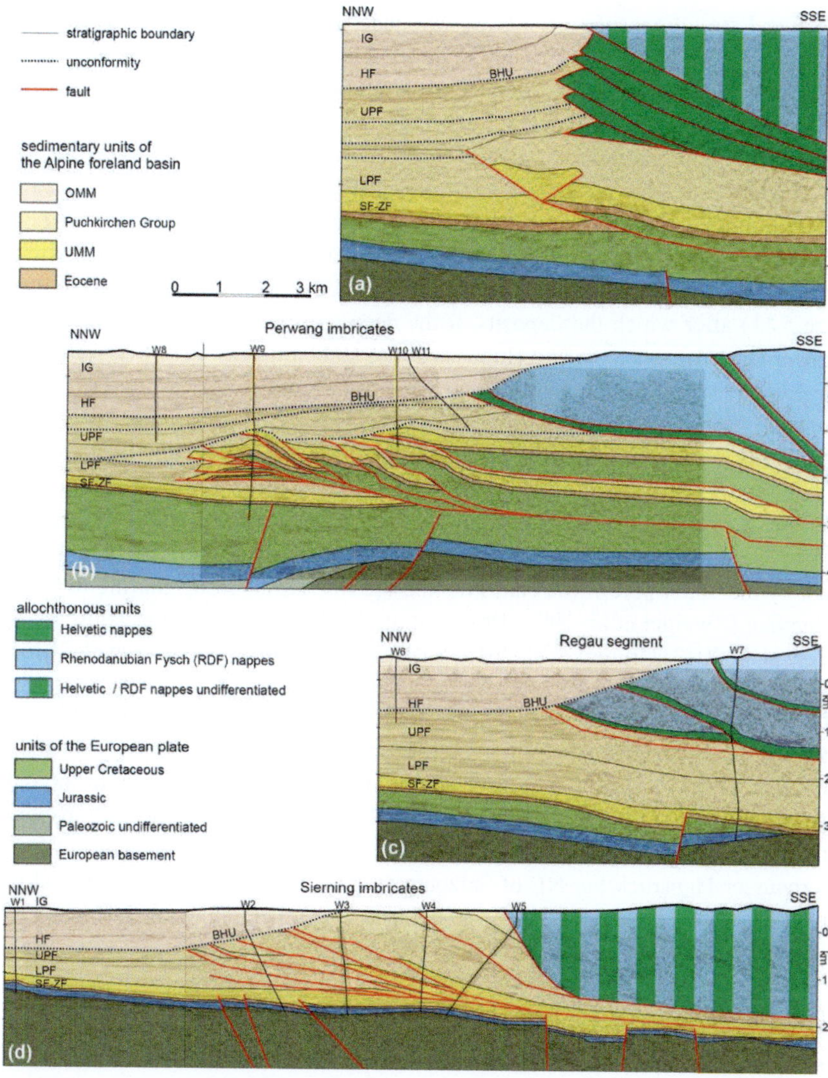

Figure 5.12. *Four interpreted seismic sections in the area of eastern Bavaria, Salzburg and Upper Austria showing the deformation style in different segments. A) Western end of Perwang segment, B) central Perwang segment, C) Regau segment and D) Sierning segment (all sections modified after Hinsch, 2013). Location of the sections is indicated in Figure 5.10. Reference to wells W1–W11 given in Table 5.1. OMM = Upper Marine Molasse, UMM = Lower Marine Molasse, IG = Innviertel Group, HF = Hall Fm., UPF = Upper Puchkirchen Fm., LPF = Lower Puchkirchen Fm., ZF = Zupfing Fm., SF = Schöneck Fm., BHU = Base Hall Unconformity*

During Late Oligocene to the Early Miocene time, sediment distribution in this 500–1500 m deep basin was strongly controlled by an axial, partly sinuous channel belt (Linzer 2001; de Ruig 2003; de Ruig and Hubbard 2006; Sharman et al. 2018). Overbank deposits are dominated by pelagic shales, while the channel deposits are comprised of conglomeratic debris flows, turbiditic sands and shales (de Ruig and Hubbard 2006; Bernhardt et al. 2012; Masalimova et al. 2015). Lithostratigraphically most of the corresponding sediments are part of the Puchkirchen Group (Grunert et al. 2015; Hülscher et al. 2019). The Puchkirchen Group is topped by a basin-wide unconformity (Base Hall Unconformity (BHU); Figure 5.11) after which the deposits of the deep marine channel system prevail in the lower Hall Fm. (de Ruig and Hubbard 2006; Hubbard et al. 2009). Traditionally, the age of the top of the Puchkirchen Group was assumed to correspond to the Top Aquitanian (Early Miocene). However, the Upper Puchkirchen Fm. may extend into the Lower Burdigalian (Figure 5.11; Piller et al. 2004; Grunert et al. 2015; Hülscher et al. 2019). The BHU is covered by redeposited sediments reworked from the Bavarian shelf to the west (Grunert et al. 2013; Hülscher et al. 2019). A period of strongly increasing sedimentation rates led to the final infill of the Puchkirchen trough by the upper part of the Hall Fm. and the Innviertel Group starting in Middle Burdigalian (Zweigel et al. 1998; Brügel et al. 2003; Borowski 2006; Hinsch 2008; Grunert et al. 2013; Grunert et al. 2015; Sharman et al. 2018; Hülscher et al. 2019).

OSM deposition starts on top of another unconformity (Figure 5.11). The continental deposits have a complex internal stratigraphy that has been reviewed by Rupp et al. (2011) and Freudenberger and Schwerd (1996). In contrast to the older fill of the foreland basin, the OSM continental deposits were generally transported from east to west. The coarsest deposits are found in the Munderfinger-Kobernausser-Hausruck fan NE of Salzburg (Figure 5.10). The clastic material was sourced mainly from the growing Alps, but also the Bohemian Massif (Unger 1989; Brügel 1998; Brügel et al. 2003).

5.2.3.3. *Structural architecture along the Alpine deformation front*

From west to east the sector can be subdivided into three different (sub-) segments (Figure 5.10). The Perwang imbricates, the Regau overthrust area and the Sierning imbricates (Hinsch 2013). The Perwang imbricates represent a deep marine accretionary wedge in front of the (Penninic) RDF wedge and imbricated slivers of Ultrahelvetic to Helvetic marls and thin limestones (Figure 5.12(B)). The Perwang imbricates have a basal detachment in Upper Cretaceous marls (Figure 5.12(A) and (B)) and include Upper Cretaceous to Lower Puchkirchen Fm. sediments. The number of imbricates within this salient decreases toward its fringes (Figure 5.12(A); see also Figure 8 in Hinsch 2013 and Figure 1 in Gusterhuber et al. 2014). The deformed wedge is onlapped by syntectonic Lower Puchkirchen sediments in the front and finally covered by Upper Puchkirchen sediments (Covault et al. 2009). The geometry of the stacked units of the frontal central wedge indicates frontal accretion which was followed by back-stepping imbrication (Figure 5.12(B)). At the western lateral fringe of the Perwang lobe, several stacked imbricates of Helvetic marls and limestones have been interpreted in seismic data (Figure 5.12(A)) and from wells (Hinsch 2013). Additional structural wedging must have occurred after the Burdigalian, as witnessed by tilted sediments (Figure 5.12(A); Hinsch 2013).

In the Regau segment, the autochthonous Mesozoic and younger Molasse sediments, which are about 2 km thick, are overthrust by the RDF wedge (Figure 5.12(C)). There is no basal detachment developed in the imaged frontal part of this segment. Molasse sediments are only imbricated in small and thin thrust slices of Upper Puchkirchen Fm. at the base of the overthrust wedge. The thrust wedge is onlapped by strata of the Hall Fm. The onlapping reflectors are tilted today toward the north, indicating differential uplift toward the south caused by internal thickening of the wedge after deposition of the Hall Fm. and some minor additional long-wavelength differential uplift to the south (see also Figure 5 in Hinsch 2013). Internal thickening of the wedge would require a roof thrust close to or at the location of the unconformity.

Similar to the Perwang imbricate, the Sierning imbricates represent an accretionary wedge of Molasse sediments in front of the RDF wedge. The interface to the RDF is relatively steep and was refolded/steepened in a piggy-back position (Figure 5.12(D); Hinsch 2013). The number of thrust sheets in the Molasse sediments increases from one to seven from west to east (Figures 5.10 and 5.12(D)). The detachment is located close to the base of the Molasse sediments, most likely in the lowermost shales of the Zupfing Fm. (Figure 5.12(D)). The increasing number of thrust sheets correlates with a decreasing stratigraphic thickness of the pre-kinematic Molasse sediments in the same direction (Hinsch 2013). The Sierning imbricates comprise sediments of Zupfing Fm. and Puchkirchen Group. The frontal parts of the Sierning imbricate are partly eroded and the unconformity is, similar to the Regau segment, successively onlapped by Hall Fm. sediments. Dips increase toward the south, which could be a result of the internal thickening of the wedge, differential compaction and long-wavelength tilt due to erosion and uplift in the south.

Figure 5.12(D) shows that the basement and the lowermost covering sediments (thin Cretaceous and lowermost Molasse sediments) are dissected by normal and reverse faults. Normal faults can be interpreted as a result of flexural bending and forebulge migration of the Lower European Plate under the load of the advancing Alpine wedge, with reactivation of inherited faults or the formation of new normal faults. Reverse faults can be associated with Alpine convergence, leading to minor inversion of especially E–W trending faults. An example of an inversion feature is the Voitsdorf oil field (Figure 5.10), located in fault-bounded Eocene sandstones just north of the tip of the imbricates, about 5 km west of the section in Figure 5.12(D). Growth strata in Upper Puchkirchen Fm. indicate that the inversion is slightly older than the frontal imbrication (Hinsch 2013).

5.2.4. *Imbricated European margin of Lower Austria*

5.2.4.1. *Structural geometry and Lower Miocene thrust sequence*

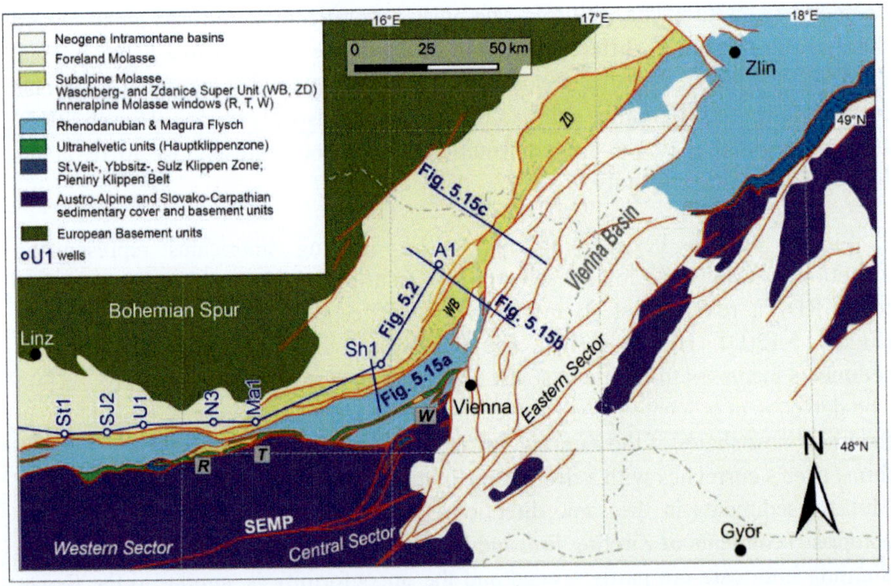

Figure 5.13. *Geological map of east Upper Austria and Lower Austria. Compiled from Lexa et al. (2000), Schnabel et al. (2002) and Krenmayr et al. (2006). R, T, W = Molasse Windows of Rogatsboden, Texing and Wolfsgraben. WB, ZD = Waschberg and Zdanice Super Units. Sector sub-division following Beidinger and Decker (2014). Blue lines refer to cross-section traces. See Table 5.1 for full names of wells and references*

The large-scale geometry of the Subalpine Molasse in Lower Austria is characterized by a change in the strike of the orogen from an E–W toward NE–SW in the area of the Vienna basin (Figures 5.1 and 5.13). Likewise, the internal nappe architecture, as well as the Lower Miocene thrust sequence of the Subalpine Molasse, are subject to changes from west to east. Using data from 65 deep exploration wells, supplemented by 2D seismic reflection profiles and published cross-sections, Beidinger and Decker (2014) describe three major along-strike differences, which are:

1) the different timing of propagation of thrusting from the RDF units into the Subalpine Molasse;

2) the different ages of youngest in-sequence thrust at the orogenic front and hence the time of termination of foreland-propagating thrusting;

3) the different amounts of Oligocene to Lower Miocene in-sequence and out-of-sequence thrust shortening.

The diachronic termination of Lower Miocene foreland-propagating thrusting and structural differences along strike led Beidinger and Decker (2014) to subdivide the Subalpine Molasse of eastern Upper and Lower Austria into a Western, a Central and an Eastern Sector (Figure 5.13). Whereas the structural architecture of the Western Sector is described in the previous section, the paragraphs below focus on the Central and Eastern Sector, both located to the east of the so-called Bohemian Spur. In order to allow for a detailed description of the thrust sequence, the provided age information refers to the stages of the Miocene Central Paratethys (Figure 5.14; e.g. Piller et al. 2007).

The seismic section to the west of the Vienna basin depicts the RDF units overthrusting two nappes of the Subalpine Molasse (Figure 5.15(A)). The latter include Molasse sediments of Egerian to Eggenburgian age and Eggenburgian to Lower Ottnangian age respectively. At the orogenic front, the imbricate Molasse overthrusts autochthonous Lower Miocene Molasse sediments of the Late Ottnangian age. The Alpine floor thrust is located within either Egerian or Eggenburgian Molasse sediments. No autochthonous Mesozoic cover units are evident in this sector resulting in the Upper Oligocene and Lower Miocene autochthonous Molasse resting directly on top of the European Crystalline Basement.

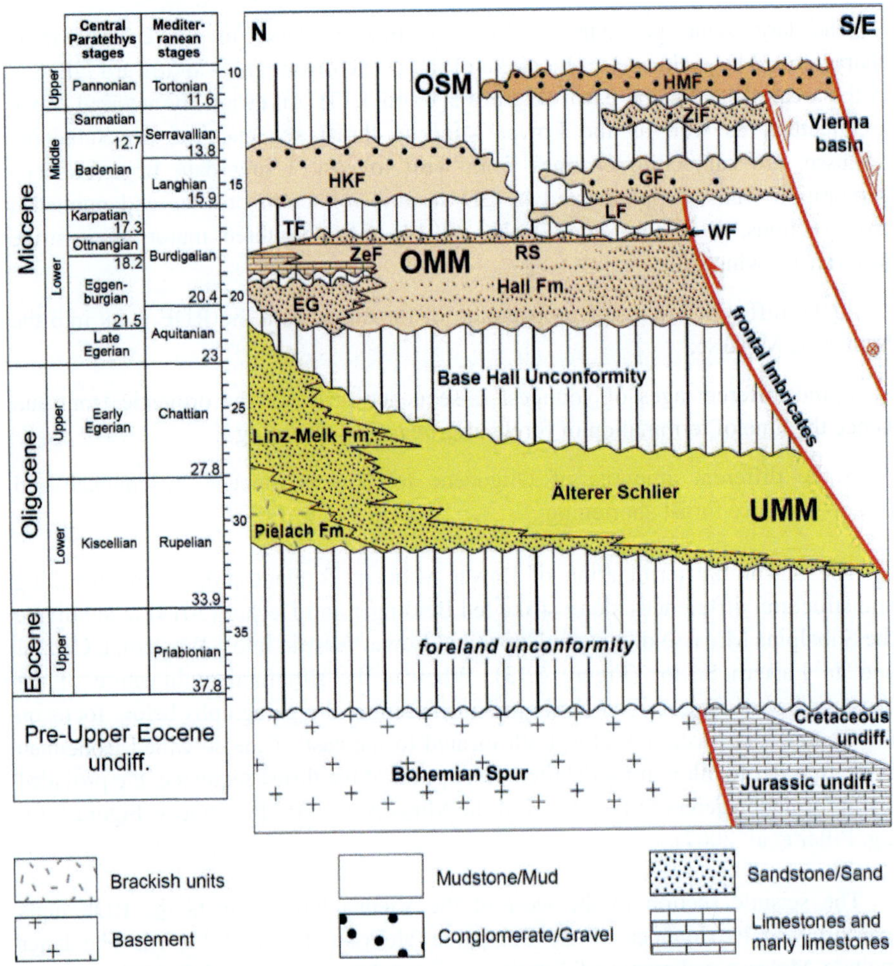

Figure 5.14. *Simplified lithostratigraphic chart of the Foreland Molasse north of the Alpine thrust front. Adapted from Krenmayr (1999), following Gebhardt et al. (2013) and Palzer-Khomenko et al. (2018). The correlation between Central Paratethys stages and Mediterranean stages follows Hilgen et al. (2012), absolute ages are from Cohen et al. (2013; updated). The correlation to the UMM, OMM and OSM of the foreland west of the Bohemian Spur is used tentatively for comparison. EG = Eggenburg Group, GF = Grund Fm., HKF = Hollabrunn-Karlstetten Fm., HMF = Hollabrunn-Mistelbach Fm., LF = Laa Fm., RS = Robulusschlier, TF = Traisen Fm. (Oncophora Beds), WF = Wildendürnbach Fm., ZeF = Zellerndorf Fm., ZiF = Ziersdorf Fm. The youngest frontal thrusts, found NW of the Vienna basin are of Late Karpatian age and cut the Karpatian Laa Fm*

The Egerian to Eggenburgian age of the autochthonous Molasse underneath the RDF units indicates that thrusting propagated from the RDF units into the imbricate Molasse during the Eggenburgian. The ages of the youngest overthrust sediments of the imbricate Molasse show that the nappe structure of the Lower Miocene of this sector was formed by foreland propagating thrusting, which lasted from the Eggenburgian until the Late Ottnangian (Figure 5.15(A)). No out-of-sequence thrusts are apparent within the Lower Miocene imbricate Molasse along with this sector. Instead, out-of-sequence thrusts are evident in the hinterland from the presence of horses and duplexes consisting of sedimentary rocks of Ultrahelvetic origin (mostly Buntmergel Fm. of the Gresten Klippen Zone) and Oligocene (UMM) foreland deposits within the RDF (Beidinger and Decker 2014). The frontal thrusts of these horses are characterized by thrust contacts with younger sediments in the hanging wall over older sediments in the footwall. Prominent occurrences of such Ultrahelvetic rocks are the Hauptklippenzone within the RDF (e.g. Figure 5.13; Wessely et al. 2006) and the Molasse Windows of Rogatsboden and Texing further west (Figure 5.13; Prey 1980). The Molasse sediments provide a maximum age of Oligocene for these out-of-sequence thrusts (Prey 1957, 1980; Beidinger and Decker 2014). The direction of Paleogene and Lower Miocene in-and out-of-sequence thrusting in this sector is toward N to NNW (Decker et al. 1993; Beidinger and Decker 2016).

In contrast to the Central Sector (Figure 5.15(A)), autochthonous Mesozoic cover units overlie the European Basement units in the area of the present-day Vienna basin (Figures 5.15(B) and (C); Wessely 1987). Both sections show that the Alpine floor thrust below the Subalpine Molasse is located within the so-called Mergelsteinserie (Mikulov Fm.) of the Upper Jurassic age. This leads to the incorporation of Jurassic to Paleogene sedimentary rocks into the so-called Waschberg Super Unit and the development of a thrust wedge at the base of the Foreland Molasse (e.g. Krhovsky et al. 2001; Wessely et al. 2006) covered by Ottnangian to Karpatian growth strata (Figure 5.15(B) and (C)). The deformation front is thus hidden (compare section 5.2.3).

In the two profiles in Figure 5.15(B) and (C), the RDF units are depicted to overthrust the Waschberg Super Unit. The nappes of the Waschberg Super Unit overthrust the Eggenburgian to Karpatian Foreland Molasse, which is further deformed by a set of frontal blind thrusts. A Lower Miocene wedge top basin, which includes a sedimentary succession ranging from the Eggenburgian to the Karpatian developed on top of the Alpine thrust belt in the hinterland (Kapounek et al. 1965; Hölzel et al. 2010; Harzhauser et al. 2017).

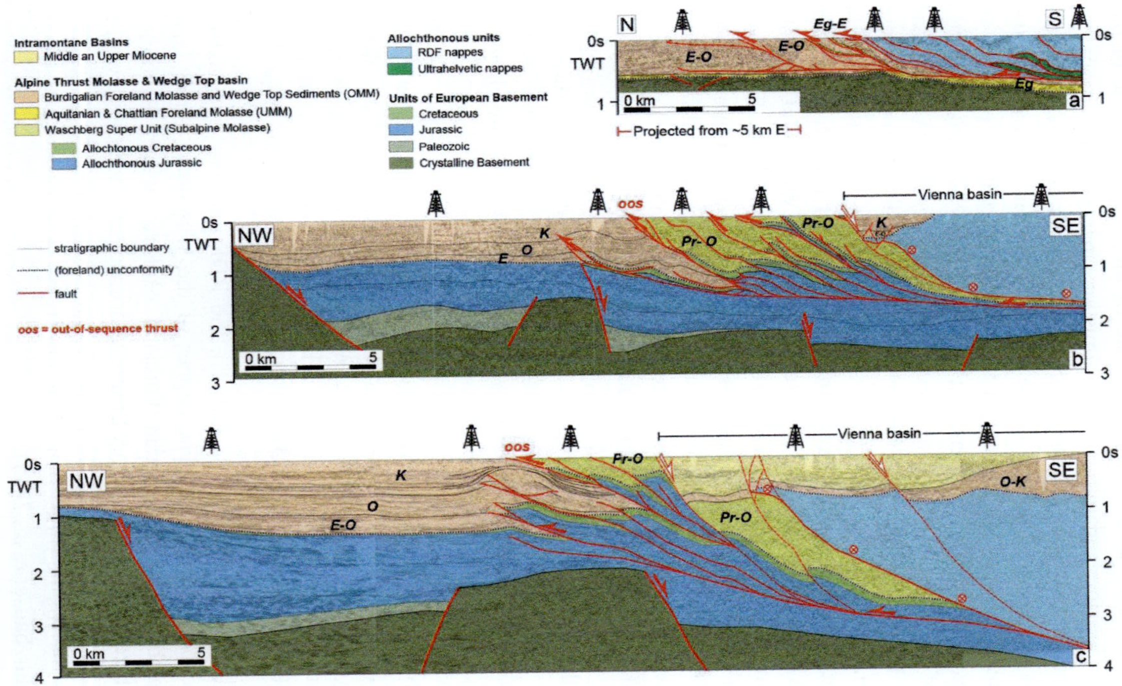

Figure 5.15. *Selection of cross-sections through the Lower Austrian imbricate European margin. (A) Interpreted seismic section, Central Sector–west of the Vienna basin. (B) Interpreted seismic section, Vienna basin area–Eastern Sector. (C) Interpreted seismic section, Eastern Sector–Vienna basin area. Sections modified from Beidinger and Decker (2014) and (C) adapted from Hölzel et al. (2010) and Beidinger and Decker (2014). See Figure 5.13 for the location of sections. Pr = Priabonian, Eg = Egerian, E = Eggenburgian, O = Ottnangian, K = Karpatian*

Ages of youngest overthrust sedimentary rocks show that the nappes of the Waschberg Super Unit and the Molasse have formed by a generally foreland-propagating thrust sequence lasting from the Eggenburgian to the Karpatian. The Alpine floor thrust propagated from the RDF units into the Waschberg Super Unit during the Eggenburgian. Further propagation of thrusting from the Waschberg Super Unit into the frontal blind thrusts below occurs during the Late Ottnangian to Karpatian times. Karpatian growth strata, related to the frontal blind thrusts mark the youngest in-sequence thrust deformation at the orogenic front (Hölzel et al. 2010; Beidinger and Decker 2014).

Out-of-sequence thrusts within the imbricate Molasse with offsets of more than 3 km occurred during the late stage of foreland propagating thrusting in the Karpatian (oos of Figure 5.15(C); Zámolyi et al. 2008). Comparable Lower Miocene (Ottnangian–Karpatian) out-of-sequence thrusts are documented from the RDF units and the Northern Calcareous Alps in the hinterland (Hölzel et al. 2010; Beidinger and Decker 2014; Beidinger et al. 2019). Apart from the described frontal late-stage in- and out-of-sequence thrusting, Granado et al. (2016) showed Lower Miocene inversion of Mesozoic normal faults in the European basement NW of the Vienna basin. Likewise, a reactivation of Variscan basement thrusts during a late stage of thrust shortening is reported from below the Outer West Carpathian Flysch units further to the NE (Beidinger and Decker 2016).

Whereas Paleogene thrusting in the Eastern Sector is approximately directed toward the N, the direction of Lower Miocene thrusting within the Subalpine Molasse is approximately directed toward the NW (Beidinger and Decker 2016). Lower Miocene out-of-sequence thrusts within the tectonically higher units in the hinterland are directed both, toward N and NW (Beidinger and Decker 2016; Beidinger et al. 2019).

5.2.4.2. *Thrust distances and the interplay between in- and out-of-sequence thrusting*

An assessment of minimum Late Oligocene and Early Miocene minimum thrust distances is available from palinspastic reconstructions, which cover a 250 km wide strike segment of the Eastern Alps extending from eastern Upper Austria in the west to the area of the present-day Vienna basin in the east (Beidinger and Decker 2014). The reconstructions include the three sections shown in Figure 5.15. While in the Eastern Sector, minimum Eggenburgian to Karpatian shortening is 21 km (Figure 5.15(B) and (C)), minimum post-Egerian shortening further to the west is only about 15 km (5.15(A)). Post-Egerian in-sequence shortening decreases further toward the west with no in-sequence thrusting documented in eastern Upper Austria after the Egerian (Beidinger and Decker 2014; see also Figure 5.12(D)). In this area, a minimum of 26 km of Egerian thrusting can be determined based on the well Molln 1 (Hamilton 1989; Beidinger and Decker 2014). The total cumulative amount of Egerian to Karpatian in-sequence thrusting may be estimated with a minimum of about 47 and 41 km for the sections in Figure 5.15 (Beidinger and Decker 2014).

Beidinger and Decker (2014) argue that the along-strike variations of both thrust sequences and in-sequence thrust distances are governed by the topography of the underlying European Plate. They propose basement buttressing as the main mechanism for the diachronous termination of in-sequence thrusting at the orogenic front and consequently also for the along-strike variation of the Early Miocene in-sequence thrust distances. A major recess of the so-called Bohemian Spur (e.g. Wessely 1987) toward the NE in the area of the present-day Vienna basin is thought to result in longer-lasting in-sequence thrusting and larger in-sequence thrust distances in the east when compared to the Western Sector.

Out-of-sequence thrusting in the Molasse basin NW of the Vienna basin only occurred during the latest stage of thrust shortening. Seismic data from the area indicate Ottnangian to Late Karpatian out-of-sequence thrusts at the orogenic front and in the internal units of the Eastern Alps (Figure 5.13; Zámolyi et al. 2008; Hölzel et al. 2010; Beidinger and Decker 2014; Beidinger et al. 2019).

The difference between the thrust displacements compensated by in-sequence thrusting in the east (47 km) and further west (26 km) is accommodated by out-of-sequence thrusting in the western parts (Beidinger and Decker 2014). Out-of-sequence thrusting there is evident from the numerous occurrences of Molasse and Ultrahelvetic units in windows within the RDF (Figure 5.13; Prey 1957, 1980; Egger 1987; Braunstingl 1988; Wagner 1996; Wessely et al. 2006). Beidinger and Decker (2014) show an increase in the cumulative amount of out-of-sequence thrust distance from east to west, which supposedly balances the decrease of Early Miocene in-sequence thrust distances.

Other interpretations of Early Miocene out-of-sequence thrusting west of the Vienna basin are based on apatite fission-track data from the RDF (Trautwein et al. 2001). The data indicate the burial of some of the RDF nappes of up to more than 6 km and exhumation of these nappes commencing during the Late Oligocene. A similar burial depth has independently been derived earlier from vitrinite reflectance by Gmach (1999) (see also review in Egger and Ćorić (2017)). Based on the data from the Wienerwald area, Trautwein et al. (2001) propose a Late Oligocene to Early Miocene age range for out-of-sequence thrusting, which is an earlier onset than proposed by Beidinger and Decker (2014). Comparable N-directed out-of-sequence thrusting within the RDF units, older than the timing of the termination of in-sequence thrusting at the orogenic front has recently been described in thrust units below the Vienna basin, too (Beidinger et al. 2019). Such earlier onset of out-of-sequence thrusts may again be related to the shape and geometry of the underlying basement units. The far SE-reaching Bohemian Spur (Tari 2005; Wessely et al. 2006) exhibits its highest topography in the Central Sector, which may have favored both, a relatively early onset of out-of-sequence thrusting and/or a relatively early exhumation as well as erosion of higher Austroalpine nappes.

5.3. Discussion

5.3.1. *Timing and sequence of thrusting*

The timing and sequence of thrusting at the northern deformation front of the European Alps have been constrained by stratigraphic data including syntectonic sediments, as well as low-temperature thermochronometry. It is important to note that the different tectonic slices present along the strike of the Subalpine Molasse have different sedimentary contents and thus may correspond to distinct paleogeographic sub-domains. Discrete topographic barriers may arise from early fault-related folding leading to syntectonic sedimentation in small sub-basins. This close link between tectonics and sedimentation is witnessed in the partly complex stratigraphy of the foreland basin deposits, as reviewed above. Where preserved, these complexities in turn provide constraints on the tectonic sequence of the basin. In the following, we review the sequence of thrusting from west to east.

5.3.1.1. *Western and the Central Alps*

5.3.1.1.1. Subalpine Molasse and JFTB

In the western part of the foreland, the southernmost parts of the Molasse Basin became incorporated into the orogenic wedge shortly after deposition in Late Oligocene times, when large bodies of USM conglomerates were thrusted onto the foreland. Syn-tectonic unconformities have been described in the Salève area (Beck et al. 1998) suggesting Late Oligocene–Early Miocene (Chattian–Aquitanian) (USM) folding and thrusting. Re-deposition of fan conglomerates and progressive foreland-directed migration of alluvial fans in response to the advancing thrust sheets possibly indicate in-sequence thrusting during the period between ca. 24–20 Ma (Schlunegger et al. 1997; Kempf et al. 1999). However, internal blind thrusts may be active at the same time (Bonnet et al. 2007). After ca. 20 Ma, the thrust front advanced toward the foreland at an average rate of approximately 1.4 km/Myr, as witnessed by the linear northward decrease of stratigraphic ages exposed in the Subalpine Molasse (von Hagke et al. 2012; von Hagke et al. 2014b). Approximately at 16 Ma, a triangle zone evolved at the front of the Subalpine Molasse, between

Lake Thun and Schongau, south of Munich (Figures 5.3 and 5.8). The formation of the triangle zone is associated with a regional-scale erosional unconformity (Schlunegger et al. 1997; Kempf et al. 1999) as well as deposition of growth strata at the southern end of the (undeformed) Molasse (Kempf et al. 1999; Ortner et al. 2015). In Early and Mid-Miocene times the triangle zone constitutes the northern deformation front of the entire Central Alps and reaches as far east as the western Tauern Window. West of Lake Thun similarly, the leading edge of deformation presumably also was in a similar position at that time, though it may have initiated earlier in a more meridional position (Burkhard and Sommaruga, 1998). In Late Miocene times, thrusting stepped back into internal parts of the Subalpine Molasse, as has been shown by low-temperature apatite (U-Th-Sm)/He thermochronometric studies from the Lake Thun, Entlebuch, Rigi and Bregenzerach area. (von Hagke et al. 2012; von Hagke et al. 2014b; Mock et al. 2019). Here, thermochronological data indicate thrust-driven exhumation pulses between ca. 12–4 Ma Based on the observation that the youngest movements in the Subalpine Molasse east of lake Thun must be younger than erosional features of Tortonian age, Haus (1935) already recognized that the Subalpine Molasse was subject to Late Miocene thrusting. While Late Miocene foreland deformation in the eastern non-detached Molasse Basin was mostly restricted to the Subalpine Molasse, large-scale strain partitioning occurred at that time in the western wedge-top (i.e. detached) part of the Molasse Basin and Alpine foreland (Figure 5.5).

In the west, deformation was partially translated along with the intra-Triassic décollement level northward, across the Molasse Basin, into the JFTB. It was proposed early that the basal décollement of the JFTB is connected to a basal thrust in the more internal Alps and that a distant push is exerted by the northward emplacement of the Alpine nappe stack (Laubscher 1974). In Late Miocene times and possibly even in the Early Miocene (Stutenbecker et al. 2019), when the main deformation of the JFTB–Subalpine Molasse system occurred, the ECMs had already started exhuming and the overlying Helvetic main thrust was folded (e.g. Burkhard 1990; Leloup et al. 2005; Egli and Mancktelow 2013; Egli et al. 2017b; Herwegh et al. 2019). Consequently, despite evidence for early deformation in Oligocene times (Aubert 1958; Liniger 1967), the time span for the main thin-skinned deformation phase of the JFTB is classically constrained to ca. 12–4 Ma (e.g. Becker 2000), which coincides with the climax of the ECM exhumation and northward translation. Earlier deformation would require the initial basal décollement to be connected to the basal thrust of the Helvetic nappe system or possibly even the Pennine thrust, thus rooting south of the ECM. This solution may apply to the earliest thrusting in Subalpine Molasse imbricates, as also proposed in analog modeling (Bonnet et al. 2008), but would lead to kinematic and timing incompatibilities if applied to the JFTB basal décollement or post 10 Ma thrusting in the Subalpine Molasse. Thus early deformation in Oligocene times may be related to the flexural bending of the European Plate in front of the Alpine orogen or the opening of the Rhine–Bresse Graben system. The main deformation in the JFTB and the Subalpine Molasse in late Miocene times shows strain partitioning between the JFTB and the Subalpine Molasse in the wedge-top part of the basin (west), while in the non-detached part of the basin (east) thrusting in the foreland was restricted to the proximal realm, i.e. the Subalpine Molasse (Figures 5.5 and 5.6; Burkhard 1990).

Though most of the deformation in the JFTB is linked to the thin-skinned development, recent deformation, post-dating the main deformation event, possibly since 4 Ma, has been linked to reactivation of inherited basement faults thus suggesting a recent thick-skinned style of deformation (Mosar 1999; Giamboni et al. 2004a; Giamboni et al. 2004b; Ustaszewski and Schmid 2006; Madritsch et al. 2008; Madritsch et al. 2010). Also, recent studies on Pleistocene river gravels deposited along the northern edge of the Jura Mountains south of the Rhine Graben and the Vosges mountains have shown that these gravel layers are uplifted or even deformed (Giamboni et al. 2004a, 2004b; Ustaszewski and Schmid 2007a; Madritsch et al. 2010). These would be the most recent traces of Jura folding, associated with sediments. Changes in the river network north of the Jura suggest that active deformation has not been completed (Giamboni et al. 2004a). This deformation appears to be linked to the reactivation of basement faults in the transfer zone between the Rhine and Bresse Grabens as well as of the JFTB main detachment (Ustaszewski and Schmid 2007a). Although this is corroborated by seismicity in the basement (Diehl et al. 2015), it cannot be excluded that thin-skinned deformation continues to be active given the fact that much seismicity continues to occur in the detached cover series of the Alpine whole foreland (Thouvenot et al. 1998; Courboulex et al. 1999; Mosar et al. 2011; Vouillamoz et al. 2016, 2017; Abednego et al. 2017).

This coeval deformation of the JFTB and the Subalpine Molasse is also mirrored in the cumulative estimates for Late Miocene horizontal shortening, which remain more or less constant along strike the western North Alpine Foreland between Lake Thun and Lake Constance (von Hagke et al. 2014b). In the JFTB shortening decreases from a maximum of ca. 25–30 km in its centre to zero at its eastern tip (Laubscher 1965; Mugnier and Vialon 1986; Burkhard 1990; Guellec et al. 1990; Philippe et al. 1996; Affolter and Gratier 2004; Leloup et al. 2005; Bellahsen et al. 2014; Malz et al. 2016). This decreasing trend is compensated by an increase in shortening in the Subalpine Molasse from ca. 10 km in the west (Lake Geneva) to ca. 20 km farther east (Lake Constance) (Figure 5.6) (Burkhard and Sommaruga 1998; Bellahsen et al. 2014; von Hagke et al. 2014b; Ortner et al. 2015; Mock et al.

2019). It is important to note that the shortening estimates in the Subalpine Molasse are minimum estimates, as the hangingwall cut-offs of individual thrust sheets are commonly not preserved. Moreover, the most proximal part of the Subalpine Molasse is hidden below the basal thrusts of the frontal Helvetic nappes and Penninic Klippen units (Figure 5.4), complicating the determination of the maximum amount of shortening (e.g. Menkveld 1995; Ortner et al. 2015).

5.3.1.1.2. Relation of shortening at the Alpine front to the ECMs

Late Miocene deformation in the Subalpine Molasse and the JFTB are considered to be kinematically linked to the exhumation of the ECMs (e.g. Boyer and Elliott 1982; Burkhard 1990; Pfiffner et al. 1997; Burkhard and Sommaruga 1998; Mosar 1999; Leloup et al. 2005; Bellahsen et al. 2014). Consequently, shortening in the fold-thrust belt is partly witnessed in the exhumation signal of the ECMs (von Hagke et al. 2014b). This implies that the amount of thrust displacement within the Subalpine Molasse plus the thrust displacement in the Jura Mountains are balanced by northward displacement of the ECMs on their basal detachment and along with fault arrays documented at their northern front (Bellahsen et al. 2014; Herwegh et al. 2017; Herwegh et al. 2019). For the Late Miocene, i.e. the time frame for which exhumation estimates for the ECMs are available, the Aar Massif in Central Switzerland (Michalski and Soom 1990; Reinecker et al. 2008; Glotzbach et al. 2010; Valla et al. 2012) and the Mont Blanc and Aiguilles Rouges Massifs in western Switzerland and France (Leloup et al. 2005; Bellahsen et al. 2014; Boutoux et al. 2016; Egli et al. 2017a) show rather a continuous exhumation. This can be translated into displacement along with the thrust systems at its northern front. The resulting shortening estimates are similar to estimates obtained at the deformation front in the same time frame (von Hagke et al. 2012; von Hagke et al. 2014a; Herwegh et al. 2017; Mock et al. 2019). However, Late Miocene thrusting is not restricted to the foreland of the ECMs. Although the easternmost ECM, the Aar Massif, disappears in eastern Switzerland, Late Miocene thrusting can be observed along the entire Subalpine Molasse until ca. Salzburg. Hence, the non-cylindrical configuration of the ECMs in the hinterland is only partly reflected in the foreland deformation with respect to the amount of shortening. The observation that shortening occurs almost simultaneously within the entire Subalpine Molasse over such long distances indicates that plate convergence or deep-seated processes are the main drivers of shortening (Mosar 1999; Bonnet et al. 2007; Bonnet et al. 2008; von Hagke et al. 2012; von Hagke et al. 2014b; Mock et al. 2019).

The Subalpine Molasse has been subject to late reactivation (see above) and has itself been overridden by the Penninic and Helvetic nappes (e.g. Menkveld 1995; Herwegh et al. 2017). The timing of this internal thrusting is not yet known and may be either seen as an early event, suggesting the Subalpine Molasse has been imbricated as early as Oligocene times. Alternatively, emplacement of the Penninic Klippen and Helvetic units on top of the Subalpine Molasse may have occurred later, suggesting reactivation of their basal thrusts. These two models are not mutually exclusive and mixtures between the two end-member scenarios are possible.

5.3.1.2. *Eastern Alps west of the Bohemian Spur*

5.3.1.2.1. Subalpine Molasse

In Bavaria and western Austria, timing and sequence of thrusting are constrained by growth strata in the OMM and OSM in the northern limb of the most external structure between Lake Constance and Salzburg (Eberhard 1986; Ortner et al. 2015). This means that the frontal anticline initiated in the Eggenburgian and grew at least until the Pannonian, growth outlasting the youngest preserved deposits. The deformation front did not propagate further into the foreland since the Eggenburgian (Ortner et al. 2015).

Cross-cutting relationships between thrusts in the cross-sections of Figure 5.9 provide further clues on the relative age of thrusting. The passive backthrust of the frontal triangle zone is truncated by the most external thrust (Salmas thrust of Figure 5.9(A); see also Figure 5.9(C)). The most external thrust is systematically steeper than most internal thrusts (e.g. Horn thrust of Figure 5.9(A)), suggesting that tightening of the frontal structure predated initiation of the more internal thrusts. There are no geometric indications of the relative age of the thrusts at the base of the Murnau, Steinberg and Hausham slices. However, cross-cutting relationships show that these thrusts postdate folds, where those are present (e.g. thrust at the base of Murnau and Hausham slices crosscuts folds formed above a thrust below; Figure 5.9(C) and (D)). Low-temperature thermochronology supports a hinterland-breaking thrust sequence (see also below). Thus, a foreland-propagating thrust sequence ended in the Eggenburgian, presumably as a result of processes in the internal part of the orogen, but the thrusts continued to be active at least until the Pannonian. Eventually, the thrusts were abandoned and the strain was accommodated along with more internal thrusts.

Line length balancing of nine Subalpine Molasse cross-sections by Ortner et al. (2015) and Pomella et al. (2015) demonstrate that shortening decreases from 45 km in the west near the Rhine Valley (section of Figure 5.9(A)) to 8 km in the TRANSALP section (section of Figure 5.9(D)). While it is difficult to separate post- and pre-Eggenburgian shortening in the west, it is clear in the TRANSALP section that the frontal thrust has more offset at depth within the UMM (15 km) than at the surface within the Cyrena Beds (4.3 km). This suggests that part of the offset predates deposition of the Cyrena Beds, is of Oligocene age and is related to the initial emplacement of the Helvetic thrust wedge (2 of Figure 5.9(D)). Post-Eggenburgian shortening entirely disappears at the deformation front east of the TRANSALP section and shifts into the more internal part of the orogen. As a consequence, Upper Chattian to Burdigalian sediments bury the retreating Alpine wedge between Salzburg and Linz at the southern margin of the Foreland Molasse and suffered only a slight basinward tilt as the buried thrust wedge thickens (section 5.2.3.3; Hinsch 2013).

5.3.1.2.2. Perwang, Sierning and Regau imbricates

The timing of frontal accretion of the Perwang accretionary wedge is constrained by the strata in and around the wedge. Within the given resolution of lithostratigraphic correlation, sediments belonging to the Lower Puchkirchen Fm. are present within the wedge, but are also found onlapping the front of the wedge. The small-scale imbrications in the frontmost part of the central wedge indicate back-stepping deformation and the resulting morphological high is covered by onlapping Upper Puchkirchen Fm. sediments (Figure 5.12(B)). Consequently, all deformation observed in the frontal part of the Perwang imbricates can be considered to have happened during the time of deposition of the Lower Puchkirchen Fm., i.e. Late Chattian to Early Aquitanian (Grunert et al. 2015; Hülscher et al. 2019). However, along with the western fringe of the Perwang lobe, younger deformation up to at least Burdigalian age is recorded at the steep interface of Helvetic imbricates to slightly deformed Molasse sediments (Figure 5.12(A)).

The termination of the frontal advance of the over-thrust wedge in the Regau segment is constrained by the Burdigalian age of the onlapping Hall Fm. The slight tilting of the Hall Fm. indicates minor post-depositional internal deformation (Late or post-Burdigalian). The geometry of the thrust slices in the Sierning segment indicates a dominant in-sequence frontal accretion. The BHU is sealing the deformation and constraining the time of accretion to be younger than the Early Burdigalian age of the unconformity (19.6–19.0 Ma; Hülscher et al. 2019). However, similar to the Regau segment there is some minor successive tilting of the beds above the unconformity into northern directions, which indicates post-depositional processes, e.g. internal structural thickening of the wedge or differential compaction.

Balancing of the sections indicates that the amount of shortening varies across the frontal structures (Hinsch 2013). The Sierning imbricates have about 6 km of minimum shortening while the Perwang imbricates accumulate 18.5 km of shortening. For both sections, the amount of overthrusting of the RDF and Helvetic wedge above the Molasse still needs to be added. Additional shortening in all sections is accommodated by out-of-sequence deformation, as demonstrated by the presence of Ultrahelvetic marls in narrow tectonic windows within the RDF wedge (see section 5.2.3.1 above; “Streifenfenster”; Prey 1980). Due to missing growth strata and poor seismic data, it is difficult to estimate the amount of shortening accommodated more internally. However, wells in the hinterland of the Regau and Sierning imbricates drilled Molasse sediments below the RDF and Helvetic wedge which (Figure 5.10) allows estimating that the amount of advance of the orogenic wedge was laterally similar.

Beidinger and Decker (2014) and Hinsch (2013) estimate about 30 km of northward progression of the northern boundary of the Helvetic and RDF wedge from the Chattian to the present-day position. In the Regau and Sierning segments, the propagation of the deformation front is likely relatively gradual in Chattian, Aquitanian and Earliest Burdigalian. In the Perwang segment, in contrast, the deformation propagated close to the deformation front already in Chattian to earliest Aquitanian and then remained in that position. Additional shortening was accommodated by back-stepping the deformation into the orogenic wedge. With the given resolution and constraints, it can be estimated that the total shortening in all segments is relatively constant. However, the mode and the location of the deformation varies (i.e. frontal accretion versus overthrusting versus internal thrusting in the Helvetic and RDF units). Consequently, the frontal-most structures do not represent the youngest shortening deformation in the three segments.

5.3.1.3. *Eastern Alps east of the Bohemian spur*

The Subalpine Molasse reappears at the surface east of Linz. In the internal Alps east of the Tauern Window, the crust is thinned toward the Pannonian basin as a consequence of lateral extrusion (see section 5.3.2) and thrusting should predate this crustal thinning, as thin and thus weak crust would localize thrusting and consume shortening in the internal Alps. However, the process of thinning the orogenically thickened crust needs time. An older Chattian to Karpatian part of extrusion tectonics and related lateral movements (e.g. SEMP fault; Figure 5.13) is contemporaneous with Subalpine Molasse thrusting. A younger, post-Karpatian part is connected to the opening of the Vienna basin and postdates thrusting and the Vienna basin strike-slip fault system and linked normal faults overprint and truncate the Subalpine Molasse thrusts (Figures 5.13 and 5.15; Beidinger and Decker 2014). Thrusting within the Subalpine Molasse ends at the end of the Early Miocene in the Vienna basin area, but already at the end of the Egerian to Eggenburgian on top of the Bohemian Spur (Beidinger and Decker 2014). The main controlling parameter may be the topography of the Bohemian Spur, which forms a barrier for thrusting and forces shortening into the hinterland (Beidinger and Decker 2014).

The most external part of the Alpine wedge was probably affected by differential exhumation related to changes in the flexural behavior of the Lower European Plate after the end of the Early Miocene (see section 5.3.4). Right on top of the Bohemian Spur, Molasse Windows below Ultrahelvetic and RDF nappes related to out-of-sequence thrusts occur (R, T and W of Figure 5.13; Richter 1950; Prey 1957; Beidinger and Decker 2014). Such windows are not known anywhere else further west or east and allow the observation of the deepest part of the Subalpine Molasse wedge on top of the Bohemian Spur.

5.3.2. *Relation of shortening at the Alpine front to the internal Eastern Alps*

Similar to the Western and Central Alps, shortening at the deformation front is closely related to processes in the internal part of the Alpine orogen. However, no ECMs exist in the Eastern Alps. The Engadine and Tauern Windows are located more internally (Figure 5.1) and Alpine metamorphism is much higher as compared to the ECMs (e.g. Miller et al. 1980; Selverstone et al. 1984; Bousquet et al. 1999; Bousquet et al. 2012). Nevertheless, final exhumation of the Engadine Window (e.g. Mattmüller 1991; Pfiffner and Hitz 1997) and the Tauern Window (e.g. Fügenschuh et al. 1997; Luth and Willingshofer 2008) has been interpreted to be of Miocene age and should therefore be connected to Subalpine Molasse shortening (see e.g. cross-sections in Ortner et al. 2006; Rosenberg and Kissling 2013; Rosenberg et al. 2015).

The exhumation of the Tauern Window is different from the ECMs and the Engadine Window, as crustal thickening was accompanied by horizontal stretching (Lammerer and Weger 1998; Rosenberg and Garcia 2011; Scharf et al. 2013) caused by east-directed extrusion tectonics north of the Adriatic indenter (Ratschbacher et al. 1991; Linzer et al. 2002; Rosenberg et al. 2004; van Gelder et al. 2017; Rosenberg et al. 2018). Horizontal stretching and eastward movement of blocks is coeval with the growth of the Subalpine Molasse thrust belt and counteracts thickening within the wedge. E-directed movement between faults accommodates part of the shortening (Linzer et al. 2002; Rosenberg et al. 2018) and progressively less shortening arrives at the deformation front (Figure 5.1). Thus, it is no coincidence that the Subalpine Molasse dies out east of Salzburg where crustal thinning is significant in present-day W–E cross-sections across the internal Eastern Alps (Ratschbacher et al. 1991; Rosenberg et al. 2018). Additionally and potentially linked to the location of the E-directed movement, the termination of the Subalpine Molasse east of Salzburg coincides with the extent of the present-day downgoing European slab (Mock et al. 2019; Kästle et al. 2020). This implies that the boundary between the Eastern and Western Alps as witnessed at the surface within the orogen (i.e. roughly south of Lake Constance) is not expressed in the deep slab configuration (Mock et al. 2019). Instead, the Subalpine Molasse is more directly linked to deep-seated processes.

5.3.3. *Alongstrike changes in structural geometry*

The main reason for along-strike changes in structural geometry are (1) changes in mechanical stratigraphy of the lower plate, from which the units of the Subalpine Molasse are detached and (2) changes in the sedimentary fill of the foreland basin itself. The thrusts stripping off units from the European margin cut up-section from the Helvetic units, which are detached within the sedimentary succession of the European margin, to the Subalpine Molasse that is detached within the foreland succession (Figure 5.5). The northern pinch-out of the UMM in the Alpine foreland disappears below the Alpine wedge to the west (Figure 5.1) and demonstrates the oblique orientation of the Oligocene foreland flexure as compared to the Middle Miocene to the present-day deformation front. This has been related to differential shortening within the Subalpine Molasse and Helvetic thrust belts (Ortner et al. 2015).

The basal décollement of the Subalpine Molasse lies within Rupelian marls (Tonmergel) close to the base of the foreland sequence in Switzerland, western Austria and southern Germany, in the deepest part of the foreland basin (Figures 5.2, 5.7). In Switzerland west of Zurich, a second, deeper décollement below the Molasse Plateau lies within evaporites of the Triassic succession (Figure 5.2). At the faulted southwestern boundary of the Bohemian Massif, the décollement jumps between Cretaceous and Rupelian marls (Figure 5.2; Hinsch 2013). SW of the Central Swell a half-graben with thick (overpressured) Upper Cretaceous shales and marls are present, allowing the propagation of the Perwang imbricates on a weak décollement. The shape of the Perwang salient is thus controlled by the geometry of the half-graben and the pinching Cretaceous at the Central Swell (Hinsch 2013). Similar controls have been observed for salients in other fold-and-thrust belts (Boyer 1995; Mitra, 1997). The stacked Helvetic imbricates near the western end of the Perwang imbricates (Figure 5.12(A)) indicate continuous out-of-sequence deformation. Hinsch (2013) relates this to the sediment deposition rate in the foreland basin and absolute thickness in this part of the basin in front of the entry point of the paleo-Inn Valley. Thrust-wedge analog modeling with syntectonic sedimentation has shown that high sedimentation rates can inhibit frontal propagation (e.g. Storti and McClay 1995).

The Sierning imbricates develop in the Bad Hall Basin (Figure 5.10). Several factors are considered to locate the décollement dominantly in the shales of the Zupfing Fm.: Cretaceous sediments are usually relatively thin and the facies is much sandier than in the Salzach Basin (Nachtmann and Wagner 1987). Furthermore, the pre-Molasse and Early Molasse layers are strongly dissected and offset by faults, destroying a potential continuous detachment layer. The very shaly Zupfing Fm. is considered to level most of the fault-related morphology and thus is suitable as a relatively continuous detachment layer.

The Regau overthrust segment is roughly located above the Central Swell (Figures 5.10 and 5.12(C)). There, no Cretaceous marls are present. Furthermore, Eocene to lowermost Oligocene strata is dissected by faults similar to the Sierning segment. The shales of the Rupelian, that host the detachment in the Sierning imbricates further to the east are considered to have a higher content of turbiditic sands. Therefore, a weak décollement is missing. Considering the relatively high friction in the foreland succession, deformation in the Regau segment can be considered to be an effect of (cyclic) overthrusting, following Gutscher et al. (1996, 1998), who observed cyclic overthrusting in scaled sandbox experiments, where high friction favors this behavior.

On top of and east of the Bohemian Spur, the Foreland Molasse is very narrow and shallow, as compared to the western part of the foreland basin (Figures 5.1 and 5.2). The foreland sequence starts with the Egerian, older foreland deposits are only observed in the Molasse Windows within the Alpine wedge (R, T and W of Figure 5.13). This all suggests that much of the foredeep has been filled by the advancing thrust wedge. The basal décollement lies at the base of the Jurassic succession in the Ultrahelvetic Hauptklippenzone of the continental slope (Figure 5.15(A)). Only where the floor thrust cuts across the rifted continental margin, Jurassic–Cretaceous sediments detached from horsts of the rifted margin are piled up together with Cenozoic sediments of the Subalpine Molasse ("klippen"; Figures 5.15(B) and (C)).

5.3.4. *(Late) Early Miocene changes*

In the foreland basin on the western slope of the Bohemian Massif, post-OMM west-directed Karpatian tilting has been documented (Gusterhuber et al. 2012), also known as the formation of the Amstetten high (Lemcke, 1972, 1973, 1984). Seismic facies interpretation of Burdigalian reflectors on a regional W–E strike line through the Molasse zone of Upper Austria indicates a west to east infill of the Puchkirchen trough in the Hall Fm. However, the present-day position of the top of the Hall Fm. is 500 m higher on the eastern side of the section than in the western part, corresponding to a westward slope of approx. 0.5% (Figures 4 and 5 of Gusterhuber et al. 2012; see also Figure 5.2, 560–640 km). Lemcke (1972, 1973) deduced this tilting from an east-directed onlap of the OSM onto west-tilted and eroded OMM NE of Salzburg (MKH of Figure 5.10). During the tilting, the foreland basin further west subsided, while the Bohemian Massif was uplifted (Freudenberger and Schwerd 1996). The tilting is not related to the activity of the fault bundle at the SW margin of the Bohemian Massif, which affects mainly Mesozoic and older rocks (Figures 5.1, 5.2 and 5.10) and ends within Upper Egerian deposits (Nachtmann and Wagner 1987).

The tilting may be related to changes in flexure of the European Plate (Ortner et al. 2015). In N–S cross-sections across the foreland west of the Bohemian Massif, OMM and OSM dip shallowly toward the orogen in cross-sections north of the Subalpine Molasse thrust belt (Figures 5.6(A) and 5.16(A); Abele et al. 1955). There, the orogenic load continued to move closer to the foreland, but not into the foreland. This redistribution of loads caused a hiatus and angular unconformity at the end of the Egerian (Lemcke 1984, 1988) related to the ongoing flexure of the lower plate (Sinclair et al. 1991; Pfiffner et al. 2002). On the Central Swell and the Bohemian Massif, the bedding of the OMM and OSM is essentially horizontal and the OMM covers the thrust belt (Figures 5.12 and 5.16(B); e.g. Brix and Götzinger 1964; Wagner 1996) indicating a rapid retreat of shortening into the more internal Alps and thus the end of foreland flexure. Visco-elastic relaxation after the end of flexure (Zweigel et al. 1998; Hülscher et al. 2019) caused steepening of the lower plate and thus the pronounced BHU angular unconformity.

In this scenario, differential uplift and tilting in the E–W section are related to the combined effects of (1) ongoing flexure-related subsidence west of the Bohemian Spur and (2) forebulge formation within the Bohemian Massif at the margin to the subsiding basin at an oblique but high angle to the general basin long axis (Figure 5.1). This effect is probably more continuous than suggested by Gusterhuber et al. (2012), as clastic material from the Bohemian Massif started to appear during the Ottnangian, but supply culminated in the Pannonian (Freudenberger and Schwerd 1996). Uplift of the Bohemian Massif and the foreland basin on top of it (Amstetten high) had important consequences for paleogeography. While throughout the Egerian, sediment was transported toward the east, the OSM fluvial system drained the foreland to the west (e.g. Füchtbauer 1967; Lemcke 1984; Freudenberger and Schwerd 1996). This west-directed drainage persisted until the thrusting of the JFTB and the Subalpine Chains closed the west-directed sediment routing system around 10 Ma and the present-day eastbound Danube catchment was established (Ziegler and Fraefel 2009; Winterberg and Willett 2019).

5.3.5. *Exhumation of the foreland in the Late Miocene*

A large number of studies have contributed to the understanding of the exhumation history of the Alps (e.g. Frisch et al. 1998; Champagnac et al. 2009; Vernon et al. 2009; Glotzbach et al. 2010; Valla et al. 2012; Scharf et al. 2013; von Hagke et al. 2014b; Schlunegger and Kissling 2015) and their foreland (Pfiffner 1986; Allen et al. 1991; Sinclair and Allen 1992; Schlunegger et al. 1997; Kuhlemann, 2000; Mazurek et al. 2006; Cederbom et al. 2011; von Hagke et al. 2012; Gusterhuber et al. 2012; Hinsch 2013; Ortner et al. 2015; Mock et al. 2019). The stratigraphic record shows that the entire Molasse Basin has experienced broad Cenozoic uplift, however to a variable extent along strike the basin. For instance, near the village of Kiffis (in Switzerland, southwest of Basel) a syncline in the eastern part of the JFTB contains sediments of the OMM at +600 m elevation (Laubscher 1974; Berger et al. 2005a) and it has been reconstructed that sea level during OMM deposition was similar to today. Additionally, thermochronologic data shows 2–3 km of exhumation since approximately 10 Ma in large parts of the basin (Mazurek et al. 2006; Cederbom et al. 2011; von Hagke et al. 2012; von Hagke et al. 2014b). Schegg and Leu (1998) estimate the total amount of post-Serravallian erosion of Molasse sediments between 1.5 and 3 km in the western part of the foreland basin. Cederbom et al. (2011) and Ziegler and Fraefel (2009) argue for the onset of this major erosional phase during the Pliocene which may have been controlled by a reorganization of the regional drainage system (Petit et al. 1996; Madritsch et al. 2010; Willett and Schlunegger 2010; Schlunegger and Mosar 2011; Yanites et al. 2013). In the eastern part of the basin (i.e. east of Lake Constance), only very limited exhumation data from the foreland and the fold-thrust belt exists (Gusterhuber et al. 2012). However, the stratigraphic record indicates there has been an erosion in the Late Miocene (e.g. Genser et al. 2007). The magnitude of erosion is partly constrained and a general increase in the amount of erosion toward the west seems plausible. Furthermore, the amount of exhumation that can be related to thrusting in the foreland fold-thrust belt is well constrained in Switzerland (Schlunegger and Mosar 2011; von Hagke et al. 2012; von Hagke et al. 2014b; Mock et al. 2019). However, the timing of the basin-wide exhumation signal is only poorly defined. New thermal models show that there is substantial freedom for inferring exhumation scenarios. Possible exhumation scenarios range from fast exhumation in Plio-Pleistocene times to long-lasting constant exhumation initiating in the Late Miocene (von Hagke et al. 2015). The reason for this ambiguity of exhumation estimates lies in the uncertainty of the thermochronological data and the poorly constrained paleo-heat flow estimates.

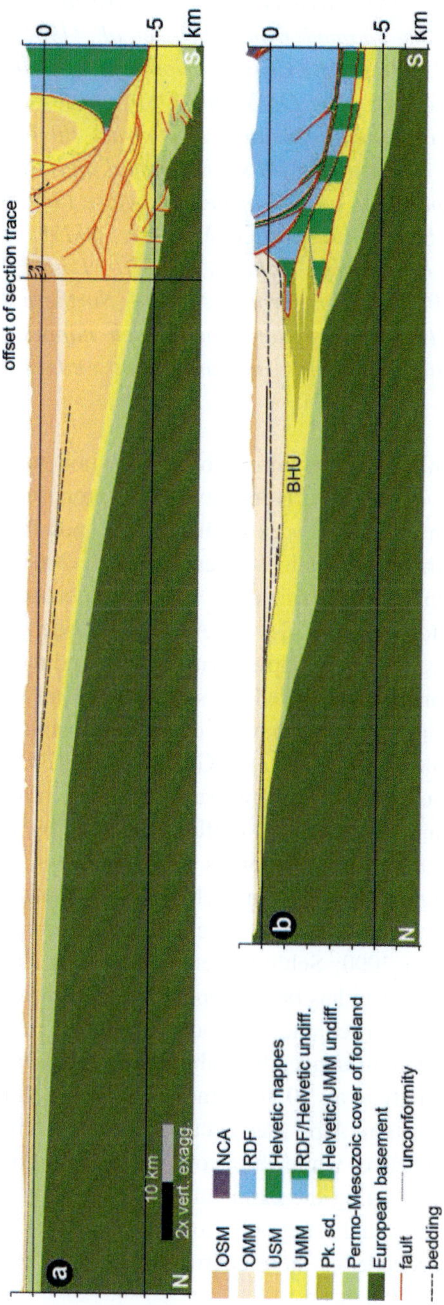

Figure 5.16. *Two sections across the Alpine foreland basin from the thrust belt to the present-day northern basin margin*

COMMENTS ON FIGURE 5.16.– *(A) Cross-section of the Bavarian part of the foreland, north of section trace offset adapted from Abele et al. (1955), south of offset redrawn from Ortner et al. (2015). (B) Cross-section in the Salzburg area adapted from Wagner (1996). In (A), the base OMM and OSM dip toward the thrust belt, therefore foreland flexure outlasted deposition. In contrast, the base of OSM in (B) is horizontal and bedding within the OMM subhorizontal and draping the foreland basin topography, which points to an end of flexure before the onset of the OMM. Abbreviations: OSM = Upper Freshwater Molasse, OMM = Upper Marine Molasse, USM = Lower Freshwater Molasse, UMM = Lower Marine Molasse, Pk. sd. = sands of the Puchkirchen Group, NCA = Northern Calcareous Alps (Austroalpine nappes), RDF = Rhenodanubian Flysch nappes. Undifferentiated RDF/Helvetic und Helvetic/UMM units indicate thrust wedges composed of the respective units.*

The uncertainty of timing and magnitude of exhumation makes it difficult to assess the underlying driver. Three drivers have been proposed: (1) It has been suggested that exhumation is driven by exogenic forcing, possibly associated with the isostatic rebound of the lower plate (Cederbom et al. 2004; Cederbom et al. 2011). Alternatively, drainage reorganization may have caused incision of the Molasse sediments (Yanites et al. 2013). (2) Analogue modeling has shown that changes inside a mechanical wedge, coupled with erosion/sedimentation rates, may explain the transition from an underfilled Flysch basin to an overfilled Molasse Basin as well as the major cyclic behavior of the Cenozoic Molasse deposits (Bonnet et al. 2007; Bonnet et al. 2008). (3) Genser et al. (2007) and von Hagke et al. (2012) notice that much of the foreland basin exhumation may also be caused by deep-seated drivers and can only be partly explained by climate events or thrusting. The deep (mantle-related) tectonics of the Alps is complex and slab dynamics may differ along strike (e.g. Lippitsch et al. 2003; Kissling et al. 2006). This variable slab dynamics cause differential exhumation in the adjacent foreland basin (Wortel and Spakman 2000; Schlunegger and Kissling 2015). Whereas the western part of the Molasse basin has been proposed to have witnessed exhumation due to slab breakoff or rollback (Schlunegger and Kissling 2015), the Eastern Alpine Molasse basin may have been influenced by slab polarity reversal and lateral escape of the Tauern Window (Handy et al. 2010; Ortner et al. 2015). However, whether a polarity flip below the Eastern Alps has occurred remains unclear, as existing geophysical data is inconclusive, while the geological record does not necessarily require such a reversal. Untangling the exhumation signal related to slab dynamics from thrusting along the deformation front or climate drivers requires high-resolution as well as extensive thermochronological age dating of the frontal structures as well as the foreland basin.

5.4. Conclusion

The position of the deformation front of the Alps depends on different controls on various scales. Part of the controls is related to the subducting Lower European Plate. These are (1) lower plate flexure, that responds to changes in orogenic loading parallel to the deformation front. Post-Early Miocene loading is controlled by the termination of frontal thrusting and extrusion-related crustal thinning east of Salzburg and the Tauern Window. (2) Mechanical stratigraphy of the Mesozoic passive margin and the Cenozoic foreland basin determines the position of the basal décollement, which is either located within passive margin deposits or within the foreland basin succession. This is in turn determined by the (3) width of the foreland, which increases from west to east toward the Bohemian Massif margin. The foreland is very narrow on top of the Bohemian Spur, then the width increases to the east. Width directly correlates with foreland basin depth at the Alpine deformation front (Figure 5.2) and forces the basal décollement into the Mesozoic where the décollement layer in the Cenozoic foreland succession pinches out (west of Zurich and east of Salzburg; Figures 5.1 and 5.2). Narrowing and shallowing of the foreland is related to the large-scale obliquity of the Alpine deformation front to foreland flexure and thus to, e.g. the UMM pinch-out (Figure 5.1).

In terms of timing of thrust transport, a Western and Eastern Sector can be distinguished. Between the Geneva and Salzburg, structural and sedimentological evidence indicates post-Early to, at least, Late Miocene thrust transport into the foreland in the Subalpine Molasse thrust belt and the salient of the JFTB in the west. Thrusting is directly related to the growth of the ECMs that sit on the same detachment. East of Salzburg, thrust transport ends during the Middle Miocene at the deformation front and the Molasse slices and the frontal Alpine wedge are covered by younger, mostly undeformed sediments. Differential uplift of the Bohemian Spur in relation to the subsiding foreland to the west exhumes some of the thrusts to the surface and causes a recess in the Alpine deformation front. Thrusts root below the Tauern Window in this part of the Alpine system, comparable to the ECMs, but the Alps east of Munich are strongly affected by extrusion tectonics. Shortening was partitioned between forelandward thrusting and E-directed transport between strike-slip faults in the extrusion corridor. Strain partitioning and buttressing effects of the Bohemian Spur ended in-sequence propagation of the thrusts into the foreland in the Eggenburgian west, in the Karpatian east of the Bohemian Spur. Shortening did not reach the Alpine deformation front afterward, causing out-of-sequence thrusting within the wedge that ends at the end of the Karpatian. The extrusion corridor decoupled the Alpine deformation front from processes in the more internal and Southern Alps.

5.5. Acknowledgments

We thank OMV, who provided the seismograms shown in Figures 5.9(C) and 5.15 and Ewald Lüschen, who provided the TRANSALP seismogram in Figure 5.9(D). We thank the reviewers Herfried Madritsch and Kurt Decker for their constructive criticism and Claudio L. Rosenberg and Nicolas Bellahsen for editorial handling.

5.6. References

Abednego, M., Blascheck, P., Schefer, S., Nussbaum, C., Joswig, M., Bossart, P., Mosar, J. (2017). Seismotectonic analysis around the Mont Terri rock laboratory (Switzerland): A pilot study. *Swiss Journal of Geosciences*, 110(1), 233–251.

Abele, G., Beschoren, B., Dehm, R., Erb, L., Fuchs, B., Ganss, O., Kiderlen, H., Nathan, H., Neumaier, F., Schmidt-Thomé, P. et al. (1955). Erläuterungen zur Geologischen Übersichtskarte der Süddeutschen Molasse 1:300.000. Map, Bayerisches Geologisches Landesamt, Munich.

Affolter, T. and Gratier, J.-P. (2004). Map view retrodeformation of an arcuate fold-and-thrust belt: The Jura case. *J. Geophys. Res.*, 109(B3), B03404.

Allen, P.A., Homewood, P., Williams, G.D. (1986). Foreland basins – An introduction. In *Spec. Publ. Int. Ass. Sediment., Foreland basins*, Allen, P.A. and Homewood, P. (eds). Blackwell, Oxford.

Allen, P.A., Crampton, S.L., Sinclair, H.D. (1991). The inception and early evolution of the North Alpine foreland basin. *Basin Res.*, 3, 143–163.

Aubert, D. (1958). Sur l'existence d'une ride de plissement oligocène dans le Jura vaudois. *Bull. la Société Neuchâteloise des Sci. Nat.*, 81, 47–57.

Bachmann, G.H., Dohr, G., Müller, M. (1982). Exploration in a classic thrust belt and its foreland: Bavarian Alps, Germany. *AAPG Bull.*, 66, 2529–2542.

Bachmann, G.H., Müller, M., Weggen, K. (1987). Evolution of the Molasse basin (Germany, Switzerland). *Tectonophysics*, 137, 77–92.

Beck, C., Deville, E., Blanc, E., Philippe, Y., Tardy, M. (1998). Horizontal shortening control of middle Miocene marine siliciclastic accumulation (Upper Marine Molasse) in the southern termination of the Savoy Molasse Basin (northwestern Alps/southern Jura). In *Cenozoic Foreland Basins of Western Europe*, Mascle, A., Puigdefábegras, C., Luterbacher, H.-P., Fernández, M. (eds). Geological Society Special Publication, London.

Becker, A. (2000). The Jura Mountains – An active foreland fold-and-thrust belt? *Tectonophysics*, 321, 381–406.

Beidinger, A. and Decker, K. (2014). Quantifying Early Miocene in-sequence and out-of-sequence thrusting at the Alpine-Carpathian junction. *Tectonics*, 33, 222–252.

Beidinger, A. and Decker, K. (2016). Paleogene and Neogene kinematics of the Alpine-Carpathian fold-thrust belt at the Alpine-Carpathian transition. *Tectonophysics*, 690, 263–287.

Beidinger, A., Schuller, V., Decker, K. (2019). High-risk exploration in a hypermature hydrocarbon basin: An example from the Vienna pull-apart Basin, Austria. *Abu Dhabi International Petroleum Exhibition & Conference*, Abu Dhabi.

Bellahsen, N., Mouthereau, F., Boutoux, A., Bellanger, M., Lacombe, O., Jolivet, L., Rolland, Y. (2014). Collision kinematics in the western external Alps. *Tectonics*, 33(6), 1055–1088.

Berger, J.-P., Reichenbacher, B., Becker, D., Grimm, M., Grimm, K., Picot, L., Storni, A., Pirkenseer, C., Derer, C., Schaefer, A. (2005a). Paleogeography of the Upper Rhine Graben (URG) and the Swiss Molasse Basin (SMB) from Eocene to Pliocene. *Int. J. Earth. Sci.*, 94(4), 697–710.

Berger, J.-P., Reichenbacher, B., Becker, D., Grimm, M., Grimm, K., Picot, L., Storni, A., Pirkenseer, C., Schaefer, A. (2005b). Eocene-Pliocene time scale and stratigraphy of the Upper Rhine Graben (URG) and the Swiss Molasse Basin (SMB). *Int. J. Earth. Sci.*, 94, 711–731.

Bergerat, F. (1987). Stress fields in the European platform at the time of Africa – Eurasia collision. *Tectonics*, 6(2), 99–132.

Bergerat, F. and Geyssant, J. (1983). Fracturation tertiaire et évolution des contraintes en Bavière orientale : le Jura franconien et la forêt bavaroise (R.F.A.). *Geol. Rundsch.*, 72(3), 935–953.

Bernhardt, A., Stright, L., Lowe, D.R. (2012). Channelized debris-flow deposits and their impact on turbidity currents: The Puchkirchen axial channel belt in the Austrian Molasse Basin. *Sedimentology*, 59(7), 2042–2070.

Bläsi, H.-R., Jordan, P., Laubscher, H.P., Reber, D., Burkhalter, R. (2018). Geologischer Atlas der Schweiz 1:25.000, Blatt 1088 Hauenstein. Map, swisstopo, Wabern.

Bonnet, C., Malavieille, J., Mosar, J. (2007). Interactions between tectonics, erosion, and sedimentation during the recent evolution of the Alpine orogen: Analogue modeling insights. *Tectonics*, 26, TC6016.

Bonnet, C., Mosar, J., Malavieille, J. (2008). Surface processes versus kinematics of thrust belts: Impact on rates of erosion, sedimentation, and exhumation – Insights from analogue models. *Bulletin de la Société Géologique de France*, 179(3), 297–314.

van der Boon, A., Beniest, A., Ciurej, A., Gaździcka, E., Grothe, A., Sachsenhofer, R.F., Langereis, C.G., Krijgsman, W. (2018). The Eocene–Oligocene transition in the North Alpine Foreland Basin and subsequent closure of a Paratethys gateway. *Glob. Planet. Change*, 162, 101–119.

Borowski, K. (2006). Lithofacies and depositional environment of the upper Hall Formation, Alpine Molasse basin, Upper Austria. Diploma TU Bergakademie Freiberg, Freiberg.

Bousquet, R., Oberhänsli, R., Goffé, B., Jolivet, L., Vidal, O. (1999). High pressure-low temperature metamorphism and deformation in the Bündnerschiefer of the Engadine window: Implications for the regional evolution of the eastern Central Alps. *Journal of Metamorphic Geology*, 17, 657–674.

Bousquet, R., Oberhänsli, R., Schmid, S., Zeilinger, G., Moeller, A., Berger, A., Wiederkehr, M., Rosenberg, C.L., Koller, F. (2012). Metamorphic framework of the Alps (1:1.000.000) 2nd edition. Commission for the Geological Map of the World, Paris.

Boutoux, A., Bellahsen, N., Nanni, U., Pik, R., Verlaguet, A., Rolland, Y., Lacombe, O. (2016). Thermal and structural evolution of the external Western Alps: Insights from (U–Th–Sm)/He thermochronology and RSCM thermometry in the Aiguilles Rouges/Mont Blanc massifs. *Tectonophysics*, 683, 109–123.

Boyer, S.E. (1995). Sedimentary basin taper as a factor controlling the geometry and advance of thrust belts. *American Journal of Science*, 295, 1220–1254.

Boyer, S.E. and Elliott, D. (1982). Thrust systems. *AAPG Bull.*, 66, 1196–1230.

Braunstingl, R. (1988). Die Flyschzone südwestlich von Steyr (Oberösterreich): Geologischer Bau und Überlegungen zum Ultrahelvetikum. *Jb. Geol. Bundesanst.*, 131(2), 231–243.

Brix, F. and Götzinger, K. (1964). Die Ergebnisse der Aufschlußarbeiten der ÖMV AG in der Molassezone Niederösterreichs in den Jahren 1957–1963. *Erdoel-Zeitschrift*, 80, 57–76.

Brix, F. and Schultz, O. (1993). Erdöl und Erdgas in Österreich. Veröffentlichungen aus dem Naturhistorischen Museum in Wien, 19 (2. Auflage). Berger, Horn.

Brügel, A. (1998). Provenances of alluvial conglomerates from the Eastalpine foreland: Oligo-/Miocene denudation history and drainage evolution of the Eastern Alps. *Tübinger Geowiss. Arb., A.*, 40, 168.

Brügel, A., Dunkl, I., Frisch, W., Kuhlemann., A., Balogh, K. (2003). Geochemistry and geochronology of gneiss pebbles from foreland molasse conglomerates: Geodynamic and paleogeographic implications for the Oligo-Miocene evolution of the Eastern Alps. *Jour. Geol.*, 111, 543–563.

Burkhard, M. (1990). Aspects of large-scale Miocene deformation in the most external part of the Swiss Alps (Subalpine Molasse to Jura fold belt). *Eclogae Geol. Helv.*, 83, 559–583.

Burkhard, M. and Sommaruga, A. (1998). Evolution of the western Swiss Molasse basin: Structural relations with the Alps and the Jura belt. In *Cenozoic Foreland Basins of Western Europe*, Mascle, A., Puigdefábegras, C., Luterbacher, H.-P., Fernández, M. (eds). Geological Society Special Publication, London.

Buxtorf, A. (1907). Zur Tektonik des Kettenjura. *Bericht der Versammlung des Oberrheinischen Geologischen Vereins*, 40, 79–111.

Caron, C., Homewood, P., Wildi, W. (1989). The original Swiss flysch: A reappraisal of the type deposits in the Swiss Prealps. *Earth-Science Reviews*, 26, 1–45.

Cederbom, C., Sinclair, H.D., Schlunegger, F., Rahn, M. (2004). Climate-induced rebound and exhumation of the European Alps. *Geology*, 32, 709–712.

Cederbom, C., van der Beek, P., Schlunegger, F., Sinclair, H.D., Oncken, O. (2011). Rapid extensive erosion of the North Alpine foreland basin at 5-4 Ma. *Basin Res.*, 23(5), 528–550.

Champagnac, J.-D., Schlunegger, F., Norton, K., von Blanckenburg, F., Abbühl, L.M., Schwab, M. (2009). Erosion-driven uplift of the modern Central Alps. *Tectonophysics*, 174(1), 236–249.

Chauve, P., Martin, J., Petitjean, E., Sequeiros, F. (1988). Le chevauchement du Jura sur la Bresse. Données nouvelles et réinterprétation des sondages. *Bulletin de la Société géologique de France*, 8(4–5), 861–870.

Cohen, K.M., Finney, S.C., Gibbard, P.L., Fan, J.-X. (2013). The ICS International Chronostratigraphic Chart. *Episodes*, 36, 199–204.

Courboulex, F., Deichmann, N., Gariel, J.-C. (1999). Rupture complexity of a moderate intraplate earthquake in the Alps: The 1996 M5 Epagny–Annecy earthquake. *Geophysical Journal International*, 139(1), 152–160.

Covault, J.A., Hubbard, S.M., Graham, S.A., Hinsch, R., Linzer, H.-G. (2009). Turbidite-reservoir architecture in complex foredeep-margin and wedge-top depocenters, tertiary Molasse foreland basin system, Austria. *Mar. Pet. Geol.*, 26, 379–396.

Decker, K., Meschede, M., Ring, U. (1993). Fault slip analysis along the northern margin of the Eastern Alps (Molasse, Helvetic nappes, North and South penninic flysch, and the Northern Calcareous Alps). *Tectonophysics*, 223, 291–312.

Diebold, P. and Noack, T. (1997). Late Paleozoic troughs and Tertiary structures in the eastern folded Jura. In *Results of NRP20*, Pfiffner, O.A., Lehner, P., Heitzmann, P., Mueller, S., Steck, A. (eds). Birkhäuser, Basel.

Diehl, T., Deichmann, N., Clinton, J., Kästli, P., Cauzzi, C., Kraft, T., Behr, Y., Edwards, B., Guilhem, A., Korger, E. et al. (2015). Earthquakes in Switzerland and surrounding regions during 2014. *Swiss Journal of Geosciences*, 108(2), 425–443.

Eberhard, M. (1986). Litho- und Biostratigraphie im Oberen Süßwassermolasse-Fächer der Adelegg (Südbayern). *Jb. Geol. Bundesanst.*, 129, 5–39.

Egger, H. (1987). Die Geologie der Rhenodanubischen Flyschzone südöstlich von Steyr (Oberösterreich, Niederösterreich). *Jb. Geol. Bundesanst.*, 130(2), 139–151.

Egger, H. and Ćorić, S. (2017). Erläuterungen zur Geologischen Karte der Republik Österreich 1:50.000, Blatt 56 St. Pölten. Geologische Karte der Republik Österreich 1:50.000. Map, Geologische Bundesansalt, Vienna.

Egli, D. and Mancktelow, N. (2013). The structural history of the Mont Blanc massif with regard to models for its recent exhumation. *Swiss Journal of Geosciences*, 106(3), 469–489.

Egli, D., Mancktelow, N., Spikings, R. (2017a). Constraints from 40Ar/39Ar geochronology on the timing of Alpine shear zones in the Mont Blanc–Aiguilles Rouges region of the European Alps. *Tectonics*, 36(4), 730–748.

Egli, D., Mosar, J., Ibele, T., Madritsch, H. (2017b). The role of precursory structures on Tertiary deformation in the Black Forest–Hegau region. *Int. J. Earth. Sci.*, 106(7), 2297–2318.

Freitag, U. (2003). Abwicklung eines Strukturprofils über den Nordrand der Alpen (Molasse bis Nördliche Kalkalpen) im Bereich der Bohrungen Legau und Hindelang. Diploma University of Jena, Jena.

Freudenberger, W. and Schwerd, K. (1996). Erläuterungen zur geologischen Karte von Bayern 1:500 000. Map, Bayerisches Geologisches Landesamt, Munich.

Frieling, D., Aehnelt, M., Scholz, H., Reichenbacher, B. (2009). Sequence stratigraphy of an alluvial fan-delta in the Upper Marine Molasse (Pfänder area, Late Burdigalian, Miocene). *Zeitschr. Dt. Geol. Ges.*, 160, 333–357.

Frisch, W., Kuhlemann, J., Dunkl, I., Brügel, A. (1998). Palinspastic reconstruction and topographic evolution of the Eastern Alps during late Tertiary tectonic extrusion. *Tectonophysics*, 297, 1–15.

Füchtbauer, H. (1967). Die Sandsteine in der Molasse nördlich der Alpen. *Geol. Rundsch.*, 56, 266–300.

Fügenschuh, B., Seward, D., Mancktelow, N. (1997). Exhumation in a convergent orogen: The western Tauern window. *Terra Nova*, 9, 213–217.

Ganss, O. (1977). Erläuterungen zur geologischen Karte von Bayern, 1:25.000, Blatt Nr. 8140 Prien a. Chiemsee und zum Blatt Nr. 8141 Traunstein. Map, Bayerisches Geologisches Landesamt, Munich.

Ganss, O. (1983). Erläuterungen zur geologischen Karte von Bayern, 1:25.000, Blatt Nr. 8040 Eggstätt. Map, Bayerisches Geologisches Landesamt, Munich.

Ganss, O. and Schmidt-Thomé, P. (1955). Die gefaltete Molasse am Alpenrand zwischen Bodensee und Salzach. *Zeitschr. Dt. Geol. Ges.*, 105(1953), 402–495.

Gebhardt, H., Ćorić, S., Krenmayr, H.G., Steininger, H., Schweigl, J. (2013). Neudefinition von lithostratigraphischen Einheiten des oberen Ottnangium (Unter- miozän) in der alpin-karpatischen Vortiefe Niederösterreichs: Pixendorf-Gruppe, Traisen-Formation und Dietersdorf-Formation. *Jb. Geol. Bundesanst.*, 153, 15–32.

Geissler, P. (1975). Zur Geologie im Ostfeld des Kohlenbergwerks Peißenberg. *Geol. Bav.*, 73, 55–57.

van Gelder, I.E., Willingshofer, E., Sokoutis, D., Cloetingh, S.A.P.L. (2017). The interplay between subduction and lateral extrusion: A case study for the European Eastern Alps based on analogue models. *Earth and Planetary Science Letters*, 472, 82–94.

Genser, J., Cloething, S.A.P.L., Neubauer, F. (2007). Late orogenic rebound and oblique Alpine convergence: New constraints from subsidence analysis of the Austrian Molasse basin. *Glob. Planet. Change*, 58, 214–233.

Giamboni, M., Ustaszewski, K., Schmid, S.M., Schumacher, M.E., Wetzel, A. (2004a). Plio-Pleistocene transpressional reactivation of Paleozoic and Paleogene structures in the Rhine-Bresse transform zone (northern Switzerland and eastern France). *Int. J. Earth. Sci.*, 93(2), 207–223.

Giamboni, M., Wetzel, A., Nivière, B., Schumacher, M. (2004b). Plio-Pleistocene folding in the southern Rhinegraben recorded by the evolution of the drainage network (Sundgau area; northwestern Switzerland and France). *Eclogae Geol. Helv.*, 97(1), 17–31.

Gillitzer, G. (1955). Geologische Neuaufnahme des Peißenberger Kohlereviers. *Geol. Bav.*, 23, 1–64.

Glotzbach, C., Reinecker, J., Danisik, M., Rahn, M., Frisch, W., Spiegel, C. (2010). Thermal history of the central Gotthard and Aar massifs, European Alps: Evidence for steady state, long-term exhumation. *J. Geophys. Res.*, 115(F3), F03017.

Gmach, H. (1999). Thermische Reife der Flyschzone im Bereich des Wienerwaldes und die Flyschanteile nördlich der Donau und deren Einfluß auf die Kohlenwasserstoffgenese am Rand des Wiener Beckens. Diploma Montanuniv. Leoben.

Grad, M., Tiira, T., ESC Working Group (2009). The Moho depth map of the European Plate. *Geophysical Journal International*, 176(1), 279–292.

Granado, P., Thöny, W., Carrera, N., Gratzer, O., Strauss, P., Muñoz, J.A. (2016). Basement-involved reactivation in foreland fold-and-thrust belts: The Alpine–Carpathian Junction (Austria). *Geological Magazine*, 153(5–6), 1110–1135.

Grill, R. and Waldmann, L. (1949). Zur Kenntnis des Untergrundes der Molasse in Österreich. *Jb. Geol. Bundesanst.*, 94, 1–40.

Gruber, M. (2017). Structural investigations of the Western Swiss Molasse Basin – From 2D seismic interpretation to a 3D geological Model. *GeoFocus*, 41, 1–190.

Grunert, P., Hinsch, R., Sachsenhofer, R.F., Bechtel, A., Ćorić, S., Harzhauser, M., Piller, W.E., Sperl, H. (2013). Early Burdigalian infill of the Puchkirchen Trough (North Alpine Foreland Basin, Central Paratethys): Facies development and sequence stratigraphy. *Mar. Pet. Geol.*, 39(1), 164–186.

Grunert, P., Auer, G., Harzhauser, M., Piller, W.E. (2015). Stratigraphic constraints for the upper Oligocene to lower Miocene Puchkirchen Group (North Alpine Foreland Basin, Central Paratethys). *Newsletters on Stratigraphy*, 48, 111–133.

Guellec, S., Mugnier, J.L., Tardy, M., Roure, F. (1990). Neogene evolution of the western Alpine foreland in the light of ECORS data and balanced cross sections. In *Deep Structure of the Alps 1*, Roure, F., Heitzmann, P., Polino, R. (eds). Mém. Soc. géol. suisse, Zurich.

Gusterhuber, J., Dunkl, I., Hinsch, R., Linzer, H.-G., Sachsenhofer, R.F. (2012). Neogene uplift and erosion in the Alpine foreland basin (Upper Austria and Salzburg). *Geologica Carpathica*, 63(4), 295–305.

Gusterhuber, J., Hinsch, R., Sachsenhofer, R.F. (2014). Evaluation of hydrocarbon generation and migration in the Molasse fold and thrust belt (Central Eastern Alps, Austria) using structural and thermal basin models. *AAPG Bull.*, 98(2), 253–277.

Gutscher, M.-A., Kukowski, N., Malavieille, J., Lallemand, S. (1996). Cyclical behavior of thrust wedges: Insights from high basal friction sandbox experiments. *Geology*, 24(2), 135–138.

Gutscher, M.-A., Kukowski, N., Malavieille, J., Lallemand, S. (1998). Episodic imbricate thrusting and underthrusting: Analog experiments and mechanical analysis applied to the Alaskan Accretionary Wedge. *Journal of Geophysical Research: Solid Earth*, 103(B5), 10161–10176.

von Hagke, C., Cederbom, C.E., Oncken, O., Stöckli, D.F., Rahn, M.K., Schlunegger, F. (2012). Linking the northern Alps with their foreland: The latest exhumation history resolved by low-temperature thermochronology. *Tectonics*, 31(5), TC5010.

von Hagke, C., Oncken, O., Evseev, S. (2014a). Critical taper analysis reveals lithological control of variations in detachment strength: An analysis of the Alpine basal detachment (Swiss Alps). *Geochemistry, Geophysics, Geosystems*, 15(1), 176–191.

von Hagke, C., Oncken, O., Ortner, H., Cederbom, C.E., Aichholzer, S. (2014b). Late Miocene to present deformation and erosion of the Central Alps – Evidence for steady state mountain building from thermokinematic data. *Tectonophysics*, 632, 250–260.

von Hagke, C., Lujiendjik, E., Ondrak, R., Lindow, J. (2015). Quantifying erosion rates in the Molasse basin using a high resolution data set and a new thermal model. *Geotectonic Research*, 97(1), 94–97.

Hamilton, W. (1989). Geologische Ergebnisse von Tiefbohrungen im Flysch und Kalkalpin zwischen Wien und Salzburg. *Führer zur Exkursion der Österreichischen Geologischen Gesellschaft*, 12, 1–57.

Handy, M.R., Schmid, S.M., Bousquet, R., Kissling, E., Bernoulli, D. (2010). Reconciling plate-tectonic reconstructions of Alpine Tethys with the geological–geophysical record of spreading and subduction in the Alps. *Earth-Science Reviews*, 102(3), 121–158.

Harzhauser, M., Theobalt, D., Strauss, P., Mandic, O., Carnevale, G., Piller, W.E. (2017). Miocene biostratigraphy and paleoecology of the Mistelbach Halfgraben in the northwestern Vienna Basin (Lower Austria). *Jb. Geol. Bundesanst.*, 157, 57–108.

Haus, H. (1935). Über alte Erosionserscheinungen am Südrand der miocaenen Nagelfluh des oberen Emmentals und deren Bedeutung für die Tektonik des Alpenrandes. *Eclogae Geol. Helv.*, 28(2), 667–677.

Herwegh, M., Berger, A., Baumberger, R., Wehrens, P., Kissling, E. (2017). Large-scale crustal-block-extrusion during late alpine collision. *Scientific Reports*, 7(1), 413.

Herwegh, M., Berger, A., Glotzbach, C., Wangenheim, C., Mock, S., Wehrens, P., Baumberger, R., Egli, D., Kissling, E. (2019). Late stages of continent–continent collision: Timing, kinematic evolution, and exhumation of the Northern Rim (Aar Massif) of the Alps. *Earth-Science Reviews*, 200, 102959.

Heuberger, S., Roth, P., Zingg, O., Naef, H., Meier, B.P. (2016). The St. Gallen Fault Zone: A long-lived, multiphase structure in the North Alpine Foreland Basin revealed by 3D seismic data. *Swiss Journal of Geosciences*, 109(1), 83–102.

Heyng, A. (2012). Lithostratigraphie der Adelholzen Formation (Eozän, Lutetium) im Raum Siegsdorf (Oberbayern). *Documenta Naturae*, 186, 1–105.

Hilgen, F.J., Lourens, L.J., van Dam, J.A. (2012). The Neogene period. In *The Geologic Time Scale 2012*, Gradstein, F.M., Ogg, J.G., Schmitz, M., Ogg, G. (eds). Elsevier, Amsterdam.

Hiltmann, W., Kuckelkorn, K., Wehner, H. (1999). Das Inkohlungsprofil der Bohrung Grambach 1- erster Hinweis auf eine Ölküche im Molassebecken. *Erdöl Erdgas Kohle*, 115, 294–297.

Hinsch, R. (2008). New insights into the Oligocene to Miocene geological evolution of the Molasse basin of Austria. *Oil and Gas European Magazine*, 34(3), 138–143.

Hinsch, R. (2013). Laterally varying structure and kinematics of the Molasse fold and thrust belt of the Central Eastern Alps: Implications for exploration. *AAPG Bull.*, 97(10), 1805–1831.

Hofmann, R. (1962). Die Tektonik des Bayerischen Pfahls. *Geol. Rundsch.*, 52(1), 332–346.

Hölzel, M., Decker, K., Zámolyi, A., Strauss, P., Wagreich, M. (2010). Lower Miocene structural evolution of the central Vienna Basin (Austria). *Mar. Pet. Geol.*, 27, 666–681.

Homewood, P. (1983). Paleogeography of Alpine flysch. *Paleogeogr. Paleoclimatol. Paleoecol.*, 44, 169–184.

Homewood, P. and Lateltin, O. (1988). Classic Swiss clastics (flysch and molasse) – The Alpine connection. *Geodinamica Acta*, 2(1), 1–11.

Homewood, P., Allen, P.A., Williams, G.D. (1986). Dynamics of the Swiss Molasse Basin. In *Foreland Basins*, Allen, P.A. and Homewood, P. (eds). Spec. Publ. Int. Ass. Sediment., Blackwell, Oxford.

Hooker, J.J. and Weidmann, M. (2000). The Eocene mammal faunas of Mormont, Switzerland: Systematic revision and resolution of dating problems. *Mémoires suisses de Paléontologie*, 120, 1–141.

Hubbard, S.M., de Ruig, M.J., Graham, S.A. (2009). Confined channel-levee complex development in an elongate depo-center: Deep-water Tertiary strata of the Austrian Molasse basin. *Mar. Pet. Geol.*, 26, 85–112.

Huber, K. and Schwerd, K. (1995). Das geologische Profil der Bohrung Hindelang 1 (Allgäuer Alpen). *Geol. Bav.*, 100, 23–54.

Hülscher, J., Fischer, G., Grunert, P., Auer, G., Bernhardt, A. (2019). Selective recording of tectonic forcings in an Oligocene/Miocene submarine channel system: Insights from new age constraints and sediment volumes from the Austrian Northern Alpine Foreland Basin. *Frontiers in Earth Science*, 7, 302.

Ibele, T. (2011). Tectonics of the western Swiss Molasse Basin during Cenozoic time. *GeoFocus*, 27, 1–148.

Jenny, J., Burri, J.-P., Muralt, R. (1995). Le forage géothermique de Thônex (Canton de Genève) : aspects stratigraphiques, tectoniques, diagénétiques, géophysiques et hydrogéologiques. *Eclogae Geol. Helv.*, 88(2), 365–396.

Jost, J., Kempf, O., Kälin, D. (2016). Stratigraphy and palaeoecology of the Upper Marine Molasse (OMM) of the central Swiss Plateau. *Swiss Journal of Geosciences*, 109(2), 149–169.

Kapounek, J., Kröll, A., Papp, A., Turnovsky, K. (1965). Die Verbreitung von Oligozän, Unter- und Mittelmiozän in Niederösterreich. *Erdöl Erdgas Kohle*, 81, 109–116.

Kästle, E.D., Rosenberg, C.L., Boschi, L., Bellahsen, N., Meier, T., El-Sharkawy, A. (2020). Slab break-offs in the Alpine subduction zone. *Int. J. Earth. Sci.*, 109(2), 587–603.

Keller, B. (1989). Fazies und Stratigraphie der oberen Meeresmolasse (unteres Miozän) zwischen Napf und Bodensee. PhD Thesis, University of Bern, Bern.

Kempf, O. and Pross, J. (2005). The lower marine to lower freshwater Molasse transition in the northern Alpine foreland basin (Oligocene; central Switzerland–south Germany): Age and geodynamic implications. *Int. Jour. Earth Sci.*, 94, 160–171.

Kempf, O., Matter, A., Burbank, D.W., Mange, M. (1999). Depositional and structural evolution of a foreland basin margin in a magnetostratigraphic framework: The eastern Swiss Molasse Basin. *Int. J. Earth Sci. (Geol. Rundsch.)*, 88, 253–275.

Kissling, E., Schmid, S.M., Lippitsch, R., Ansorge, J., Fügenschuh, B. (2006). Lithosphere structure and tectonic evolution of the Alpine arc: New evidence from high-resolution teleseismic tomography. In *European Lithosphere Dynamics*, Gee, D.G., Stephenson, R.A. (eds). Memoir, Geological Society, London.

Kittler, G. and Neumayer, R. (1983). Austrian Molasse Basin. In *Seismic Expression of Structural Styles, a Picture and Work Atlas*, Bally, A.W. (ed.). Studies in Geology, AAPG, Tulsa.

Kohl, H. and Krenmayr, H.-G. (1997). Erläuterungen zu Blatt 49 Wels. Geologische Karte der Republik Österreich. Map, Geologische Bundesanstalt, Vienna.

Krenmayr, H.G. (1999). The Austrian sector of the North Alpine Molasse: A classic foreland basin. *Berichte der Geologischen Bundesanstalt*, 49, 22–26.

Krenmayr, H.G., Schnabel, W., Reitner, J.M., Van Husen, D., Finger, F., Linner, M., Roetzel, R., Rupp, C., Bryda, G., Mandl, G.W. et al. (2006). Geologische Karte von Oberösterreich 1:200.000 mit Legende und Kurzerläuterungen. Map, Geologische Bundesanstalt, Vienna.

Krhovsky, J., Rögl, F., Hamrsmid, B. (2001). Stratigraphic correlation of the Late Eocene to Early Miocene of the Waschberg Unit (Lower Austria) with the Zdanice and Pouzdrany Units (South Moravia), in Paleogene of the Eastern Alps. In *Paleogene in Austria*, Piller, W. and Rasser, M. (eds). Schriftenreihe der Erdwissenschaftlichen Kommissionen, Österreichische Akademie der Wissenschaften, Vienna.

Kröll, A. and Wessely, G. (2001). Molassezone Niederösterreich und angrenzende Gebiete 1:200.000: Geologische Karte der Molassebasis. Map, Geologische Bundesanstalt, Vienna.

Kröll, A., Wagner, L., Wessely, G., Zych, D. (2006). Molassezone Salzburg – Oberösterreich 1:200.000: Strukturkarte der Molassebasis. Map, Verlag der Geologischen Bundesanstalt (GBA), Vienna.

Kuhlemann, J. (2000). Post-collisional sediment budget of circum-Alpine basins (Central Europe). *Memorie di Scienze Geologiche, Padova*, 52(1), 1–91.

Kuhlemann, J. and Kempf, O. (2002). Post-Eocene evolution of the North Alpine Foreland Basin and its response to Alpine tectonics. *Sediment. Geol.*, 152, 45–78.

Kuhlemann, J., Dunkl, I., Brügel, A., Spiegel, C., Frisch, W. (2006). From source terrains of the Eastern Alps to the Molasse Basin: Detrital record of non-steady-state exhumation. *Tectonophysics*, 413, 301–316.

Lahusen, P.H. and Wyss, R. (1995). Erdöl- und Erdgasexploration in der Schweiz: Ein Rückblick. *Bull. Ver. schweiz. Petroleum-Geol. u. -Ing.*, 62(141), 43–62.

Lammerer, B. and Weger, M. (1998). Footwall uplift in an orogenic wedge: The Tauern Window in the Eastern Alps of Europe. *Tectonophysics*, 285, 213–230.

Lange, H. (1985). Tiefbohrungen. In *Erläuterungen zur geologischen Karte von Bayern 1:25.000, Blatt Nr. 8036 Otterfing und zum Blatt Nr. 8136 Holzkirchen*, Grottenthaler, W. (ed.). Bayerisches Geologisches Landesamt, Munich.

Laubscher, H.P. (1961). Die Fernschubhypothese der Jurafaltung. *Eclogae Geol. Helv.*, 54, 221–280.

Laubscher, H.P. (1965). Ein kinematisches Modell der Jurafaltung. *Eclogae Geol. Helv.*, 58, 232–318.

Laubscher, H.P. (1974). Basement uplift and decollement in the Molasse Basin. *Eclogae Geol. Helv.*, 67(3), 531–537.

Leloup, P.H., Arnaud, N., Sobel, E.R., Lacassin, R. (2005). Alpine thermal and structural evolution of the highest external crystalline massif: The Mont Blanc. *Tectonics*, 24(4), 4002.

Lemcke, K. (1972). Die Lagerung der jüngsten Molasse im nördlichen Alpenvorland. *Bull. Ver. schweiz. Petroleum-Geol. u. -Ing.*, 39, 29–41.

Lemcke, K. (1973). Zur nachpermischen Geschichte des Alpenvorlandes. *Geol. Bav.*, 69, 5–48.

Lemcke, K. (1983). Indications of a large eustatic sea-level fall at the Rupelian/Chattian boundary in the German Molasse Basin. *Bull. Ver. Schweiz. Petroleum-Geol. u. -Ing.*, 49(117), 57–60.

Lemcke, K. (1984). Geologische Vorgänge in den Alpen ab Obereozän im Spiegel vor allem der deutschen Molasse. *Geol. Rundsch.*, 73, 371–397.

Lemcke, K. (1988). Geologie von Bayern I – Das bayerische Alpenvorland vor der Eiszeit. Schweizerbart, Stuttgart.

Lemcke, K., Büchi, U.P., Wiener, G. (1968). Einige Ergebnisse der Erdölexploration auf die mittelländische Molasse der Zentralschweiz. *Bull. Ver. schweiz. Petroleum-Geol. u. -Ing.*, 35(87), 15–34.

Lexa, J., Bezák, V., Elecko, J., Mello, J., Polák, M., Vozár, J. (2000). Geological map of Western Carpathians and adjacent areas, 1:500.000. Map, Ministry of Environment of Slovak Republic and ŠGÚDŠ, Bratislava.

Liniger, H. (1967). Pliozän und Tektonik des Juragebirges. *Eclogae Geol. Helv.*, 60(2), 407–490.

Linzer, H.-G., Decker, K., Peresson, H., Dell'Mour, R., Frisch, W. (2002). Balancing lateral orogenic float of the Eastern Alps. *Tectonophysics*, 354, 211–237.

Lippitsch, R., Kissling, E., Ansorge, J. (2003). Upper Mantle structure beneath the Alpine orogen from high-resolution teleseismic tomography. *J. Geophys. Res.*, 108(B8), 2376.

Lüschen, E., Lammerer, B., Gebrande, H., Millhan, K., Nicolich, R., Group, T.W. (2004). Orogenic structure of the Eastern Alps, Europe from TRANSALP deep seismic reflection profiling. *Tectonophysics*, 388, 85–102.

Luth, S.W. and Willingshofer, E. (2008). Mapping of the post-collisional cooling history of the Eastern Alps. *Swiss Journal of Geosciences*, 101(1), 207–223.

Madritsch, H. (2015). Outcrop-scale fracture systems in the Alpine foreland of central-northern Switzerland: Kinematics and tectonic context. *Swiss Journal of Geosciences*, 108(2), 155–181.

Madritsch, H., Schmid, S.M., Fabbri, O. (2008). Interactions between thin- and thick-skinned tectonics at the northwestern front of the Jura fold-and-thrust belt (eastern France). *Tectonics*, 27, TC5005.

Madritsch, H., Kounov, A., Schmid, S.M., Fabbri, O. (2009). Multiple fault reactivations within the intra-continental Rhine–Bresse Transfer Zone (La Serre Horst, eastern France). *Tectonophysics*, 471(3), 297–318.

Madritsch, H., Preusser, F., Fabbri, O., Bichet, V., Schlunegger, F., Schmid, S.M. (2010). Late Quaternary folding in the Jura Mountains: Evidence from syn-erosional deformation of fluvial meanders. *Terra Nova*, 22(2), 147–154.

Madritsch, H., Naef, H., Meier, B., Franzke, H.J., Schreurs, G. (2018). Architecture and kinematics of the Constance-frick trough (Northern Switzerland): Implications for the formation of post-Variscan Basins in the foreland of the Alps and scenarios of their Neogene reactivation. *Tectonics*, 37(7), 2197–2220.

Malz, A., Madritsch, H., Meier, B., Kley, J. (2016). An unusual triangle zone in the external northern Alpine foreland (Switzerland): Structural inheritance, kinematics and implications for the development of the adjacent Jura fold-and-thrust belt. *Tectonophysics*, 670, 127–143.

Malzer, O., Rögl, F., Seifert, P., Wagner, L., Wessely, G., Brix, F. (1993). Die Molassezone und deren Untergrund. In *Erdöl und Erdgas in Österreich*, Brix, F. and Schultz, O. (eds). Naturhistorisches Museum, Vienna.

Masalimova, L.U., Lowe, D.R., Mchargue, T., Derksen, R. (2015). Interplay between an axial channel belt, slope gullies and overbank deposition in the Puchkirchen Formation in the Molasse Basin, Austria. *Sedimentology*, 62(6), 1717–1748.

Mattmüller, C.R. (1991). Überlegungen zur Deckenkinematik im Engadiner Fenster. *Jb. Geol. Bundesanst.*, 134(2), 319–328.

Maurer, H. (1983). Sedimentpetrographische Analysen an Molasseabfolgen der Westschweiz. *Jb. Geol. Bundesanst.*, 126(1), 23–69.

Maurer, H. and Nabholz, W. (1980). Sedimentpetrographie in der Molasse-Abfolge der Bohrung Romanens 1 und in der benachbarten subalpinen Molasse (Kt. Fribourg). *Eclogae Geol. Helv.*, 73(1), 205–222.

Maurer, H., Buchner, E., Seyfried, H. (2002). Wie marin ist die Untere Meeresmolasse? Überlegungen zur Entstehung der Deutenhausener Schichten (basale Untere Meeresmolasse) im westlichen Oberbayern. Zeitschr. *Dt. Geol. Ges.*, 153(1), 77–91.

Mazurek, M., Hurford, A.J., Leu, W. (2006). Unravelling the multi-stage burial history of the Swiss Molasse Basin: Integration of apatite fission track, vitrinite reflectance and biomarker isomerisation analysis. *Basin Res.*, 18(1), 27–50.

Menkveld, J.-W. (1995). Der geologische Bau des Helvetikums der Innerschweiz. PhD Thesis, University of Bern.

Menkveld-Gfeller, U., Kempf, O., Funk, H. (2016). Lithostratigraphic units of the Helvetic Palaeogene: Review, new definition, new classification. *Swiss Journal of Geosciences*, 109(2), 171–199.

Michalski, I. and Soom, M. (1990). The Alpine thermotectonic evolution of the Aar and Gotthard massifs, central Switzerland: Fission track ages on zircon and apatite and K-Ar mica ages. *Schweizerische mineralogische und petrographische Mitteilungen*, 70, 373–387.

Michel, P., Appert, G., Lavigne, J., Lefavrais, A., Bonte, A., Lienhardt, G., Ricour, J. (1953). Le contact Jura – Bresse dans la région de Lons-le-Saunier. *Bulletin de la Société géologique de France*, 6(3), 861–870.

Miller, C., Satir, M., Morteani, G. (1980). High-pressure Metamorphism in the Tauern window. *Mitt. Österr. Geol. Ges.*, 71–72, 89–97.

Mitra, G. (1997). Evolution of salients in a fold-and-thrust belt: The effects of sedimentary basin geometry, strain distribution and critical taper. In *Evolution of Geological Structures in Micro- to Macro-scales*, Sengupta, S. (ed.). Chapman and Hall, London.

Mock, S. and Herwegh, M. (2017). Tectonics of the central Swiss Molasse Basin: Post-Miocene transition to incipient thick-skinned tectonics? *Tectonics*, 36(9), 1699–1723.

Mock, S., von Hagke, C., Schlunegger, F., Dunkl, I., Herwegh, M. (2019). Late Miocene thrusting in the North Alpine foreland: Driven by a deep-seated process and shaped by the local mechanical stratigraphy. *Solid Earth Discussions*, 1–32.

Moeck, I. and Mraz, E. (2015). Multiphase fossil normal faults in geothermal systems: Insights from the Bavarian Molasse Basin. *Geotectonic Research*, 97(1), 70–72.

Mosar, J. (1999). Present-day and future tectonic underplating in the western Swiss Alps: Reconciliation of basement/wrench-faulting and décollement folding of the Jura and Molasse basin in the Alpine foreland. *EPSL*, 173, 143–155.

Mosar, J., Abednego, M., Ibele, T., Matzenauer, E., Meier, B., Sommaruga, A., Sprecher, C. H., Vouillamoz, N. (2011). Du Jura central aux Préalpes romandes – Une tectonique active dans l'avant-pays des Alpes. *Géochronique*, 117, 52–55.

Mugnier, J.L. and Vialon, P. (1986). Deformation and displacement of the Jura cover on its basement. *J. Struct. Geol.*, 8(3), 373–387.

Mühlberg, M. (1943). Temperaturmessungen in der Bohrung Tuggen in der Linthebene und einige andere Befunde in der Schweiz. *Eclogae Geol. Helv.*, 36(1), 17–34.

Müller, M. (1970). Das Ergebnis der Bohrung Staffelsee 1 als Grundlage für neue Vorstellungen über Bau und Untergrund der gefalteten Molasse. *Geol. Bav.*, 63, 86–106.

Müller, M. (1978). Schichtenverzeichnisse von Bohrungen. In *Erläuterungen zur geologischen Karte von Bayern 1:25.000, Blatt Nr. 8042 Waging a. See*, Müller, M., Ziegler, J.H. (eds). Bayerisches Geologisches Landesamt, Munich.

Müller, M. (2011). Die Geologie beiderseits des Lechs im Raum Schongau. *Geol. Bav.*, 111, 1–163.

Müller, M., Nieberding, F., Wanninger, A. (1988). Tectonic style and pressure distribution at the northern margin of the Alps between Lake Constance and the River Inn. *Geol. Rundsch.*, 77, 787–796.

Nachtmann, W. and Wagner, L. (1987). Mesozoic and Early Tertiary evolution of the Alpine foreland in Upper Austria and Salzburg, Austria. *Tectonophysics*, 137, 61–76.

Ortner, H. and Stingl, V. (2001). Facies and basin development of the Oligocene in the Lower Inn Valley, Tyrol/Bavaria. In *Paleogene in Austria*, Piller, W. and Rasser, M. (eds). Schriftenreihe der Erdwissenschaftlichen Kommissionen, Österreichische Akademie der Wissenschaften, Vienna.

Ortner, H., Reiter, F., Brandner, R. (2006). Kinematics of the Inntal shear zone–sub-Tauern ramp fault system and the interpretation of the TRANSALP seismic section, Eastern Alps, Austria. *Tectonophysics*, 414, 241–258.

Ortner, H., Aichholzer, S., Zerlauth, M., Pilser, R., Fügenschuh, B. (2015). Geometry, amount and sequence of thrusting in the Subalpine Molasse of Western Austria and Southern Germany, European Alps. *Tectonics*, 34 (1), 1–30.

Palzer-Khomenko, M., Wagreich, M., Knierzinger, W., Meszar, M., Gier, S., Kallanxhi, M.-E., Soliman, A. (2018). A calcite crisis unravelling Early Miocene (Ottnangian) stratigraphy in the North Alpine–Carpathian Foreland Basin: A litho- and chemostratigraphic marker for the Rzehakia Lake System. *Geologica Carpathica*, 69(4), 315–334.

Paulus, B. (1963). Zur Stratigraphie und Fazies der oligozänen und miozänen Molasse im südlichen Oberbayern. *Bull. Ver. schweiz. Petroleum-Geol. u. -Ing.*, 30, 53–97.

Paulus, B. (1981). Eine Revision des Bohrprofils Tölz 1 und der Nachweis der bisher unbekannten Kirchsee-Mulde am Nordrand der Faltenmolasse zwischen Isar und Mangfall. *Zeitschr. Dt. Geol. Ges.*, 132, 253–276.

Paulus, B. and Brockert, M. (1968). Die Tiefbohrung Feilnbach 1. In *Erläuterungen zur geologischen Karte von Bayern, 1:25.000, Blatt Nr. 8237 Miesbach*, Pflaumann, U. and Stephan, W. (eds). Bayerisches Geologisches Landesamt, Munich.

Petit, C., Campy, M., Chaline, J., Bonvalot, J. (1996). Major palaeohydrographic changes in Alpine foreland during the Pliocene – Pleistocene. *Boreas*, 25, 131–143.

Pfiffner, O.A. (1986). Evolution of the north Alpine foreland basin in the Central Alps. In *Foreland Basins*, Allen, P.A. and Homewood, P. (eds). Spec. Publ. Int. Ass. Sediment., Blackwell, Oxford.

Pfiffner, O.A. (1993). The structure of the Helvetic nappes and its relation to mechanical stratigraphy. *J. Struct. Geol.*, 15, 511–521.

Pfiffner, O.A. (2009). *Geologie der Alpen*. Haupt, Stuttgart.

Pfiffner, O.A. and Hitz, L. (1997). Geologic interpretation of the seismic profiles of the Eastern Traverse (lines E1-E3, E7-E9): Eastern Swiss Alps. In *Results of NRP20*, Pfiffner, O.A., Lehner, P., Heitzmann, P., Mueller, S., Steck, A. (eds). Birkhäuser, Basel.

Pfiffner, O.A., Erard, P.-F., Stäuble, M. (1997). Two cross-sections through the Swiss Molasse basin (line E4-E6, W1, W7-W10). In *Results of NRP20*, Pfiffner, O.A., Lehner, P., Heitzmann, P., Mueller, S., Steck, A. (eds). Birkhäuser, Basel.

Pfiffner, O.A., Schlunegger, F., Buiter, S.J.H. (2002). The Swiss Alps and their peripheral foreland basin: Stratigraphic response to deep crustal processes. *Tectonics*, 21, 1009.

Philippe, Y., Coletta, B., Deville, E., Mascle, A. (1996). The Jura fold-and-thrust belt: A kinematic model based on map-balancing. In *Peri-Tethys Memoir 2: Structure and Prospects of Alpine Basins and Forelands*, Ziegler, P.A. and Horvath, F. (eds). Mémoire du Muséum national d'Histoire naturelle, Paris.

Picha, F.J., Stránik, Z., Krejči, O. (2006). Geology and hydrocarbon resources of the outer Western Carpathians and their foreland, Czech Republic. In *The Carpathians and their Foreland: Geology and Hydrocarbon Resources*, Golonka, J. and Picha, F.J. (eds). Memoir, AAPG, Tulsa.

Piller, W.E., Egger, H., Erhart, C.W., Gross, M., Harzhauser, M., Hubmann, B., van Husen, D., Krenmayr, H.-G., Krystyn, L., Lein, R. et al. (2004). Die stratigraphische Tabelle von Österreich 2004 (sedimentäre Schichtfolgen). Kommission für die paläontologische und stratigraphische Erforschung Österreichs der Österreichische Akademie der Wissenschaften und Österreichische Stratigraphische Kommission.

Piller, W.E., Harzhauser, M., Mandic, O. (2007). Miocene Central Paratethys stratigraphy – Current status and future directions. *Stratigraphy*, 4(2/3), 151–168.

Pilser, R. (2008). Kinematische Entwicklung der tektonischen Großeinheiten am Alpennordrand östlich des Schliersees. Diploma University of Innsbruck.

Plöchinger, B. and Prey, S. (1968). Profile durch die Windischgarstener Störungszone im Raume Windischgarsten – St. Gallen. *Jb. Geol. Bundesanst.*, 111, 175–211.

Pomella, H., Ortner, H., Zerlauth, M., Fügenschuh, B. (2015). The Alpine nappe stack in western Austria: A crustal-scale cross-section. *Int. J. Earth. Sci.*, 104, 733–745.

Prey, S. (1953). Flysch, Klippenzone und Kalkalpenrand im Almtal bei Scharnstein und Grünau (O.Ö.). *Jb. Geol. Bundesanst.*, 96, 301–343.

Prey, S. (1957). Ergebnisse der bisherigen Forschungen über das Molassefenster von Rogatsboden (NÖ). *Jb. Geol. Bundesanst.*, 100, 299–358.

Prey, S. (1980). Helvetikum, Flysche und Klippenzonen von Salzburg bis Wien. In *Der geologische Aufbau Österreichs*, Oberhauser, R. (ed.). Springer, Vienna.

Ratschbacher, L., Frisch, W., Linzer, G., Merle, O. (1991). Lateral extrusion in the Eastern Alps, Part 2: Structural analysis. *Tectonics*, 10, 257–271.

Reinecker, J., Danisík, M., Schmid, C., Glotzbach, C., Rahn, M., Frisch, W., Spiegel, C. (2008). Tectonic control on the late stage exhumation of the Aar Massif (Switzerland). Constraints from apatite fission track and (U-Th)/He data. *Tectonics*, 27, TC6009.

Richter, M. (1950). Die Molassefenster in der Flyschzone von Niederösterreich. N. *Jb. Geol. Paläont., Abh.*, 92, 31–46.

Rögl, F., Hochuli, P., Müller, C. (1979). Oligocene – Early Miocene stratigraphic correlations in the Molasse Basin of Austria. *Ann. Geol. Pays Helleniques. Tome Hors Series*, 30, 1045–1049.

Rosenberg, C.L. and Garcia, S. (2011). Estimating displacement along the Brenner Fault and orogen-parallel extension in the Eastern Alps. *Int. J. Earth. Sci.*, 100(5), 1129–1145.

Rosenberg, C.L. and Kissling, E. (2013). Three-dimensional insight into Central-Alpine collision: Lower-plate or upper-plate indentation? *Geology*, 41(12), 1219–1222.

Rosenberg, C.L., Brun, J.-B., Gapais, D. (2004). Indentation model of the Eastern Alps and the origin of the Tauern Window. *Geology*, 32, 997–1000.

Rosenberg, C.L., Berger, A., Bellahsen, N., Bousquet, R. (2015). Relating orogen width to shortening, erosion, and exhumation during Alpine collision. *Tectonics*, 34(6), 1306–1328.

Rosenberg, C.L., Schneider, S., Scharf, A., Bertrand, A., Hammerschmidt, K., Rabaute, A., Brun, J.P. (2018). Relating collisional kinematics to exhumation processes in the Eastern Alps. *Earth-Science Reviews*, 176, 311–344.

de Ruig, M.J. (2003). New insights from 3D seismic data – Deep marine sedimentation and gas reservoir distribution in Upper Austria. *Oil and Gas European Magazine*, 119, 64–73.

de Ruig, M.J. and Hubbard, S.M. (2006). Seismic facies and reservoir characteristics of a deep-marine channel belt in the Molasse foreland basin, Puchkirchen Formation, Austria. *AAPG Bull.*, 90, 735–752.

Rupp, C. (2008). Erläuterungen zu Blatt 47 Ried im Innkreis. Geologische Karte der Republik Österreich. Map, Geologische Bundesanstalt, Vienna.

Rupp, C., Linner, M., Mandl, G.W., Atzenhofer, B., Berning, B., Bieber, G., Draxler, I., Egger, H., Finger, F., Heinrich, M. et al. (2011). Geologische Karte von Oberösterreich 1:200 000: Erläuterungen. Map, Geologische Bundesanstalt, Vienna.

Sachsenhofer, R.F. and Schulz, H.-M. (2006). Architecture of Lower Oligocene source rocks in the Alpine Foreland Basin: A model for syn- and post-depositional source-rock features in the Paratethyan realm. *Petroleum Geoscience*, 12(4), 363–377.

Schaad, W., Keller, B., Matter, A. (1992). Die Obere Meeresmolasse (OMM) am Pfänder: Beispiel eines Gilbert-Deltakomplexes. *Eclogae Geol. Helv.*, 85, 145–168.

Schardt, H. (1893). Sur l'origine des Préalpes romandes. *Archives de physique et des sciences naturelles de Genève*, 3(30), 570–583.

Scharf, A., Handy, M.R., Favaro, S., Schmid, S.M., Bertrand, A. (2013). Modes of orogen-parallel stretching and extensional exhumation in response to microplate indentation and roll-back subduction (Tauern Window, Eastern Alps). *Int. J. Earth. Sci.*, 102(6), 1627–1654.

Schegg, R. and Leu, W. (1998). Analysis of erosion events and palaeogeothermal gradients in the North Alpine Foreland Basin of Switzerland. In *Basin Modelling: Practice and Progress*, Düppenbecker, S.J. and Iliffe, J.E. (eds). Special Publication No. 141, Geological Society, London.

Schegg, R., Leu, W., Cornford, C., Allen, P.A. (1997). New coalification profiles in the Molasse Basin of Western Switzerland: Implications for the thermal and geodynamic evolution of the Alpine Foreland. *Eclogae Geol. Helv.*, 90(1), 79–96.

Schlunegger, F. and Kissling, E. (2015). Slab rollback orogeny in the Alps and evolution of the Swiss Molasse basin. *Nat. Commun.*, 6, 8605.

Schlunegger, F. and Mosar, J. (2011). The last erosional stage of the Molasse Basin and the Alps. *Int. J. Earth. Sci.*, 100(5), 1147–1162.

Schlunegger, F., Matter, A., Burbank, D.W., Klaper, E.M. (1997). Magnetostratigraphic constraints on relationships between evolution of the central Swiss Molasse basin and Alpine orogenic events. *Geol. Soc. Am. Bull.*, 109, 225–241.

Schmid, S.M., Pfiffner, O.A., Froitzheim, N., Schönborn, G., Kissling, E. (1996). Geophysical-geological transect and tectonic evolution of the Swiss-Italian Alps. *Tectonics*, 15(5), 1036–1064.

Schnabel, W., Bryda, G., Egger, H., Fuchs, G., Krenmayer, H.G., Mandl, G.W., Matura, A., Nowotny, A., Roetzel, R., Schabert, S. et al. (2002). Geologische Karte von Niederösterreich 1:200.000 mit Legende und Kurzerläuterung. Map, Geologische Bundesanstalt, Vienna.

Scholz, H. (2000). Die tertiären Grobschüttungen am Südrande des Molassebeckens im Allgäu (Südwestbayern) – eine Synopsis. *N. Jb. Geol. Paläont., Abh.*, 218, 61–84.

Schuster, R., Finger, F., Frank, W., Egger, H. (2007). Kristallingerölle aus verschiedenen tektonischen Einheiten im Bereich des Almtales (Oberösterreich) – Petrologie, Geochemie, Geochronologie. In *Arbeitstagung der Geologischen Bundesanstalt 2007*, Egger, H. and Rupp, C. (eds). Geologische Bundsanstalt, Vienna.

Schuster, R., Daurer, A., Krenmayr, H.G., Linner, M., Mandl, G.W., Pestal, G., Reitner, J.M. (2015). Rocky Austria. Geologie von Österreich – kurz und bunt. Map, Geologische Bundesanstalt, Vienna.

Schwerd, K. (1978). Über die Bausteinschichten der Allgäuer und östlichen Vorarlberger Faltenmolasse zwischen Lech und Bregenzer Ach. *Geol. Jb.*, *A*, 46, 3–24.

Schwerd, K. (1983). Helvetikum, Ultrahelvetikum, Feuerstätter Decke, Rhenodanubischer Flysch und Arosa-Zone zwischen Burgberg und Hindelang (Exkursion G am 8. April 1983). *Jahresberichte und Mitteilungen des Oberrheinischen Geologischen Vereines, Neue Folge*, 65, 99–112.

Schwerd, K. (1996). Geologische Karte von Bayern 1:500.000. Map, Bayerisches Geologisches Landesamt, Munich.

Selverstone, J., Spear, F.S., Franz, G., Morteani, G. (1984). High-pressure metamorphism in the SW Tauern window, Austria: P-T paths from Hornblende-Kyanite-Staurolite Schists. *Journal of Petrology*, 25(2), 501–531.

Sharman, G.R., Hubbard, S.M., Covault, J.A., Hinsch, R., Linzer, H.-G., Graham, S.A. (2018). Sediment routing evolution in the North Alpine Foreland Basin, Austria: Interplay of transverse and longitudinal sediment dispersal. *Basin Res.*, 30(3), 426–447.

Sinclair, H.D. (1997). Flysch to molasse transition in peripheral foreland basins: The role of the passive margin versus slab breakoff. *Geology*, 25, 1123–1126.

Sinclair, H.D. and Allen, P.A. (1992). Vertical versus horizontal motions in the Alpine orogenic wedge: Stratigraphic response in the foreland basin. *Basin Res.*, 4, 215–232.

Sinclair, H.D., Coakley, B.J., Allen, P.A., Watts, A.B. (1991). Simulation of Foreland Basin Stratigraphy using a diffusion model of mountain belt uplift and erosion: An example from the central Alps, Switzerland. *Tectonics*, 10(3), 599–620.

Sommaruga, A. (1999). Décollement tectonics in the Jura foreland fold-and-thrust belt. *Mar. Pet. Geol.*, 16, 111–134.

Sommaruga, A., Eichenberger, U., Marillier, F. (2012). Seismic atlas of the Swiss Molasse basin. Matériaux pour la Géologie de la Suisse Geophysique No. 44. swisstopo, Wabern.

Sommaruga, A., Gruber, M., Mosar, J. (2016). Structure géologique profonde du Canton de Fribourg: Études régionales Fribourg et Bulle. *Geofocus*, 39, 103.

Sommaruga, A., Mosar, J., Schori, M., Gruber, M. (2017). The role of the Triassic evaporites underneath the North Alpine foreland. In *Permo-Triassic Salt Provinces of Europe, North Africa and the Atlantic Margins*, Soto, J.I., Flinch, J.F., Tari, G. (eds). Elsevier, Amsterdam.

Stackelberg, U.V. (1960). Oberkreide und Alttertiär des Helvetikums am bayerischen Alpenrand im Westen von Tölz. *Geol. Bav.*, 41, 3–54.

Storti, F. and McClay, K. (1995). Influence of syntectonic sedimentation on thrust wedges in analog models. *Geology*, 23, 999–1002.

Strunck, P. (2001). The Molasse of Western Switzerland. Unpublished PhD Thesis, University of Bern.

Stutenbecker, L., Tollan, P.M.E., Madella, A., Lanari, P. (2019). Miocene basement exhumation in the Central Alps recorded by detrital garnet geochemistry in foreland basin deposits. *Solid Earth*, 10(5), 1581–1595.

Tari, G. (2005). The divergent continental margins of the Jurassic Proto-Pannonian Basin: Implications for the petroleum systems of the Vienna Basin and the Moesian platform. In *Petroleum Systems of Divergent Continental Margin Basins*, Post, P.J., Rosen, N.C., Olson, D.L., Palmes, S.L., Lyons, K.T., Newton, G.B. (eds). GCSSEPM Transactions, SEPM.

Tari, G. and Strauss, P. (2018). The Jurassic Gresten facies of the European margin in Austria, Hungary and Romania: A regional overview. *Berichte der Geologischen Bundesanstalt*, 128, 154.

Tesauro, M., Kaban, M.K., Cloetingh, S.A.P.L. (2008). EuCRUST-07: A new reference model for the European crust. *Geophysical Research Letters*, 35, L05313.

Thouvenot, F., Fréchet, J., Tapponnier, P., Thomas, J.-C., Le Brun, B., Ménard, G., Lacassin, R., Jenatton, L., Grasso, J.-R., Coutant, O. et al. (1998). The ML 5.3 Épagny (French Alps) earthquake of 1996 July 15: A long-awaited event on the Vuache Fault. *Geophysical Journal International*, 135(3), 876–892.

Trautwein, B., Dunkl, I., Frisch, W. (2001). Accretionary history of the Rhenodanubian flysch zone in the Eastern Alps – Evidence from apatite fission-track geochronology. *Int. J. Earth. Sci.*, 90(3), 703–713.

Trümpy, R. (1969). Die helvetischen Decken der Ostschweiz: Versuch einer palinspastischen Korrelation und Ansätze zu einer kinematischen Analyse. *Eclogae Geol. Helv.*, 62, 105–142.

Unger, H.J. (1989). Die Lithozonen der Oberen Süßwassermolasse Südostbayerns und ihre vermutlichen zeitlichen Äquivalente gegen Westen und Osten. *Geol. Bav.*, 94, 195–237.

Ustaszewski, K. and Schmid, S.M. (2006). Control of preexisting faults on geometry and kinematics in the northernmost part of the Jura fold-and-thrust belt. *Tectonics*, 25(5), TC5003.

Ustaszewski, K. and Schmid, S.M. (2007a). Latest Pliocene to recent thick-skinned tectonics at the Upper Rhine Graben – Jura Mountains junction. *Swiss Journal of Geosciences*, 100(2), 293–312.

Ustaszewski, K. and Schmid, S.M. (2007b). Neotectonic activity in the Upper Rhine Graben – Jura Mountains junction (North-Western Switzerland and adjacent France). *Bulletin angewandter Geologie*, 12(1), 3–19.

Valla, P.G., van der Beek, P.A., Shuster, D.L., Braun, J., Herman, F., Tassan-Got, L., Gautheron, C. (2012). Late Neogene exhumation and relief development of the Aar and Aiguilles Rouges massifs (Swiss Alps) from low-temperature thermochronology modeling and 4He/3He thermochronometry. *Journal of Geophysical Research: Earth Surface*, 117(F1), F01004.

Vernon, A.J., van der Beek, P.A., Sinclair, H.D., Persano, C., Foeken, J., Stuart, F.M. (2009). Variable late Neogene exhumation of the central European Alps: Low-temperature thermochronology from the Aar Massif, Switzerland, and the Lepontine Dome, Italy. *Tectonics*, 28, TC5004.

Vouillamoz, N., Abednego, M., Wust, G.H., Mosar, J. (2016). Optimizing event detection and location in low-seismicity zones using nanoseismic monitoring: Case study from Western Switzerland. *Bulletin of the Seismological Society of America*, 106(5), 2023–2036.

Vouillamoz, N., Mosar, J., Deichmann, N. (2017). Multi-scale imaging of a slow active fault zone: Contribution for improved seismic hazard assessment in the Swiss Alpine foreland. *Swiss Journal of Geosciences*, 110(2), 547–563.

Wachtel, G. and Wessely, G. (1981). Die Tiefbohrung Berndorf 1 in den östlichen Kalkalpen und ihr geologischer Rahmen. *Mitt. Österr. Geol. Ges.*, 74–75, 137–165.

Wagner, L. (1996). Stratigraphy and Hydrocarbons in the Upper Austrian Molasse Foredeep (active margin). In *Oil and Gas in Alpidic Thrustbelts and Basins of Central and Eastern Europe, Spec. Publ. No. 5*, Wessely, G. and Liebl, W. (eds). EAEG Special Publication, London.

Wagner, L. (1998). Tectono-stratigraphy and hydrocarbons in the Molasse Foredeep of Salzburg, Upper and Lower Austria. In *Cenozoic Foreland Basins of Western Europe*, Mascle, A. (ed.). Geological Society Special Publication, London.

Weidmann, M., Homewood, P., Morel, R., Berchten, J.-D., Bucher, H., Burri, J.-D., Cornioley, J.-D., Escher, P., Rück, P., Tabotta, A. et al. (1993). Atlas géologique de la Suisse 1:25.000, Feuille 1244 Châtel-St-Denis. swisstopo, Wabern.

Wessely, G. (1987). Mesozoic and Tertiary evolution of the Alpine-Carpathian foreland in eastern Austria. *Tectonophysics*, 137(1), 45–59.

Wessely, G., Draxler, I., Gangl, G., Gottschling, P., Heinrich, M., Hofmann, T., Lenhardt, W., Matura, A., Pavuza, R., Peresson, H. et al. (2006). Niederösterreich. Geologische Bundesanstalt, Vienna.

Willett, S.D. and Schlunegger, F. (2010). The last phase of deposition in the Swiss Molasse Basin: From foredeep to negative-alpha basin. *Basin Res.*, 22, 623-639.

Winterberg, S. and Willett, S.D. (2019). Greater Alpine river network evolution, interpretations based on novel drainage analysis. *Swiss Journal of Geosciences*, 112(1), 3–22.

Wortel, M.J.R. and Spakman, W. (2000). Subduction and slab detachment in the Mediterranean-Carpathian region. *Science*, 290, 1910–1917.

Yanites, B.J., Ehlers, T.A., Becker, J.K., Schnellmann, M., Heuberger, S. (2013). High magnitude and rapid incision from river capture: Rhine River, Switzerland. *Journal of Geophysical Research: Earth Surface*, 118(2), 1060–1084.

Zámolyi, A., Decker, K., Hölzel, M., Strauss, P., Wagreich, M. (2008). Kinematics and timing of deformation at the front of the Alpine–Carpathian wedge (Waschberg– Zdanice unit, Austria–Czech Republic). *Geophysical Research Abstracts*, 10, 04297.

Zerlauth, M. (2008). Kinematische Entwicklung der tektonischen Großeinheiten am Alpennordrand westlich des Schliersees. Diploma Thesis, University of Innsbruck.

Zerlauth, M. (2014). The Vorarlberg and Upper Allgäu Helvetics: Thermo-tectonic evolution of an external fold-and-thrust belt at the northern Alpine rim. PhD Thesis, University of Innsbruck.

Zerlauth, M., Ortner, H., Pomella, H., Pfiffner, A.O., Fügenschuh, B. (2014). Inherited tectonic structures controlling the deformation style – An example from the Helvetic nappes of the Eastern Alps. *Swiss Journal of Geosciences*, 107(2/3), 157–175.

Zerlauth, M., Bertrand, A., Rantitsch, G., Groß, D., Ortner, H., Pomella, H., Fügenschuh, B. (2016). Thermal history of the westernmost Eastern Alps (Penninic Rhenodanubian Flysch nappes, Helvetic nappes, and Subalpine Molasse thrust sheets). *Int. J. Earth. Sci.*, 105, 1525–1547.

Ziegler, P.A. and Dèzes, P. (2006). Crustal evolution of Western and Central Europe. In *European Lithosphere Dynamics*, Gee, D.G. and Stephenson, R.A. (eds). Geological Society, London.

Ziegler, P.A. and Fraefel, M. (2009). Response of drainage systems to Neogene evolution of the Jura fold-thrust belt and Upper Rhine Graben. *Swiss Journal of Geosciences*, 102(1), 57–75.

Ziegler, P.A. and Stampfli, G.M. (2001). Late Paleozoic–Early Mesozoic plate boundary reorganisation: Collapse of the Variscan orogen and opening of Neotethys. In *The Continental Permian of the Southern Alps and Sardinia (Italy) Regional Reports and General Correlations*, Cassinis, R. (ed.). Annali Museo Civico Scienze Naturali, Brescia, 25, 17–34.

Ziegler, P.A., Cavazza, W., Robertson, A.H.F., Crasquin-Soleau, S. (eds) (2001). PeriTethys Memoir 6: Peritethyan rift/Wrench basins and passive margins, IGCP369. *Mémoires du Museum National d'Histoire Naturelle*, 186, 259–294.

Zweigel, J., Aigner, T., Luterbacher, H. (1998). Eustatic versus tectonic controls an Alpine foreland basin fill: Sequence stratigraphy and subsidence analysis in the SE German Molasse. In *Cenozoic Foreland Basins of Western Europe*, Mascle, A., Puigdefábegras, C., Luterbacher, H.-P., Fernández, M. (eds). Geological Society Special Publication, London.

List of Authors

Peter VAN DER BEEK
CNRS, ISTerre
Université Grenoble Alpes
France
and
Institute of Geosciences
University of Potsdam
Germany

Andreas BEIDINGER
OMV Exploration & Production GmbH
Vienna
Austria

Nicolas BELLAHSEN
ISTeP
Sorbonne Université
UMR 7193
CNRS
Paris
France

Alfons BERGER
Institute of Geological Sciences
University of Bern
Switzerland

Matthias BERNET
CNRS, ISTerre
Université Grenoble Alpes
France

Christoph GLOTZBACH
Department of Geosciences
University of Tübingen
Germany

Christoph VON HAGKE
Department of Geography and Geology
University of Salzburg
Austria

Marco HERWEGH
Institute of Geological Sciences
University of Bern
Switzerland

Ralph HINSCH
OMV Exploration & Production GmbH
Vienna
Austria

Edi KISSLING
Department of Earth Sciences
Institute of Geophysics
ETH Zürich
Switzerland

Samuel MOCK
Institute of Geological Sciences
University of Bern
and
Fachstelle Gefahrenprävention
Kantonales Laboratorium Basel-Stadt
Switzerland

Giancarlo MOLLI
Department of Earth Sciences
University of Pisa
Italy

Jon MOSAR
Department of Geosciences
University of Fribourg
Switzerland

Hugo ORTNER
Institute of Geology
University of Innsbruck
Austria

Yann ROLLAND
EDYTEM
Université Savoie Mont Blanc
UMR 5204
CNRS
Le Bourget-du-Lac
France

Claudio L. ROSENBERG
ISTeP
Sorbonne Université
UMR 7193
CNRS
Paris
France

Fritz SCHLUNEGGER
Institute of Geological Sciences
University of Bern
Switzerland

Anna SOMMARUGA
Department of Geosciences
University of Fribourg
Switzerland

Pierre G. VALLA
CNRS, ISTerre
Université Grenoble Alpes
France
and
Institute of Geological Sciences &
Oeschger Center for Climate Research
University of Bern
Switzerland

Index

A, B, C

D, E

F, G, M

N, O

S, T

Summary of Volume 1

List of Authors

Index

Summary of Volume 2

Chapter 2. Paleogeography and Architecture of the Alpine Tethys Margins

Yves LAGABRIELLE

Chapter 3. One View on the Petrology and Geochemistry of the Mantle and Oceanic Crustal Remnants in the Alps, Reconciling Rifts and Ultra-slow Seafloor Spreading

Othmar MÜNTENER and Anders MCCARTHY

Chapter 4. Oceanic Subduction in the Alps: Lessons from the Rock Record

Philippe AGARD

List of Authors

Index